The AI Policy Sourcebook 2025

MARC ROTENBERG
ELENI KYRIAKIDES

Center for AI and Digital Policy
WASHINGTON DC

Center for AI and Digital Policy
1100 13th St NW
Suite 800
Washington, DC 20005

Visit the CAIDP online:
https://www.caidp.org

Edition 2025
Printed in the United States of America
All Rights Reserved

ISBN-13: 979-8-218-60661-9
ISBN-10: 8-218-60661-9

Table of Contents

Preface

When we published the first compendium of AI policies in 2019, our aim was to provide an essential reference book for those studying AI policy. At that time, we noted the rapid explosion of AI ethics frameworks, but it was still early days for AI governance. The OECD had just finalized the first AI principles endorsed by national governments. The Universal Guidelines for AI, published the year before, were gaining influence among AI policymakers. But work on the EU AI Act had not yet started. At the Council of Europe, the first steps were taken toward a global AI treaty.

A lot has happened in five years. Governments have raced to develop national AI strategies, amend current laws, and enact new laws. Remarkable progress has been made by international organizations. The early, now venerable, OECD AI Principles were recently updated to account for recent developments in AI technology. The EU AI Act was adopted in 2024, and implementation has begun. Forty countries have endorsed the Council of Europe Framework Convention on AI. At the United Nations, there is consensus on the need to ensure safe, secure, and trustworthy AI. But challenges remain in many domains – sustainability, autonomous weapons, algorithmic transparency, labor impacts, copyright, and more

Below, we discuss in more detail the significance of the AI governance frameworks we chose to include in the *2025 AI Policy Sourcebook*. This is not a comprehensive review of all AI laws and regulations. As with the original *AI Policy Sourcebook*, we aim to highlight the most influential frameworks, show the evolution of the AI governance field, and provide an essential reference text for those studying AI governance and those drafting governance frameworks.

We also offer a few words about the current state of AI governance and the road ahead.

* * *

We begin with the **Universal Guidelines for AI (2018)**, the first global framework for AI governance. Endorsed by more than 300 experts in over 40 countries, the **Universal Guidelines** provided a foundation for much of the work that followed. The **Universal Guidelines** provided clear direction on issues that were present in the early days, such as fairness, accountability, and transparency, and anticipated emerging issues, such as AI safety. The Termination Principle, for example, centered human control over AI systems and established a clear obligation to shut down autonomous AI systems.

The following year, the **OECD AI Principles (2019)** were finalized. The **OECD AI Principles** were a remarkable achievement, outlining key goals for AI policymakers and expressing a rough consensus from countries across the Americas, Europe, and parts of East Asia on AI priorities – inclusive growth, sustainable development, human rights, democratic values, security, and safety. When the G20 countries endorsed the OECD AI Principles the same year, more than 50 countries, including Brazil, China, Russia, and South Africa, had endorsed the first government-backed global framework for AI.

Among the most remarkable AI governance frameworks is the **UNESCO Recommendation on AI Ethics (2021)**, adopted unanimously by the UN member states. Work on the UNESCO Recommendation began shortly after the conclusion of the **OECD AI Principles.** The UNESCO Recommendation set out values and principles to guide AI governance and expanded our understanding of AI policy with new issues such as sustainability, gender equity, health and well-being. The **UNESCO Recommendation** also established a **Readiness Assessment Methodology** to strengthen the capacity for AI deployment and the **Ethical Impact Assessment**, a tool for evaluating the benefits and risks of an AI system in relation to the values and principles of the **UNESCO Recommendation on AI Ethics.**

The **EU AI Act (2024)** is likely the world's most widely recognized AI governance framework. It is a comprehensive, risk-based regulation designed to protect public safety and fundamental rights while creating a regulatory framework for deploying AI systems within the European Union. Among its many notable features are prohibitions on specific AI systems, a detailed structure of independent oversight, and staggered deadlines for the two-year implementation period. It remains to be seen whether the **EU AI Act** will generate the same "Brussels Effect" associated with the GDPR, Europe's comprehensive data protection regulation.

The **Council of Europe Framework Convention on Artificial Intelligence, Human Rights, Democracy, and the Rule of Law (2024)** emerged on a parallel track, over the same five-year period as the EU AI Act. The Council of Europe "AI Treaty" is the first legally binding, global treaty for the governance of AI and reflects the mandate of the Council of Europe. Council of Europe conventions are open to signature and ratification by both COE member and non-member states. At the time of publication, forty countries have endorsed the AI Treaty, including the member states of the European Union, Canada, Israel, Japan, Norway, the United Kingdom, the United States, and others. The AI Treaty will likely enter into force later this year.

At the United Nations, important resolutions were adopted in 2024, reflecting the ongoing efforts worldwide to find common ground on AI governance. The **Global Digital Compact** set out several key recommendations, including an international scientific panel for AI, similar to the International Panel on Climate Change (IPCC), that would provide evidence-based assessments of AI capabilities and social impacts. Our contribution to this work, noted in the Resources section, is a proposal for a UN Special Rapporteur for AI and Human Rights who would possess the agility, competence, and authority to advance the UN's efforts in this rapidly evolving field.

Work continues across the African continent to develop a comprehensive AI strategy. An early resolution from the African Commission on Human and Peoples' Rights, along with several other initiatives, has resulted in the **Continental Artificial Intelligence Strategy of the African Union and** the **Windhoek Statement on Artificial Intelligence in Southern Africa**. The **Windhoek Statement** addresses numerous key AI issues, including data governance, capacity building, investment and infrastructure, education, research, and gender. UNESCO continues to play a significant role in these initiatives.

In Latin America, both the **Declaration of Montevideo (2024)** and the **Santiago Declaration (2024)** have shaped regional AI policies. The **Montevideo Declaration** emphasizes "respect for and protection of human rights, fundamental freedoms, rule of law and democracy, focused on the human being, well-being and dignity of people, fostering innovation and ensuring inclusive and sustainable technological progress." The **Montevideo Declaration** sets out a Roadmap to promote responsible AI across Governance and Regulation, Talent and future of work, Protection of vulnerable groups, Environment, Sustainability and Climate Change, and Infrastructure. The **Santiago Declaration** is a regional commitment to promote responsible and ethical development and use of AI, with a focus on inclusive governance and addressing potential societal impact

The **Saudi Data & AI Authority, AI Ethics Principles (2023)** reflect many of the goals contained in the **Universal Guideline** – including Fairness, Privacy and Security, Transparency and Explainability, Accountability and Responsibility -- and have been incorporated in other governance frameworks in the Middle East and North Africa (MENA) region.

We have traced the recent development of the AI Summit declarations, beginning with the **Bletchley AI Safety Summit Declaration (2023),** and then the **Seoul Declaration for Safe, Innovative and Inclusive AI (2024)** and the **Paris Statement on Inclusive and Sustainable Artificial Intelligence for People and the Planet (2025).** More than 60 countries endorsed the Paris statement. Notable also were related statements issued at the recent AI Action Summit, including the **Paris Declaration on Maintaining Human Control in AI-enabled Weapon Systems (2025)**.

Also notable is the **Hiroshima AI Process,** launched by the G7 countries, and reflecting Japan's leading role in the development of AI policy. The **Hiroshima AI Process** aims to establish a global framework for promoting safe, secure, and trustworthy AI with guiding principles and a code of conduct for organizations developing advanced AI systems, particularly focusing on mitigating risks associated with generative AI, all while facilitating inclusive international governance on AI development and use.

For the 2025 edition of the *AI Policy Sourcebook*, we have decided to dedicate a section to the **US AI Executive Orders**. As we go to press, the US Office of Science and Technology Policy is reviewing the US National AI strategy and the various policies regulating AI systems in the federal government. Our analysis of the **AI Executive Orders** from the past several US administrations revealed a surprising level of consensus regarding fundamental goals for the US national AI strategy. We hope that making these documents available will contribute to a thoughtful, forward-looking AI plan for the US that builds on prior achievements.

* * *

Having closely observed the development of AI governance frameworks over the past several years, I would like to share a few impressions. First, the level of engagement among countries worldwide in AI policy is striking and encouraging. A topic that was esoteric less than a decade ago is now front and center for many governments. However, the public remains largely unaware of these initiatives, particularly in the United States. Reporting on AI tends to swing between the hype of overstated innovation and the doom of exaggerated concern. Thoughtful reporting on AI governance, focusing on progress and setbacks, would engage the public, strengthen democratic institutions, and promote the development of well-informed public policy.

Second, the development of AI policy should be viewed as an evolutionary process. From the **OECD Principles** to the **EU AI Act,** we can see the rapid development and sophistication in the understanding of AI policy. More issues are considered and in greater detail. New initiatives build on earlier initiatives. It is essential for those who joined the conversation after the release of ChatGPT in late 2022 to recognize the long history of efforts by governments to regulate AI. Even this collection provides only a recent timeline of a topic that goes back to the debate over autonomous weapons more than forty years ago.

Third, AI policymakers must be careful to steer a steady course, neither veering too quickly in one direction or another. The introduction of generative AI in 2022 posed new challenges, but many of the common elements for effective governance were already well known – the need for impact assessments, the establishment of supervisory authorities, the allocation of rights and responsibilities for those who use AI systems and those who design, develop, and deploy AI systems. Coordinating new AI regulations with preexisting rules for automated decision-making is not a simple task. However, it is a mistake to exclude from a modern definition of AI systems, rule-based expert systems (symbolic AI) that have defined the field from the start. A good definition of AI should remain technology-neutral.

Fourth, we are now entering a new phase of AI governance. If the period 2019-2024 could be described as Establishing Norms for AI Governance, the period 2025-2029 is about the Implementation and Enforcement of AI Governance Norms. This is when the hard work begins. Governments must spend less time articulating AI principles and more time advancing the principles they have already endorsed. There is a real risk in this movement of moving sideways or even backward. With AI governance, we do not have the luxury of time. AI is evolving rapidly. Regulation must do so as well.

Fifth, it is not too soon to ask questions about regulatory convergence and divergence, progress and setbacks, which governance strategies are working and which are in need of repair. Public policy benefits from these comparative assessments. CAIDP's comprehensive report the *AI and Democratic Values Index*, provides the basis for this work. With the *AI and Democratic Values Index*, we provide a narrative survey of AI policies and a methodology to assess AI policies and practices against democratic values. This methodology provides an opportunity to compare national AI policies at a moment in time and to analyze trends over time.

Finally, we need to underscore the urgency of AI governance. Companies and countries are rushing quickly into a future they do not fully understand. The leading AI experts caution that advanced AI systems are not reliable or trustworthy. They have urged us to pause, at least to implement the safeguards and guardrails necessary for sustainable progress. Others warn that many of the AI systems already deployed lack meaningful transparency and accountability. Bias is embedded and replicated at scale.

It is not too soon to think critically about the consequences of the AI age and how we are to maintain human control over this rapidly evolving technology. Human reason remains central to this task. With the publication of this reference book on early efforts to govern AI, we hope to advance that work. *Sapere aude!*

Marc Rotenberg
Founder, Center for AI and Digital Policy

Acknowledgments

Thanks to my colleagues and friends at the Center for AI and Digital Policy who have helped launch this wonderful organization. Merve Hickok, Christabel Randolph, Dr. Grace Thomson, and Selim Alan, you are all rock stars. It is such a pleasure to watch the growth of CAIDP and to know the role each of you played in many of the documents in this collection. And to all the members of the CAIDP Family, including the Research Group members, Team Leaders, Teaching Fellows, Policy Group participants, academic advisors, and board members, we promised you meaningful engagement in AI policy. I hope you see that we have delivered on this promise.

We thank our funders and supporters, particularly Vilas Dahr, President of the Patrick J. McGovern Foundation. Vilas stood with us at the start and later celebrated with us when Canada and Japan signed the Council of Europe AI Treaty, bringing the number of nations committed to protecting human rights, democracy, and the rule of law to almost forty. That is leadership.

We extend our sincere gratitude to the government officials we have collaborated with over the years, from Rio to Seoul, Brussels to Nairobi, Dubai to Ottawa, Tokyo to Lisbon, and Mexico City to Luxembourg. We appreciate your consideration of the perspectives shared by our research teams and policy groups. We also admire your commitment to public service.

For the publication of the *AI Policy Sourcebook* we thank the members of the Policy Group members who identified key resources and publications: Idil Ilayda Kaner, Luciana Longo, Marian Adly, Ren Bin Lee Dixon, and Rupali Lekhi. We also thank our expert advisors who provided feedback on the content and structure of the *Sourcebook*: Almudena Arpón de Mendívil, Nathalie Smuha, and Hannes Werthner.

Finally, special thanks to Eleni Kyriakides, who helped compile the first *AI Policy Sourcebook* and enabled the launch of this new, updated version. The work continues.

Marc Rotenberg

International Instruments

Public Voice, Universal Guidelines for AI
(2018)

New developments in Artificial Intelligence are transforming the world, from science and industry to government administration and finance. The rise of AI decision-making also implicates fundamental rights of fairness, accountability, and transparency. Modern data analysis produces significant outcomes that have real life consequences for people in employment, housing, credit, commerce, and criminal sentencing. Many of these techniques are entirely opaque, leaving individuals unaware whether the decisions were accurate, fair, or even about them.

We propose these Universal Guidelines to inform and improve the design and use of AI. The Guidelines are intended to maximize the benefits of AI, to minimize the risk, and to ensure the protection of human rights. These Guidelines should be incorporated into ethical standards, adopted in national law and international agreements, and built into the design of systems. We state clearly that the primary responsibility for AI systems must reside with those institutions that fund, develop, and deploy these systems.

1. Right to Transparency. All individuals have the right to know the basis of an AI decision that concerns them. This includes access to the factors, the logic, and techniques that produced the outcome.

2. Right to Human Determination. All individuals have the right to a final determination made by a person.

3. Identification Obligation. The institution responsible for an AI system must be made known to the public.

4. Fairness Obligation. Institutions must ensure that AI systems do not reflect unfair bias or make impermissible discriminatory decisions.

5. Assessment and Accountability Obligation. An AI system should be deployed only after an adequate evaluation of its purpose and

objectives, its benefits, as well as its risks. Institutions must be responsible for decisions made by an AI system.

6. Accuracy, Reliability, and Validity Obligations. Institutions must ensure the accuracy, reliability, and validity of decisions.

7. Data Quality Obligation. Institutions must establish data provenance and assure quality and relevance for the data input into algorithms.

8. Public Safety Obligation. Institutions must assess the public safety risks that arise from the deployment of AI systems that direct or control physical devices, and implement safety controls.

9. Cybersecurity Obligation. Institutions must secure AI systems against cybersecurity threats.

10. Prohibition on Secret Profiling. No institution shall establish or maintain a secret profiling system.

11. Prohibition on Unitary Scoring. No national government shall establish or maintain a general-purpose score on its citizens or residents.

12. Termination Obligation. An institution that has established an AI system has an affirmative obligation to terminate the system if human control of the system is no longer possible.

Explanatory Memorandum and References

Context

The Universal Guidelines on Artificial Intelligence (UGAI) call attention to the growing challenges of intelligent computational systems and proposes concrete recommendations that can improve and inform their design. At its core, the purpose of the UGAI is to promote transparency and accountability for these systems and to ensure that people retain control over the systems they create. Not all systems fall within the scope of these Guidelines. Our concern is with those systems that impact the rights of people. Above all else, these systems should do no harm.

The declaration is timely. Governments around the world are developing policy proposals and institutions, both public and private, are supporting research and development of "AI." Invariably, there will be an enormous impact on the public, regardless of their participation in the design and development of these systems. And so, the UGAI reflects a public perspective on these challenges.

The UGAI were announced at the 2018 International Data Protection and Privacy Commissioners Conference, among the most significant meetings of technology leaders and data protection experts in history.

The UGAI builds on prior work by scientific societies, think tanks, NGOs, and international organizations. The UGAI incorporates elements of human rights doctrine, data protection law, and ethical guidelines. The Guidelines include several well-established principles for AI governance, and put forward new principles not previously found in similar policy frameworks.

Terminology

The term "Artificial Intelligence" is both broad and imprecise. It includes aspects of machine learning, rule-based decision-making, and other computational techniques. There are also disputes regarding whether Artificial Intelligence is possible. The UGAI simply acknowledges that this term, in common use, covers a wide range of related issues and adopts the term to engage the current debate. There is no attempt here to define its boundaries, other than to assume that AI requires some degree of automated decision-making. The term "Guidelines" follows the practice of policy frameworks that speak primarily to governments and private companies.

The UGAI speaks to the obligations of "institutions" and the rights of "individuals." This follows from the articulation of fair information practices in the data protection field. The UGAI takes the protection of the individual as a fundamental goal. Institutions, public and private, are understood to be those entities that develop and deploy AI systems. The term "institution" was chosen rather than the more familiar "organization" to underscore the permanent, ongoing nature of the obligations set out in the Guidelines. There is one principle that is addressed to "national governments." The reason for this is discussed below.

Application

These Guidelines should be incorporated into ethical standards, adopted in national law and international agreements, and built into the design of systems.

The Principles

The elements of the Transparency Principle can be found in several modern privacy laws, including the US Privacy Act, the EU Data Protection Directive, the GDPR, and the Council of Europe Convention 108. The aim of this principle is to enable independent accountability for automated decisions, with a primary emphasis on the right of the individual to know the basis of an adverse determination. In practical terms, it may not be possible for an individual to interpret the basis of a particular decision, but this does not obviate the need to ensure that such an explanation is possible.

The Right to a Human Determination reaffirms that individuals and not machines are responsible for automated decision-making. In many instances, such as the operation of an autonomous vehicle, it would not be possible or practical to insert a human decision prior to an automated decision. But the aim remains to ensure accountability. Thus where an automated system fails, this principle should be understood as a requirement that a human assessment of the outcome be made.

Identification Obligation. This principle seeks to address the identification asymmetry that arises in the interaction between individuals and AI systems. An AI system typically knows a great deal about an individual; the individual may not even know the operator of the AI system. The Identification Obligation establishes the foundation of AI accountability which is to make clear the identity of an AI system and the institution responsible.

The Fairness Obligation recognizes that all automated systems make decisions that reflect bias and discrimination, but such decisions should not be normatively unfair. There is no simple answer to the question as to what is unfair or impermissible. The evaluation often depends on context. But the Fairness Obligation makes clear that an assessment of objective outcomes alone is not sufficient to evaluate an AI system. Normative consequences must be assessed, including those that preexist or may be amplified by an AI system.

The Assessment and Accountability Obligation speaks to the obligation to assess an AI system prior to and during deployment. Regarding assessment, it should be understood that a central purpose of this obligation is to determine whether an AI system should be established. If an assessment reveals substantial risks, such as those suggested by principles concerning Public Safety and Cybersecurity, then the project should not move forward.

The Accuracy, Reliability, and Validity Obligations set out key responsibilities associated with the outcome of automated decisions. The terms are intended to be interpreted both independently and jointly.

The Data Quality Principle follows from the preceding obligation.

The Public Safety Obligation recognizes that AI systems control devices in the physical world. For this reason, institutions must both assess risks and take precautionary measures as appropriate.

The Cybersecurity Obligation follows from the Public Safety Obligation and underscores the risk that even well-designed systems may be the target of hostile actors. Those who develop and deploy AI systems must take these risks into account.

The Prohibition on Secret Profiling follows from the earlier Identification Obligation. The aim is to avoid the information asymmetry that arises increasingly with AI systems and to ensure the possibility of independent accountability.

The Prohibition on Unitary Scoring speaks directly to the risk of a single, multi-purpose number assigned by a government to an individual. In data protection law, universal identifiers that enable the profiling of individuals across are disfavored. These identifiers are often regulated and in some instances prohibited. The concern with universal scoring, described here as "unitary scoring," is even greater. A unitary score reflects not only a unitary profile but also a predetermined outcome across multiple domains of human activity. There is some risk that unitary scores will also emerge in the private sector. Conceivably, such systems could be subject to market competition and government regulations. But there is not even the possibility of counterbalance with unitary scores assigned by government, and therefore they should be prohibited.

The Termination Obligation is the ultimate statement of accountability for an AI system. The obligation presumes that systems must remain within human control. If that is no longer possible, the system should be terminated.

OECD AI Principles / G20 AI Guidelines
(2019, updated 2024)

Background Information

The Recommendation on Artificial Intelligence (AI) (hereafter the "Recommendation") – the first intergovernmental standard on AI – was adopted by the OECD Council meeting at Ministerial level on 22 May 2019 on the proposal of the Digital Policy Committee (DPC, formerly the Committee on Digital Economy Policy, CDEP). The Recommendation aims to foster innovation and trust in AI by promoting the responsible stewardship of trustworthy AI while ensuring respect for human rights and democratic values. In June 2019, at the Osaka Summit, G20 Leaders welcomed the G20 AI Principles, drawn from the Recommendation.

The Recommendation was revised by the OECD Council on 8 November 2023 to update its definition of an "AI System", in order to ensure the Recommendation continues to be technically accurate and reflect technological developments, including with respect to generative AI. On the basis of the 2024 Report to Council on its implementation, dissemination and continued relevance, the Recommendation was revised by the OECD Council meeting at Ministerial level on 3 May 2024 to reflect technological and policy developments, including with respect to generative AI, and to further facilitate its implementation.

***

THE COUNCIL,
HAVING REGARD to Article 5 b) of the Convention on the Organisation for Economic Co-operation and Development of 14 December 1960;

HAVING REGARD to standards developed by the OECD in the areas of privacy, digital security, consumer protection and responsible business conduct;

HAVING REGARD to the Sustainable Development Goals set out in the 2030 Agenda for Sustainable Development adopted by the United Nations General Assembly (A/RES/70/1) as well as the 1948 Universal Declaration of Human Rights; to the important work being carried out on artificial intelligence (hereafter, "AI") in other international governmental and non-governmental fora;

RECOGNISING that AI has pervasive, far-reaching and global implications that are transforming societies, economic sectors and the world of work, and are likely to increasingly do so in the future;

RECOGNISING that AI has the potential to improve the welfare and well-being of people, to contribute to positive sustainable global economic activity, to increase innovation and productivity, and to help respond to key global challenges;

RECOGNISING that, at the same time, these transformations may have disparate effects within, and between societies and economies, notably regarding economic shifts, competition, transitions in the labour market, inequalities, and implications for democracy and human rights, privacy and data protection, and digital security;

RECOGNISING that trust is a key enabler of digital transformation; that, although the nature of future AI applications and their implications may be hard to foresee, the trustworthiness of AI systems is a key factor for the diffusion and adoption of AI; and that a well-informed whole-of-society public debate is necessary for capturing the beneficial potential of the technology, while limiting the risks associated with it;

UNDERLINING that certain existing national and international legal, regulatory and policy frameworks already have relevance to AI, including those related to human rights, consumer and personal data protection, intellectual property rights, responsible business conduct, and competition, while noting that the appropriateness of some frameworks may need to be assessed and new approaches developed;

RECOGNISING that given the rapid development and implementation of AI, there is a need for a stable policy environment that promotes a human-centric approach to trustworthy AI, that fosters research, preserves economic incentives to innovate, and that applies to all stakeholders according to their role and the context;

CONSIDERING that embracing the opportunities offered, and addressing the challenges raised, by AI applications, and empowering stakeholders to engage is essential to fostering adoption of trustworthy AI in society, and to turning AI trustworthiness into a competitive parameter in the global marketplace.

On the proposal of the Digital Policy Committee:

I. AGREES that for the purpose of this Recommendation the following terms should be understood as follows:

> –*AI system*: An AI system is a machine-based system that, for explicit or implicit objectives, infers, from the input it receives, how to generate outputs

such as predictions, content, recommendations, or decisions that can influence physical or virtual environments. Different AI systems vary in their levels of autonomy and adaptiveness after deployment.

–AI system lifecycle: An AI system lifecycle typically involves several phases that include to: plan and design; collect and process data; build model(s) and/or adapt existing model(s) to specific tasks; test, evaluate, verify and validate; make available for use/deploy; operate and monitor; and retire/decommission. These phases often take place in an iterative manner and are not necessarily sequential. The decision to retire an AI system from operation may occur at any point during the operation and monitoring phase.

–AI actors: AI actors are those who play an active role in the AI system lifecycle, including organisations and individuals that deploy or operate AI.

–AI knowledge: AI knowledge refers to the skills and resources, such as data, code, algorithms, models, research, know-how, training programmes, governance, processes, and best practices required to understand and participate in the AI system lifecycle, including managing risks.

–Stakeholders: Stakeholders encompass all organisations and individuals involved in, or affected by, AI systems, directly or indirectly. AI actors are a subset of stakeholders.

Section 1: Principles for responsible stewardship of trustworthy AI

II. RECOMMENDS that Members and non-Members adhering to this Recommendation (hereafter the "Adherents") promote and implement the following principles for responsible stewardship of trustworthy AI, which are relevant to all stakeholders.

III. CALLS ON all AI actors to promote and implement, according to their respective roles, the following principles for responsible stewardship of trustworthy AI.

IV. UNDERLINES that the following principles are complementary and should be considered as a whole.

1.1. Inclusive growth, sustainable development and well-being

Stakeholders should proactively engage in responsible stewardship of trustworthy AI in pursuit of beneficial outcomes for people and the planet, such as augmenting human capabilities and enhancing creativity, advancing inclusion of underrepresented populations, reducing economic, social, gender and other inequalities, and protecting natural environments, thus invigorating inclusive growth, well-being, sustainable development and environmental sustainability.

1.2. Respect for the rule of law, human rights and democratic values, including fairness and privacy

> a) AI actors should respect the rule of law, human rights, democratic and human-centred values throughout the AI system lifecycle. These include non-discrimination and equality, freedom, dignity, autonomy of individuals, privacy and data protection, diversity, fairness, social justice, and internationally recognised labour rights. This also includes addressing misinformation and disinformation amplified by AI, while respecting freedom of expression and other rights and freedoms protected by applicable international law.

> b) To this end, AI actors should implement mechanisms and safeguards, such as capacity for human agency and oversight, including to address risks arising from uses outside of intended purpose, intentional misuse, or unintentional misuse in a manner appropriate to the context and consistent with the state of the art.

1.3. Transparency and explainability

AI Actors should commit to transparency and responsible disclosure regarding AI systems. To this end, they should provide meaningful information, appropriate to the context, and consistent with the state of art:

> i. to foster a general understanding of AI systems, including their capabilities and limitations,

> ii .to make stakeholders aware of their interactions with AI systems, including in the workplace,

> iii. where feasible and useful, to provide plain and easy-to-understand information on the sources of data/input, factors, processes and/or logic that led to the prediction, content, recommendation or decision, to enable those affected by an AI system to understand the output, and,

> iv. to provide information that enable those adversely affected by an AI system to challenge its output.

1.4. Robustness, security and safety

a) AI systems should be robust, secure and safe throughout their entire lifecycle so that, in conditions of normal use, foreseeable use or misuse, or other adverse conditions, they function appropriately and do not pose unreasonable safety and/or security risks.

b) Mechanisms should be in place, as appropriate, to ensure that if AI systems risk causing undue harm or exhibit undesired behaviour, they can be overridden, repaired, and/or decommissioned safely as needed.

c) Mechanisms should also, where technically feasible, be in place to bolster information integrity while ensuring respect for freedom of expression.

1.5. Accountability

a) AI actors should be accountable for the proper functioning of AI systems and for the respect of the above principles, based on their roles, the context, and consistent with the state of the art.

b) To this end, AI actors should ensure traceability, including in relation to datasets, processes and decisions made during the AI system lifecycle, to enable analysis of the AI system's outputs and responses to inquiry, appropriate to the context and consistent with the state of the art.

c) AI actors, should, based on their roles, the context, and their ability to act, apply a systematic risk management approach to each phase of the AI system lifecycle on an ongoing basis and adopt responsible business conduct to address risks related to AI systems, including, as appropriate, via co-operation between different AI actors, suppliers of AI knowledge and AI resources, AI system users, and other stakeholders. Risks include those related to harmful bias, human rights including safety, security, and privacy, as well as labour and intellectual property rights.

Section 2: National policies and international co-operation for trustworthy AI

V. RECOMMENDS that Adherents implement the following recommendations, consistent with the principles in section 1, in their national policies and international co-operation, with special attention to small and medium-sized enterprises (SMEs).

2.1. Investing in AI research and development

a) Governments should consider long-term public investment, and encourage private investment, in research and development and open science, including interdisciplinary efforts, to spur innovation in trustworthy AI that focus on challenging technical issues and on AI-related social, legal and ethical implications and policy issues.

b) Governments should also consider public investment and encourage private investment in open-source tools and open datasets that are representative and respect privacy and data protection to support an environment for AI research and development that is free of harmful bias and to improve interoperability and use of standards.

2.2. Fostering an inclusive AI-enabling ecosystem

Governments should foster the development of, and access to, an inclusive, dynamic, sustainable, and interoperable digital ecosystem for trustworthy AI. Such an ecosystem includes inter alia, data, AI technologies, computational and connectivity infrastructure, and mechanisms for sharing AI knowledge, as appropriate. In this regard, governments should consider promoting mechanisms, such as data trusts, to support the safe, fair, legal and ethical sharing of data.

2.3. Shaping an enabling interoperable governance and policy environment for AI

a) Governments should promote an agile policy environment that supports transitioning from the research and development stage to the deployment and operation stage for trustworthy AI systems. To this effect, they should consider using experimentation to provide a controlled environment in which AI systems can be tested, and scaled-up, as appropriate. They should also adopt outcome-based approaches that provide flexibility in achieving governance objectives and co-operate within and across jurisdictions to promote interoperable governance and policy environments, as appropriate.

b) Governments should review and adapt, as appropriate, their policy and regulatory frameworks and assessment mechanisms as they apply to AI systems to encourage innovation and competition for trustworthy AI.

2.4. Building human capacity and preparing for labour market transformation

a) Governments should work closely with stakeholders to prepare for the transformation of the world of work and of society. They should empower people to effectively use and interact with AI systems across the breadth of applications, including by equipping them with the necessary skills.

b) Governments should take steps, including through social dialogue, to ensure a fair transition for workers as AI is deployed, such as through training programmes along the working life, support for those affected by displacement, including through social protection, and access to new opportunities in the labour market.

c) Governments should also work closely with stakeholders to promote the responsible use of AI at work, to enhance the safety of workers, the quality of jobs and of public services, to foster entrepreneurship and productivity, and aim to ensure that the benefits from AI are broadly and fairly shared.

2.5. International co-operation for trustworthy AI

a) Governments, including developing countries and with stakeholders, should actively co-operate to advance these principles and to progress on responsible stewardship of trustworthy AI.

b) Governments should work together in the OECD and other global and regional fora to foster the sharing of AI knowledge, as appropriate. They should encourage international, cross-sectoral and open multi-stakeholder initiatives to garner long-term expertise on AI.

c) Governments should promote the development of multi-stakeholder, consensus-driven global technical standards for interoperable and trustworthy AI.

d) Governments should also encourage the development, and their own use, of internationally comparable indicators to measure AI research, development and deployment, and gather the evidence base to assess progress in the implementation of these principles.

VI. INVITES the Secretary-General and Adherents to disseminate this Recommendation.

VII. INVITES non-Adherents to take due account of, and adhere to, this Recommendation.

VIII. INSTRUCTS the Digital Policy Committee, through its Working Party on AI Governance, to:

a) continue its important work on artificial intelligence building on this Recommendation and taking into account work in other international fora, and to further develop the measurement framework for evidence-based AI policies;

b) develop and iterate further practical guidance on the implementation of this Recommendation to meet evolving developments and new policy priorities;

c) provide a forum for exchanging information on AI policy and activities including experience with the implementation of this Recommendation, and to foster multi-stakeholder and interdisciplinary dialogue to promote trust in and adoption of AI; and

d) report to Council, in consultation with other relevant committees, on the implementation, dissemination and continued relevance of this Recommendation no later than five years following its revision and at least every ten years thereafter.

African Commission on Human and Peoples' Rights, ACHPR/Res. 473, Resolution on the Need to Undertake a Study on Human and Peoples' Rights and Artificial Intelligence (AI), Robotics and Other New and Emerging Technologies in Africa (2021)

The African Commission on Human and Peoples' Rights (Commission) meeting at its 31ˢᵗ Extraordinary Session held virtually, from 19 to 25 February 2021:
Recalling its mandate to promote and protect human and peoples' rights in Africa under Article 45 of the African Charter on Human and Peoples' Rights (the African Charter);

Recognising that new and emerging technologies such as Artificial Intelligence (AI), robotics and other new emerging technologies present both opportunities and perils for the promotion and protection of human and people's rights in Africa;

Reaffirming the importance of access to the internet in the digital age and its implication for the realisation of human rights provided for in the African Charter and other African Union (AU) human rights instruments;

Recalling the AU's Science, Technology and Innovation Strategy for Africa especially within the context of the Fourth Industrial Revolution;

Noting the challenges and concerns that are posed by autonomous systems that are not under meaningful human control and the use of algorithms in Google, Amazon, Facebook and Android;

Further noting the challenges from AI and digital technologies to data protection and from information disorder, particularly, disinformation, misinformation and mal-information of shallowfakes and deepfakes as well as manipulated media enabled by certain AI technologies, robotics and other new and emerging technologies;

Further Noting the rise of online hate speech in the form of racial slurs, misogyny, misandry, xenophobia in the digital age and the challenges of content take-downs by algorithms which may negatively impact on relevant evidence of human rights abuse or discriminate against particular populations;

Recognising the need for comprehensive and multidisciplinary research on the legal, ethical, safety and security opportunities and challenges raised by AI technologies, robotics and other new and emerging technologies in Africa;

Recognising that while AI companies, as well as organisations and businesses that use AI technologies, robotics and other new and emerging technologies have a significant impact on human rights protection in Africa, there is no comprehensive framework governing their operations to ensure that they comply with human rights obligations on the continent;

Noting with concern that development and uses of AI technologies, robotics and other new and emerging technologies have far-reaching consequences on human rights in general, including the right to privacy, and socio-economic rights in particular, as provided for in the African Charter and other regional instruments including the right to work, education, health, social security and for access to social services;

Recognizing the need for creating conditions for harnessing the benefits of AI and minimizing its adverse impacts and the need for instituting sufficient frameworks in Africa to ensure equal and fair distribution of AI benefits in a way that does not exacerbate inequalities;

Further recognising the need for a comprehensive governance framework on AI technologies, robotics and other new and emerging technologies in Africa in a way that enhances human rights protection on the African continent including protection of the ownership of data on individuals experience in the digital sphere;

Emphasising the importance of participation by African states and Africans in the development of international policies and governance frameworks on AI technologies, robotics and other new and emerging technologies;

Recalling the African Commission's 2015 General Comment Number 3 on Article 4 of the African Charter on Human and Peoples' Rights noting that "any machine autonomy in the selection of human targets or the use of force should be subject to meaningful human control";

Bearing in Mind the 2018 statement to the United Nations Group of Governmental Experts on Lethal Autonomous Weapon Systems by the African Group of States on Disarmament that "notions of dignity and humanity are the parents of all human rights and should govern human conduct including human inventions" and that it is "inhumane, abhorrent, repugnant, and against public conscience for humans to give up control to machines" and "that technology should be solely dedicated for the prosperity and progress of human beings in all spheres of life";

Mindful of the 2019 report by the UN Working Group of Experts on the Rights of Peoples of African Descent (A/HRC/42/59) noting that "the development of new technologies must reflect a strong commitment to human rights and human dignity" and caution about "the ongoing influence of mindsets that channel certain narratives, including racially-biased beliefs, and remain embedded in decision-making and the importance of surfacing those views to mitigate their impact, particularly in computerized algorithms that may lack reflective capacity or effective independent oversight";

Noting that the various uses and potential uses of AI technologies, robotics and other new and emerging technologies in the criminal justice system, law enforcement, immigration, border control, elections, commercial decision-making etc. have implications for various rights under the African Charter such as the right to life; right to privacy; right to dignity; right to liberty; right to equality and non-discrimination; freedom of assembly; and the right to freedom of expression;

Noting that AI technologies, robotics and other new and emerging technologies enhanced creation and sharing of information, disinformation and unsolicited information, including non-consensual sexual images and materials for radicalization and incitement of violence, can impact freedom of expression, the right to equality and non-discrimination, freedom of assembly, privacy and right to life;

Emphasising the need for sufficient consideration of African norms, ethics, values, such as *Ubuntu*, communitarian ethos, freedom from domination of one people by another, freedom from racial and other forms of discrimination in framing of global AI governance frameworks;

Recognising the need to undertake a Study on the impact of AI, robotics and other new and emerging technologies, on human and peoples' rights in Africa;

THE COMMISSION:

1. Calls on State Parties to ensure that the development and use of AI, robotics and other new and emerging technologies is compatible with the rights and duties in the African Charter and other regional and international human rights instruments, in order to uphold human dignity, privacy, equality, non-discrimination, inclusion, diversity, safety, fairness, transparency, accountability and economic development as underlying principles that guide the development and use of AI, robotics and other new and emerging technologies.

2. Urges State Parties to ensure that all AI technologies, robotics and other new and emerging technologies that are imported from other continents are made

applicable to the African context and or adjusted to fit Africa's needs, and to give serious consideration to African values and norms in formulation of AI governance frameworks to address the global epistemic injustice that currently exists.

3.   Urges State Parties to ensure transparency in the use of AI technologies, robotics and other new and emerging technologies and that decisions made in the use of AI technologies, robotics and other new and emerging technologies are easily understandable to those affected by such decisions.

4.   Calls on State Parties to work towards a comprehensive legal and ethical governance framework for AI technologies, robotics and other new and emerging technologies so as to ensure compliance with the African Charter and other regional treaties.

5.   Calls on the African Union and regional bodies to urgently place on their agendas the rapid issue of AI technologies, robotics and other new and emerging technologies with a view to develop a regional regulatory framework that ensures that these technologies respond to the needs of the people of the continent.

6.   Appeals to State Parties to ensure that all AI technologies, robotics and other new and emerging technologies which have far reaching consequences for humans must remain under meaningful human control in order to ensure that the threat that they pose to fundamental human rights is averted. The emerging norm of maintaining meaningful human control over AI technologies, robotics and other new and emerging technologiesshould be codified as a human rights principle.

7.   Commits to undertake a study in order to further develop guidelines and norms that address issues relating to AI technologies, robotics and other new and emerging technologies and their impact on human rights in Africa working together with an African Group of Experts on AI and new technologies.

Done virtually, on 25 February 2021

African Union, Continental Artificial Intelligence Strategy: Harnessing AI for Africa's Development and Prosperity [excerpt] (2021)

2. Continental AI Strategy

2.1 Vision and Mission

The Continental AI Strategy supports the AU Agenda 2063's vision. It is rooted in Africa's unique challenges and opportunities, as the continent can leverage its youth's digital native talent, natural resources, huge market and geopolitical position to develop and promote an ethical, responsible and inclusive AI that empowers people and contributes to the continent's economic growth, social progress and cultural renaissance.

Vision

A prosperous and integrated Africa where responsible, ethical and African-centric AI is the axes of inclusive growth, resilience, socio-economic development, people empowerment and positioning the continent as a key player in the global AI landscape.

Mission

The Mission of the AI Strategy is to harness AI for accelerating social and economic transformation and promoting cultural renaissance in Africa in line with the AU Agenda 2063 and the Sustainable Development Goals, to minimise the risks that AI poses to African people and countries, and to accelerate the development of the necessary AI capabilities of the AU Member States.

2.2 Guiding Principles

The following high-level principles guide the AI Strategy:

Local First - The production, development, use and assessment of AI in Africa will be foremost to address African challenges like healthcare delivery, food security, clean energy, climate change and water management and opportunities with African solutions. The growth of local talent and ecosystems is considered paramount to advancing AI solutions that are of public value and interest, serving the Continent's needs and priorities, and respecting and preserving cultural values and customs.

People-centred - AI should promote inclusive growth, sustainable development, well-being and cultural renaissance. The production, development, use and assessment of AI in Africa will be foremost to address African challenges and opportunities with African solutions. AI solutions will address rural and remote areas specific challenges, such as agriculture, climate change (droughts and floods) and healthcare needs.

Human Rights and Human Dignity - The production, development, use and assessment of AI systems in Africa will always uphold human dignity, gender equality and respect and promote all the human rights set out under the African Charter on Human and Peoples' Rights and its subsidiary instruments, as well as the Universal Declaration on Human Rights and related instruments of international human rights law.

Peace and Prosperity - The production, development, use and assessment of AI systems in Africa shall advance peaceful and prosperous African societies, where safety and security are enjoyed by all who live in them and where the natural environment is preserved and protected.

Inclusion and Diversity - The production, development, use and assessment of AI in Africa will be inclusive, non-discriminatory, leaving no one and no place behind and benefiting everyone, and respectful of the diversity of African people, cultures, languages, gender dimensions and nations. AI will not discriminate against anyone on the basis of sex, gender, race, ethnic origin, pregnancy status, economic status, age, any form of disability, language, religion, political opinion, or any other ground as contemplated under the regional and international human rights canons. In particular, the opportunities of the AI revolution will be harnessed to empower African women.

Ethics and Transparency - The Strategy should provide guidance and recommendations to enable member States to embrace a responsible AI concept. Biasness, widening inequalities, marginalisation of certain groups who are not ready to embrace AI, loss of culture and identity, and widening of social and technological gaps should all be avoided.

Cooperation and Integration - The Continental Strategy will promote regionally integrated governance approaches and mechanisms and foster regional cooperation in advancing inclusive African AI capacities and ecosystems. Member States, AUC, AU Organs, RECs, African Institutions, and International Organisations shall cooperate to create capacity to enable African countries to self-manage their data and AI and take advantage of regional initiatives and regulated data flows to govern data appropriately.

Skills Development, Public Awareness and Education - AI solutions will be supported by formal and informal AI education to equip the African population with the necessary skills for the AI-driven future.

2.3 Overall Strategic Objectives

The Strategic Objectives foreseen to be reached by 2030 comprise the following:

i. Implement robust AI governance, regulations, standards, codes of conduct and best practices to manage AI risks and promote its growth.

ii. Promote the adoption of AI in the public sector, with a view to delivering efficient services to all citizens, businesses and others.

iii. Accelerate the integration of AI in the core sectors outlined in the Digital Transformation Strategy, notably sectors with high social and economic value, including agriculture, education, health, climate change and natural resource management, and regional peace and security.

iv. Accelerate the adoption of AI by the private sector, including small and medium-sized enterprises.

v. Create an enabling environment for a vibrant and inclusive AI start-up ecosystem focused on solving development problems.

vi. Ensure the availability of high-quality and diverse datasets to underpin AI development and ensure the availability of AI infrastructure, such as data centres, computing platforms, and IoTs, is available for data storage and management.

vii. Promote information integrity, media and information literacy.

viii. Promote diversity in AI skills and AI talent to prepare Africa's workforce of tomorrow, with particular attention to women and girls.

ix. Encourage research and innovation in AI through partnerships between academia and the private and public sectors.

x. Adopt and implement AI ethical principles that respect human rights, gender equality and African people's dignity, respect diversity, inclusivity and African culture and values.

xi. Adopt and implement safety and security in the design, development and use of AI systems.

xii. Accelerate public and private investment in AI in Africa.

xiii. Promote regional cooperation and solidarity among AU Member States to maximise the benefits of AI, minimise its risks and share capabilities and resources.

xiv. Promote African participation in shaping the global AI governance system.

xv. Stimulate AI-related partnerships between Africa and the rest of the world.

2.4 Focus Areas

The Focus Areas of the AI Strategy are:

i. Maximising the benefits of AI for social and economic development and cultural renaissance.

ii. Minimising risks and safeguarding AI development and adoption from harm to African people, societies and environments.

iii. Building capabilities in infrastructure, datasets, computing, skills and education, research and innovation, and specialised AI platforms.

iv. Fostering regional and international cooperation; and

v. Accelerating AI investment.

vi. Creating an inclusive governance and regulatory framework.

Windhoek Statement on Artificial Intelligence in Southern Africa (2022)

PREAMBLE

We, the Ministers in charge of Higher Education, ICTs, Digital Technologies, Communication, Science and Innovation from the Southern Africa countries1, gathered at the UNESCO-Southern Africa sub-Regional Forum on Artificial Intelligence (SARFAI), co-hosted by the Government of Namibia and UNESCO in Windhoek, from 7 to 9 September 2022.

Recognizing that Artificial Intelligence (AI) is fast transforming the world and the future of humanity, raising complex questions and blurring traditional boundaries of the physical, digital and biological worlds, influencing human thinking, interaction and decision-making, and raising socio-economic, ethical and political risks engendered by the use of AI systems mainly produced outside our region, that often rely on low quality and nonrepresentative data, with limited knowledge of local languages and indigenous knowledge;

Guided by the Recommendation on the Ethics of Artificial Intelligence, adopted by the General Conference of UNESCO at its 41st Session (2021), and addressed to Member States, both as AI actors and as authorities responsible for its effective implementation, and the development of the policy, and legal and regulatory frameworks needed throughout the entire AI system life cycle, to respect human dignity and exploit the full potential presented by the 4th Industrial Revolution;

Conscious of the promising regional networks of AI experts in Southern Africa, and that AI technologies are being widely used in the region, thus requires policymakers, economic actors, and people to understand how AI technologies can impact their lives, and to collaboratively engage in AI policy dialogues and governance to harness the potential of AI for the common good;

Acknowledging that the use of AI in our countries can impact human lives, especially the youth, who constitute approximately 60% of the Sub-Saharan population. AI must contribute to safeguarding human dignity and human rights, as well as gender equality, social justice, economic development, and environmental and ecosystem protection, thus contributing to building a prosperous future for our young generations;

Observing significant gaps in awareness, knowledge and human resource capabilities in the different domains of AI, including on the ethical implications, within government institutions, independent oversight bodies, human rights institutions, judiciary and law enforcement bodies, as well as the private sector and the general public in Southern Africa;

Recognizing the existence of gender bias in artificial intelligence and its profound impact on the lives of women and girls, linked to the low numbers of women in AI and emerging technologies-related research, development and innovation and gender bias in data, and its impact on the whole of society;

Bearing in mind the Universal Declaration of Human Rights (1948) and other international human rights frameworks, whereby Member States have pledged themselves to achieving, in cooperation with the United Nations, the promotion of universal respect for, and observance of human rights and fundamental freedoms;

Affirming the UNESCO Operational Strategy for Priority Africa (2022-2029) with a flagship programme on "Harnessing new and emerging technologies for sustainable development in Africa, including through the implementation of the recommendation on the ethics of AI", and taking note of the findings of UNESCO's Artificial Intelligence Needs Assessment Survey in Africa (2021), and of UNESCO's Landscape Study of AI Policies and Use in Southern Africa (2022), that indicate the need to strengthen AI governance and capacities, and to invest more efforts in AI education, research and training;

Taking into account the SADC Protocol on Science, Technology and Innovation (2008), as well as the Decision of the Joint Meeting of SADC Ministers of Education and Training, Science, Technology and Innovation, in Lilongwe, Malawi (2022), "to review its implementation to align with the new policy environment in relation to STI";

Supporting the SADC Industrialization Strategy and Roadmap and the SADC Regional Indicative Strategic Development Plan 2020-2030 which call for the mobilization of financial resources for industrial revolution and substantial higher investment in infrastructure, upgrading and diversifying the capital stock and the provision of the high-technology skills necessary in modern industry, through science, technology and innovation;

Highlighting some relevant regional frameworks, such as the model laws on Data Protection, Cybercrime, e-Commerce, Electronic Transactions, as well as the SADC Cyber Infrastructure Framework, as strong and useful guiding legal frameworks that are relevant to the current and future legislation on AI;

Emphasizing the effective use of AI in disaster risk reduction and resilience-building in supporting both national and regional strategies and the Action Plans in the areas of climate change, and green, blue and circular economies, which will require investment in infrastructure; credible data collection; support of open science; international cooperation; integration of indigenous knowledge and citizen science into AI development and deployment;

Congratulate Southern African countries that have proactively put in place the relevant national advisory Task Forces and committees to ascertain their positioning in the 4th Industrial Revolution;

Reaffirm the Outcome Statement of the UNESCO Forum on Artificial Intelligence in Africa, held in Benguérir, Morocco (2018), that calls for the organization of sub-regional forums in Africa to facilitate exchanges, elaboration of strategic frameworks and action plans in view of an AI Strategy for Africa and the draft SADC Digital Transformation Strategy;

Express our gratitude to the Government of Namibia and UNESCO for organizing the Southern Africa subregional Forum on AI (SARFAI) to facilitate both technical and policy dialogues on AI; for hosting the UNESCO World Press Freedom Day International Conference (2021), which adapted the Windhoek 30+ Declaration; and the Africa Ministers of ICT Forum (2021), which discussed the need for multistakeholder approaches to AI governance and the need for transparency and accountability of digital technology and internet companies;

Recommend this Statement to advance standard-setting initiatives, foster cooperation and exchanges of expertise involving all AI stakeholders, as well as to strengthen cooperation between Southern African countries and UNESCO, as foreseen through the SADC-UNESCO Joint Programme of Action 2022-2025, including through the establishment of a Southern African coordination mechanism for the Implementation of the UNESCO Recommendation on the Ethics of AI.

RECOMMENDATIONS

AI and Data Governance

• Develop knowledge in line with African ethical values aimed at understanding the socio-technological implications of AI in order to work with different stakeholders to co-create Africa-centric governance frameworks, with a holistic approach to the sustainable and ethical development and use of AI;

• Adopt anticipatory approaches to AI policymaking and incorporate implementation plans, based on strategic national intent, expressed through well-configured strategies and regulatory frameworks;

• Review, update and develop regulatory and legal frameworks, including science, technology and innovation (STI), ICT, digital transformation, cybersecurity policies, 4IR policies, and sector-specific strategies (such as e-health, e-governance, e-finance and e-learning) in relation to AI;

• Consider legislative tools to safeguard and secure the use of high-risk AI, including facial recognition technologies, and AI use in healthcare and education;

• Facilitate policy dialogue on national, regional and global AI frameworks and strategies;

• Enact national data protection laws to create strong personal data protection safeguards, and guarantee data free flow with trust (DFFT), based on interoperable and common standards, including between the private sector, academia and governments;

• Establish Frameworks for Data Governance, with special attention to essential data sets, in national data centres of Member States, for state sovereignty in a sustainable manner;

• Promote transparency of AI algorithms to allow for public oversight and avoid embedded biases;

• Mitigate the digital divide by fostering open and competitive markets, taking into account vulnerabilities of developing countries, such as the lack of infrastructure, human capacity and regulatory frameworks;

• Promote Pan-African data sets as public assets to create platforms to foster scientific research in line with the UNESCO Recommendation on Open Science, and to protect such data sets from exploitation by third parties;

• Support the creation and use of localized and unbiased datasets for training AI models for inclusive and nondiscriminatory outcomes, and reinforcing awareness about, and protection of data ownership; and

• Encourage the development of standards for AI and digital technologies, including facilitating the participation of African experts and standard-setting bodies in global standard-setting processes.

Capacity-Building and Awareness-Raising

• Launch programmes to promote public awareness and literacy to respond to African realities related to the ethical development and use of AI, including for youth, women, and vulnerable and marginalized groups;

• Strengthen the capacities of governments, civil society and the private sector to understand and make sense of AI technologies and applications;

• Strengthen the capacities of governments, civil society and the private sector to develop, use and deploy AI ethically for the benefit of societies and people;

• Establish and enhance the capacity of policy, regulatory and enforcement institutions, also through targeted skills development and support for AI governance; and through leveraging on and including African experts in UNESCO "AI experts without borders" initiative with the support of SADC; and

• Pilot and deploy innovative tools and approaches to AI capacity-building, including UNESCO's Readiness Assessment Methodology, Ethical Impact Assessment and Competency Framework on AI and Digital Transformation for Civil Servants.

Investment and Infrastructure

• Expand investments towards infrastructure development, to address issues such as access to electricity, connectivity, spectrum, data centres, cybersecurity (CSIRTs), cloud and high-performance computing and user terminal devices, including by leveraging SADC regional mechanisms for such infrastructure; usage and affordability gap; and the unequal distribution of unpaid work and of the caring responsibilities in our societies." This session will highlight the context of women in AI in Southern Africa and discuss strategies to strengthen the contributions of women in AI and to eliminate gender biases in AI Systems.

• Build upon the regional and national high-performance computing platforms and infrastructures already in place for the role that these infrastructures play in running AI-based applications and services in relevant areas such as health, agriculture, climate change, tourism and weather forecasting;

• Develop innovative funding mechanisms that build upon models such as the Universal Service Fund (USF) in the telecom sector, to support innovation ecosystems and start-ups;

• Establish ICT, innovation and technology hubs, including in rural and economically disadvantaged areas aimed at facilitating access, use, development and deployment of digital technologies, including AI, in supporting the creation of smart communities;

• Support the establishment of AI incubation centres, digital skills and talent hubs, and attract venture capital funding to encourage digital entrepreneurship initiatives and the incubation of technology-based companies for addressing the social problem of unemployment;

• Establish well-balanced strategies dedicating adequate domestic public funds to support the ethical development and deployment of AI, and mobilize development cooperation partners to reinforce domestic priorities to enable Africa to leapfrog in science, technology and innovation; and

• Strengthen mutual understanding of the AI-related investment needs among technology experts and policymakers who determine the funding of these initiatives.

Education

• Leverage Africa's fastest growing and youngest population in the world, turning this demographic dividend into an economic opportunity by providing quality education and lifelong learning opportunities for all, through leveraging the potential of digital technologies, with a specific focus on AI technologies, especially for girls and women;

• Strengthen ongoing regional initiatives regarding digital infrastructure to support science, technology, innovation and higher education ecosystem, such as the NRENs (National Research and Education Networks) of the Member States and UbuntuNet, for its role in promoting digital transformation initiatives in education that can be reinforced by AI;

• Promote the decolonization of the design and application of AI technologies, including by decolonizing education at all levels, developing Africa-centric AI curricula and involving communities to co-design inclusive and ethical AI applications, taking into account heritage and indigenous knowledge systems;

• Invest in and promote digital, media, information literacy and AI-related skills to strengthen critical thinking and competencies and empower human capital;

• Develop transdisciplinary curricula, for instance, Science, Technology, Engineering, Arts and Mathematics (STEAM) at all levels of education, including TVET, and remove disciplinary barriers among students in the domain of AI, especially coding, also by means of streamlining the approval processes for curricula, in order to keep pace with the rapid development of AI and other emerging technologies;

• Develop teacher competencies in AI-related skills, responding to the needs of African countries, and in line with UNESCO's ICT competency framework and the UNESCO Recommendation on the Ethics of AI; and

• Establish and support African AI and emerging technology institutes, centres of excellence (CoE), centres of specialisation (CoS) and regional research clusters. This is to expand coordination and leverage synergies between national centres, and foster strong links between education, research and commercialisation.

Research, Development and Innovation

• Increase investment in ethical AI-related research, development and innovation;

• Promote context-relevant and evidence-based research and innovation to inform AI governance and the development of applications in education, health and agriculture, among others;

• Identify, nurture and retain AI talents, including through financial incentives and support to develop relevant AI solutions; and

• Facilitate academic and student exchanges by reducing barriers to regional mobility, including through innovative initiatives such as the "Africa Innovation by Bus", implemented by the South African University of Johannesburg.

Environment and Disaster Risk Reduction

• Leverage AI to combat climate change, preserve biodiversity and address complex socio-environment challenges, such as mining and extraction of natural resources and conflict between humans and the wildlife;

• Promote environmental conservation and sustainability, including by leveraging indigenous knowledge and encouraging citizen-based science;

• Conduct a comprehensive assessment and identify opportunities to use AI for risk mitigation and disaster management; and

• Promote and implement the Recommendations on the Ethics of AI and on Open Science, leveraging initiatives and networks such as the Africa Open Science Platform.

Gender

• Support the development and implementation of gender-responsive policies and programmes, including programmes that enhance girls' and women's participation in Science, Technology, Engineering, Arts and Mathematics (STEAM), encourage female entrepreneurship, leadership and management in industries and regulatory entities, as guided by the Charter of the SADC Women in Science, Engineering, and Technology Organization (SADC WISETO);

• Promote the generation and use of data sets that reflect diversity and promote gender inclusion; and

• Establish, support and facilitate civil society organisations and networks to connect and empower women around the world, such as the Global Network of Women for Ethics of AI.

Collaboration and Partnerships

• Implement multistakeholder approaches to AI governance, such as the triple, quadruple and quintuple helix models to bring together the private sector, academia, civil society and government; in particular, ensure the engagement of youth organisations, gender equality advocates and stakeholders, small and medium enterprises, persons with disabilities, environment and climate change experts and under-represented groups in multistakeholder processes;

• Create linkages between different AI fora at a technical and policy level to reinforce the Pan-African and regional knowledge networks of experts working on AI, and facilitate the participation of women and youth;

• Promote regional and sub-regional approaches, platforms and mechanisms, such as a steering committee, to coordinate the implementation of the Recommendation on the Ethics of AI (2021), in order to address the issues, gaps and challenges identified in this document; to enhance policy harmonization and coordination across national borders; and to improve resource mobilization in collaboration with SADC and the AU;

• Promote the active and mutually beneficial engagement of African countries in multilateral platforms such as the Intergovernmental Bioethics Committee, and global initiatives, including the Global Forum on the Ethics of AI, an annual high-level event to advance state-of-the-art knowledge and skills on the challenges raised by AI-related technologies;

• Support regional alignment on key policy areas, including the AUC Data Policy Framework, the AU Digital Transformation Strategy, the Draft AU AI Continental Strategy, the Draft SADC Digital Transformation Strategy, and the UNESCO Recommendation on the Ethics of AI; and

• Strengthen mutually beneficial partnerships with the private sector for ethical AI development and use.

SCOPE OF APPLICATION

The Statement, in line with the Recommendation on the Ethics of AI, refers to AI as "systems which have the capacity to process data and information in a way that resembles intelligent behaviour, and typically includes aspects of reasoning, learning, perception, prediction, planning or control".

This Statement aims to:(i) contribute to the promotion of scientific research, technical and policy dialogues, exchanges of knowledge and practices on AI among professional networks in Southern Africa; (ii) guide the Member States in their policymaking and legislation with actionable recommendations; and (iii) facilitate SADC's design and implementation of strategic orientations and policies on AI, in the framework of the SADC Regional Indicative Strategic Development Plan (RISDP 2020-2030).

This Statement shall be submitted to the 2023 Joint Meeting of SADC Ministers responsible for Education and Training and Science, Technology and Innovation, for discussion and adoption, with a clear action plan, with timelines, and a monitoring and evaluation framework that ensures our shared accountability in the execution of actions.

The implementation of the recommendations contained in this Statement shall benefit from the technical support of international cooperating partners under the framework of the SADC-UNESCO Joint Programme of Action 2022-2025, regarding AI governance and capacity-building efforts.

UNESCO, Recommendation on the Ethics of Artificial Intelligence [excerpt] (2022)

III. Values And Principles

9. The values and principles included below should be respected by all actors in the AI system life cycle, in the first place and, where needed and appropriate, be promoted through amendments to the existing and elaboration of new legislation, regulations and business guidelines. This must comply with international law, including the United Nations Charter and Member States' human rights obligations, and should be in line with internationally agreed social, political, environmental, educational, scientific and economic sustainability objectives, such as the United Nations Sustainable Development Goals (SDGs).

10. Values play a powerful role as motivating ideals in shaping policy measures and legal norms. While the set of values outlined below thus inspires desirable behaviour and represents the foundations of principles, the principles unpack the values underlying them more concretely so that the values can be more easily operationalized in policy statements and actions.

11. While all the values and principles outlined below are desirable per se, in any practical contexts, there may be tensions between these values and principles. In any given situation, a contextual assessment will be necessary to manage potential tensions, taking into account the principle of proportionality and in compliance with human rights and fundamental freedoms. In all cases, any possible limitations on human rights and fundamental freedoms must have a lawful basis, and be reasonable, necessary and proportionate, and consistent with States' obligations under international law. To navigate such scenarios judiciously will typically require engagement with a broad range of appropriate stakeholders, making use of social dialogue, as well as ethical deliberation, due diligence and impact assessment.

12. The trustworthiness and integrity of the life cycle of AI systems is essential to ensure that AI technologies will work for the good of humanity, individuals, societies and the environment and ecosystems, and embody the values and principles set out in this Recommendation. People should have good reason to trust that AI systems can bring individual and shared benefits, while adequate measures are taken to mitigate risks. An essential requirement for trustworthiness is that, throughout their life cycle, AI systems are subject to thorough monitoring by the relevant stakeholders as appropriate. As trustworthiness is an outcome of the

operationalization of the principles in this document, the policy actions proposed in this Recommendation are all directed at promoting trustworthiness in all stages of the AI system life cycle.

III. 1 Values

Respect, protection and promotion of human rights and fundamental freedoms and human dignity

13. The inviolable and inherent dignity of every human constitutes the foundation for the universal, indivisible, inalienable, interdependent and interrelated system of human rights and fundamental freedoms. Therefore, respect, protection and promotion of human dignity and rights as established by international law, including international human rights law, is essential throughout the life cycle of AI systems. Human dignity relates to the recognition of the intrinsic and equal worth of each individual human being, regardless of race, colour, descent, gender, age, language, religion, political opinion, national origin, ethnic origin, social origin, economic or social condition of birth, or disability and any other grounds.

14. No human being or human community should be harmed or subordinated, whether physically, economically, socially, politically, culturally or mentally during any phase of the life cycle of AI systems. Throughout the life cycle of AI systems, the quality of life of human beings should be enhanced, while the definition of "quality of life" should be left open to individuals or groups, as long as there is no violation or abuse of human rights and fundamental freedoms, or the dignity of humans in terms of this definition.

15. Persons may interact with AI systems throughout their life cycle and receive assistance from them, such as care for vulnerable people or people in vulnerable situations, including but not limited to children, older persons, persons with disabilities or the ill. Within such interactions, persons should never be objectified, nor should their dignity be otherwise undermined, or human rights and fundamental freedoms violated or abused.

16. Human rights and fundamental freedoms must be respected, protected and promoted throughout the life cycle of AI systems. Governments, private sector, civil society, international organizations, technical communities and academia must respect human rights instruments and frameworks in their interventions in the processes surrounding the life cycle of AI systems. New technologies need to provide new means to advocate, defend and exercise human rights and not to infringe them.

Environment and ecosystem flourishing

17. Environmental and ecosystem flourishing should be recognized, protected and promoted through the life cycle of AI systems. Furthermore, environment and ecosystems are the existential necessity for humanity and other living beings to be able to enjoy the benefits of advances in AI.

18. All actors involved in the life cycle of AI systems must comply with applicable international law and domestic legislation, standards and practices, such as precaution, designed for environmental and ecosystem protection and restoration, and sustainable development. They should reduce the environmental impact of AI systems, including but not limited to its carbon footprint, to ensure the minimization of climate change and environmental risk factors, and prevent the unsustainable exploitation, use and transformation of natural resources contributing to the deterioration of the environment and the degradation of ecosystems.

Ensuring diversity and inclusiveness

19. Respect, protection and promotion of diversity and inclusiveness should be ensured throughout the life cycle of AI systems, consistent with international law, including human rights law. This may be done by promoting active participation of all individuals or groups regardless of race, colour, descent, gender, age, language, religion, political opinion, national origin, ethnic origin, social origin, economic or social condition of birth, or disability and any other grounds.

20. The scope of lifestyle choices, beliefs, opinions, expressions or personal experiences, including the optional use of AI systems and the co-design of these architectures should not be restricted during any phase of the life cycle of AI systems.

21. Furthermore, efforts, including international cooperation, should be made to overcome, and never take advantage of, the lack of necessary technological infrastructure, education and skills, as well as legal frameworks, particularly in LMICs, LDCs, LLDCs and SIDS, affecting communities.

Living in peaceful, just and interconnected societies

22. AI actors should play a participative and enabling role to ensure peaceful and just societies, which is based on an interconnected future for the benefit of all, consistent with human rights and fundamental freedoms. The value of living in peaceful and just societies points to the potential of AI systems to contribute throughout their life cycle to the interconnectedness of all living creatures with each other and with the natural environment.

23. The notion of humans being interconnected is based on the knowledge that every human belongs to a greater whole, which thrives when all its constituent parts are enabled to thrive. Living in peaceful, just and interconnected societies requires an organic, immediate, uncalculated bond of solidarity, characterized by a permanent search for peaceful relations, tending towards care for others and the natural environment in the broadest sense of the term.

24. This value demands that peace, inclusiveness and justice, equity and interconnectedness should be promoted throughout the life cycle of AI systems, in so far as the processes of the life cycle of AI systems should not segregate, objectify or undermine freedom and autonomous decision-making as well as the safety of human beings and communities, divide and turn individuals and groups against each other, or threaten the coexistence between humans, other living beings and the natural environment.

III. 2 Principles

Proportionality and Do No Harm

25. It should be recognized that AI technologies do not necessarily, per se, ensure human and environmental and ecosystem flourishing. Furthermore, none of the processes related to the AI system life cycle shall exceed what is necessary to achieve legitimate aims or objectives and should be appropriate to the context. In the event of possible occurrence of any harm to human beings, human rights and fundamental freedoms, communities and society at large or the environment and ecosystems, the implementation of procedures for risk assessment and the adoption of measures in order to preclude the occurrence of such harm should be ensured.

26. The choice to use AI systems and which AI method to use should be justified in the following ways: (a) the AI method chosen should be appropriate and proportional to achieve a given legitimate aim; (b) the AI method chosen should not infringe upon the foundational values captured in this document, in particular, its use must not violate or abuse human rights; and (c) the AI method should be appropriate to the context and should be based on rigorous scientific foundations. In scenarios where decisions are understood to have an impact that is irreversible or difficult to reverse or may involve life and death decisions, final human determination should apply. In particular, AI systems should not be used for social scoring or mass surveillance purposes.

Safety and security

27. Unwanted harms (safety risks), as well as vulnerabilities to attack (security risks) should be avoided and should be addressed, prevented and eliminated throughout

the life cycle of AI systems to ensure human, environmental and ecosystem safety and security. Safe and secure AI will be enabled by the development of sustainable, privacy protective data access frameworks that foster better training and validation of AI models utilizing quality data.

Fairness and non-discrimination

28. AI actors should promote social justice and safeguard fairness and non-discrimination of any kind in compliance with international law. This implies an inclusive approach to ensuring that the benefits of AI technologies are available and accessible to all, taking into consideration the specific needs of different age groups, cultural systems, different language groups, persons with disabilities, girls and women, and disadvantaged, marginalized and vulnerable people or people in vulnerable situations. Member States should work to promote inclusive access for all, including local communities, to AI systems with locally relevant content and services, and with respect for multilingualism and cultural diversity. Member States should work to tackle digital divides and ensure inclusive access to and participation in the development of AI. At the national level, Member States should promote equity between rural and urban areas, and among all persons regardless of race, colour, descent, gender, age, language, religion, political opinion, national origin, ethnic origin, social origin, economic or social condition of birth, or disability and any other grounds, in terms of access to and participation in the AI system life cycle. At the international level, the most technologically advanced countries have a responsibility of solidarity with the least advanced to ensure that the benefits of AI technologies are shared such that access to and participation in the AI system life cycle for the latter contributes to a fairer world order with regard to information, communication, culture, education, research and socio-economic and political stability.

29. AI actors should make all reasonable efforts to minimize and avoid reinforcing or perpetuating discriminatory or biased applications and outcomes throughout the life cycle of the AI system to ensure fairness of such systems. Effective remedy should be available against discrimination and biased algorithmic determination.

30. Furthermore, digital and knowledge divides within and between countries need to be addressed throughout an AI system life cycle, including in terms of access and quality of access to technology and data, in accordance with relevant national, regional and international legal frameworks, as well as in terms of connectivity, knowledge and skills and meaningful participation of the affected communities, such that every person is treated equitably.

Sustainability

31. The development of sustainable societies relies on the achievement of a complex set of objectives on a continuum of human, social, cultural, economic and environmental dimensions. The advent of AI technologies can either benefit sustainability objectives or hinder their realization, depending on how they are applied across countries with varying levels of development. The continuous assessment of the human, social, cultural, economic and environmental impact of AI technologies should therefore be carried out with full cognizance of the implications of AI technologies for sustainability as a set of constantly evolving goals across a range of dimensions, such as currently identified in the Sustainable Development Goals (SDGs) of the United Nations.

Right to Privacy, and Data Protection

32. Privacy, a right essential to the protection of human dignity, human autonomy and human agency, must be respected, protected and promoted throughout the life cycle of AI systems. It is important that data for AI systems be collected, used, shared, archived and deleted in ways that are consistent with international law and in line with the values and principles set forth in this Recommendation, while respecting relevant national, regional and international legal frameworks.

33. Adequate data protection frameworks and governance mechanisms should be established in a multi-stakeholder approach at the national or international level, protected by judicial systems, and ensured throughout the life cycle of AI systems. Data protection frameworks and any related mechanisms should take reference from international data protection principles and standards concerning the collection, use and disclosure of personal data and exercise of their rights by data subjects while ensuring a legitimate aim and a valid legal basis for the processing of personal data, including informed consent.

34. Algorithmic systems require adequate privacy impact assessments, which also include societal and ethical considerations of their use and an innovative use of the privacy by design approach. AI actors need to ensure that they are accountable for the design and implementation of AI systems in such a way as to ensure that personal information is protected throughout the life cycle of the AI system.

Human oversight and determination

35. Member States should ensure that it is always possible to attribute ethical and legal responsibility for any stage of the life cycle of AI systems, as well as in cases of remedy related to AI systems, to physical persons or to existing legal entities.

Human oversight refers thus not only to individual human oversight, but to inclusive public oversight, as appropriate.

36. It may be the case that sometimes humans would choose to rely on AI systems for reasons of efficacy, but the decision to cede control in limited contexts remains that of humans, as humans can resort to AI systems in decision-making and acting, but an AI system can never replace ultimate human responsibility and accountability. As a rule, life and death decisions should not be ceded to AI systems.

Transparency and explainability

37. The transparency and explainability of AI systems are often essential preconditions to ensure the respect, protection and promotion of human rights, fundamental freedoms and ethical principles. Transparency is necessary for relevant national and international liability regimes to work effectively. A lack of transparency could also undermine the possibility of effectively challenging decisions based on outcomes produced by AI systems and may thereby infringe the right to a fair trial and effective remedy, and limits the areas in which these systems can be legally used.

38. While efforts need to be made to increase transparency and explainability of AI systems, including those with extra-territorial impact, throughout their life cycle to support democratic governance, the level of transparency and explainability should always be appropriate to the context and impact, as there may be a need to balance between transparency and explainability and other principles such as privacy, safety and security. People should be fully informed when a decision is informed by or is made on the basis of AI algorithms, including when it affects their safety or human rights, and in those circumstances should have the opportunity to request explanatory information from the relevant AI actor or public sector institutions. In addition, individuals should be able to access the reasons for a decision affecting their rights and freedoms, and have the option of making submissions to a designated staff member of the private sector company or public sector institution able to review and correct the decision. AI actors should inform users when a product or service is provided directly or with the assistance of AI systems in a proper and timely manner.

39. From a socio-technical lens, greater transparency contributes to more peaceful, just, democratic and inclusive societies. It allows for public scrutiny that can decrease corruption and discrimination, and can also help detect and prevent negative impacts on human rights. Transparency aims at providing appropriate information to the respective addressees to enable their understanding and foster trust. Specific to the AI system, transparency can enable people to understand how

each stage of an AI system is put in place, appropriate to the context and sensitivity of the AI system. It may also include insight into factors that affect a specific prediction or decision, and whether or not appropriate assurances (such as safety or fairness measures) are in place. In cases of serious threats of adverse human rights impacts, transparency may also require the sharing of code or datasets.

40. Explainability refers to making intelligible and providing insight into the outcome of AI systems. The explainability of AI systems also refers to the understandability of the input, output and the functioning of each algorithmic building block and how it contributes to the outcome of the systems. Thus, explainability is closely related to transparency, as outcomes and sub-processes leading to outcomes should aim to be understandable and traceable, appropriate to the context. AI actors should commit to ensuring that the algorithms developed are explainable. In the case of AI applications that impact the end user in a way that is not temporary, easily reversible or otherwise low risk, it should be ensured that the meaningful explanation is provided with any decision that resulted in the action taken in order for the outcome to be considered transparent.

41. Transparency and explainability relate closely to adequate responsibility and accountability measures, as well as to the trustworthiness of AI systems.

Responsibility and accountability

42. AI actors and Member States should respect, protect and promote human rights and fundamental freedoms, and should also promote the protection of the environment and ecosystems, assuming their respective ethical and legal responsibility, in accordance with national and international law, in particular Member States' human rights obligations, and ethical guidance throughout the life cycle of AI systems, including with respect to AI actors within their effective territory and control. The ethical responsibility and liability for the decisions and actions based in any way on an AI system should always ultimately be attributable to AI actors corresponding to their role in the life cycle of the AI system.

43. Appropriate oversight, impact assessment, audit and due diligence mechanisms, including whistle-blowers' protection, should be developed to ensure accountability for AI systems and their impact throughout their life cycle. Both technical and institutional designs should ensure auditability and traceability of (the working of) AI systems in particular to address any conflicts with human rights norms and standards and threats to environmental and ecosystem well-being.

Awareness and literacy

44. Public awareness and understanding of AI technologies and the value of data should be promoted through open and accessible education, civic engagement, digital skills and AI ethics training, media and information literacy and training led jointly by governments, intergovernmental organizations, civil society, academia, the media, community leaders and the private sector, and considering the existing linguistic, social and cultural diversity, to ensure effective public participation so that all members of society can take informed decisions about their use of AI systems and be protected from undue influence.

45. Learning about the impact of AI systems should include learning about, through and for human rights and fundamental freedoms, meaning that the approach and understanding of AI systems should be grounded by their impact on human rights and access to rights, as well as on the environment and ecosystems.

Multi-stakeholder and adaptive governance and collaboration

46. International law and national sovereignty must be respected in the use of data. That means that States, complying with international law, can regulate the data generated within or passing through their territories, and take measures towards effective regulation of data, including data protection, based on respect for the right to privacy in accordance with international law and other human rights norms and standards.

47. Participation of different stakeholders throughout the AI system life cycle is necessary for inclusive approaches to AI governance, enabling the benefits to be shared by all, and to contribute to sustainable development. Stakeholders include but are not limited to governments, intergovernmental organizations, the technical community, civil society, researchers and academia, media, education, policy-makers, private sector companies, human rights institutions and equality bodies, anti-discrimination monitoring bodies, and groups for youth and children. The adoption of open standards and interoperability to facilitate collaboration should be in place. Measures should be adopted to take into account shifts in technologies, the emergence of new groups of stakeholders, and to allow for meaningful participation by marginalized groups, communities and individuals and, where relevant, in the case of Indigenous Peoples, respect for the self-governance of their data.

IV Areas of Policy Action

48. The policy actions described in the following policy areas operationalize the values and principles set out in this Recommendation. The main action is for Member States to put in place effective measures, including, for example, policy

frameworks or mechanisms, and to ensure that other stakeholders, such as private sector companies, academic and research institutions, and civil society adhere to them by, among other actions, encouraging all stakeholders to develop human rights, rule of law, democracy, and ethical impact assessment and due diligence tools in line with guidance including the United Nations Guiding Principles on Business and Human Rights. The process for developing such policies or mechanisms should be inclusive of all stakeholders and should take into account the circumstances and priorities of each Member State. UNESCO can be a partner and support Member States in the development as well as monitoring and evaluation of policy mechanisms.

49. UNESCO recognizes that Member States will be at different stages of readiness to implement this Recommendation, in terms of scientific, technological, economic, educational, legal, regulatory, infrastructural, societal, cultural and other dimensions. It is noted that "readiness" here is a dynamic status. In order to enable the effective implementation of this Recommendation, UNESCO will therefore: (1) develop a readiness assessment methodology to assist interested Member States in identifying their status at specific moments of their readiness trajectory along a continuum of dimensions; and (2) ensure support for interested Member States in terms of developing a UNESCO methodology for Ethical Impact Assessment (EIA) of AI technologies, sharing of best practices, assessment guidelines and other mechanisms and analytical work.

POLICY AREA 1: ETHICAL IMPACT ASSESSMENT

50. Member States should introduce frameworks for impact assessments, such as ethical impact assessment, to identify and assess benefits, concerns and risks of AI systems, as well as appropriate risk prevention, mitigation and monitoring measures, among other assurance mechanisms. Such impact assessments should identify impacts on human rights and fundamental freedoms, in particular but not limited to the rights of marginalized and vulnerable people or people in vulnerable situations, labour rights, the environment and ecosystems and ethical and social implications, and facilitate citizen participation in line with the values and principles set forth in this Recommendation.

51. Member States and private sector companies should develop due diligence and oversight mechanisms to identify, prevent, mitigate and account for how they address the impact of AI systems on the respect for human rights, rule of law and inclusive societies. Member States should also be able to assess the socioeconomic impact of AI systems on poverty and ensure that the gap between people living in wealth and poverty, as well as the digital divide among and within countries, are not increased with the massive adoption of AI technologies at present and in the future.

In order to do this, in particular, enforceable transparency protocols should be implemented, corresponding to the access to information, including information of public interest held by private entities. Member States, private sector companies and civil society should investigate the sociological and psychological effects of AI-based recommendations on humans in their decision-making autonomy. AI systems identified as potential risks to human rights should be broadly tested by AI actors, including in real-world conditions if needed, as part of the Ethical Impact Assessment, before releasing them in the market.

52. Member States and business enterprises should implement appropriate measures to monitor all phases of an AI system life cycle, including the functioning of algorithms used for decision-making, the data, as well as AI actors involved in the process, especially in public services and where direct end-user interaction is needed, as part of ethical impact assessment. Member States' human rights law obligations should form part of the ethical aspects of AI system assessments.

53. Governments should adopt a regulatory framework that sets out a procedure, particularly for public authorities, to carry out ethical impact assessments on AI systems to predict consequences, mitigate risks, avoid harmful consequences, facilitate citizen participation and address societal challenges. The assessment should also establish appropriate oversight mechanisms, including auditability, traceability and explainability, which enable the assessment of algorithms, data and design processes, as well as include external review of AI systems. Ethical impact assessments should be transparent and open to the public, where appropriate. Such assessments should also be multidisciplinary, multi-stakeholder, multicultural, pluralistic and inclusive. The public authorities should be required to monitor the AI systems implemented and/or deployed by those authorities by introducing appropriate mechanisms and tools.

POLICY AREA 2: ETHICAL GOVERNANCE AND STEWARDSHIP

54. Member States should ensure that AI governance mechanisms are inclusive, transparent, multidisciplinary, multilateral (this includes the possibility of mitigation and redress of harm across borders) and multi-stakeholder. In particular, governance should include aspects of anticipation, and effective protection, monitoring of impact, enforcement and redress.

55. Member States should ensure that harms caused through AI systems are investigated and redressed, by enacting strong enforcement mechanisms and remedial actions, to make certain that human rights and fundamental freedoms and the rule of law are respected in the digital world and in the physical world. Such mechanisms and actions should include remediation mechanisms provided by

private and public sector companies. The auditability and traceability of AI systems should be promoted to this end. In addition, Member States should strengthen their institutional capacities to deliver on this commitment and should collaborate with researchers and other stakeholders to investigate, prevent and mitigate any potentially malicious uses of AI systems.

56. Member States are encouraged to develop national and regional AI strategies and to consider forms of soft governance such as a certification mechanism for AI systems and the mutual recognition of their certification, according to the sensitivity of the application domain and expected impact on human rights, the environment and ecosystems, and other ethical considerations set forth in this Recommendation. Such a mechanism might include different levels of audit of systems, data, and adherence to ethical guidelines and to procedural requirements in view of ethical aspects. At the same time, such a mechanism should not hinder innovation or disadvantage small and medium enterprises or start-ups, civil society as well as research and science organizations, as a result of an excessive administrative burden. These mechanisms should also include a regular monitoring component to ensure system robustness and continued integrity and adherence to ethical guidelines over the entire life cycle of the AI system, requiring re-certification if necessary.

57. Member States and public authorities should carry out transparent self-assessment of existing and proposed AI systems, which, in particular, should include the assessment of whether the adoption of AI is appropriate and, if so, should include further assessment to determine what the appropriate method is, as well as assessment as to whether such adoption would result in violations or abuses of Member States' human rights law obligations, and if that is the case, prohibit its use.

58. Member States should encourage public entities, private sector companies and civil society organizations to involve different stakeholders in their AI governance and to consider adding the role of an independent AI Ethics Officer or some other mechanism to oversee ethical impact assessment, auditing and continuous monitoring efforts and ensure ethical guidance of AI systems. Member States, private sector companies and civil society organizations, with the support of UNESCO, are encouraged to create a network of independent AI Ethics Officers to give support to this process at national, regional and international levels.

59. Member States should foster the development of, and access to, a digital ecosystem for ethical and inclusive development of AI systems at the national level, including to address gaps in access to the AI system life cycle, while contributing to international collaboration. Such an ecosystem includes, in particular, digital technologies and infrastructure, and mechanisms for sharing AI knowledge, as appropriate.

60. Member States should establish mechanisms, in collaboration with international organizations, transnational corporations, academic institutions and civil society, to ensure the active participation of all Member States, especially LMICs, in particular LDCs, LLDCs and SIDS, in international discussions concerning AI governance. This can be through the provision of funds, ensuring equal regional participation, or any other mechanisms. Furthermore, in order to ensure the inclusiveness of AI fora, Member States should facilitate the travel of AI actors in and out of their territory, especially from LMICs, in particular LDCs, LLDCs and SIDS, for the purpose of participating in these fora.

61. Amendments to the existing or elaboration of new national legislation addressing AI systems must comply with Member States' human rights law obligations and promote human rights and fundamental freedoms throughout the AI system life cycle. Promotion thereof should also take the form of governance initiatives, good exemplars of collaborative practices regarding AI systems, and national and international technical and methodological guidelines as AI technologies advance. Diverse sectors, including the private sector, in their practices regarding AI systems must respect, protect and promote human rights and fundamental freedoms using existing and new instruments in combination with this Recommendation.

62. Member States that acquire AI systems for human rights sensitive use cases, such as law enforcement, welfare, employment, media and information providers, health care and the independent judiciary system should provide mechanisms to monitor the social and economic impact of such systems by appropriate oversight authorities, including independent data protection authorities, sectoral oversight and public bodies responsible for oversight. 63. Member States should enhance the capacity of the judiciary to make decisions related to AI systems as per the rule of law and in line with international law and standards, including in the use of AI systems in their deliberations, while ensuring that the principle of human oversight is upheld. In case AI systems are used by the judiciary, sufficient safeguards are needed to guarantee inter alia the protection of fundamental human rights, the rule of law, judicial independence as well as the principle of human oversight, and to ensure a trustworthy, public interest-oriented and human-centric development and use of AI systems in the judiciary. 64. Member States should ensure that governments and multilateral organizations play a leading role in ensuring the safety and security of AI systems, with multistakeholder participation. Specifically, Member States, international organizations and other relevant bodies should develop international standards that describe measurable, testable levels of safety and transparency, so that systems can be objectively assessed and levels of compliance determined. Furthermore, Member States and business enterprises should continuously support strategic research on potential safety and security risks

of AI technologies and should encourage research into transparency and explainability, inclusion and literacy by putting additional funding into those areas for different domains and at different levels, such as technical and natural language. 65. Member States should implement policies to ensure that the actions of AI actors are consistent with international human rights law, standards and principles throughout the life cycle of AI systems, while taking into full consideration the current cultural and social diversities, including local customs and religious traditions, with due regard to the precedence and universality of human rights. 66. Member States should put in place mechanisms to require AI actors to disclose and combat any kind of stereotyping in the outcomes of AI systems and data, whether by design or by negligence, and to ensure that training data sets for AI systems do not foster cultural, economic or social inequalities, prejudice, the spreading of disinformation and misinformation, and disruption of freedom of expression and access to information. Particular attention should be given to regions where the data are scarce.

67. Member States should implement policies to promote and increase diversity and inclusiveness that reflect their populations in AI development teams and training datasets, and to ensure equal access to AI technologies and their benefits, particularly for marginalized groups, both from rural and urban zones.

68. Member States should develop, review and adapt, as appropriate, regulatory frameworks to achieve accountability and responsibility for the content and outcomes of AI systems at the different phases of their life cycle. Member States should, where necessary, introduce liability frameworks or clarify the interpretation of existing frameworks to ensure the attribution of accountability for the outcomes and the functioning of AI systems. Furthermore, when developing regulatory frameworks, Member States should, in particular, take into account that ultimate responsibility and accountability must always lie with natural or legal persons and that AI systems should not be given legal personality themselves. To ensure this, such regulatory frameworks should be consistent with the principle of human oversight and establish a comprehensive approach focused on AI actors and the technological processes involved across the different stages of the AI system life cycle.

69. In order to establish norms where these do not exist, or to adapt the existing legal frameworks, Member States should involve all AI actors (including, but not limited to, researchers, representatives of civil society and law enforcement, insurers, investors, manufacturers, engineers, lawyers and users). The norms can mature into best practices, laws and regulations. Member States are further encouraged to use mechanisms such as policy prototypes and regulatory sandboxes to accelerate the development of laws, regulations and policies, including regular reviews thereof, in

line with the rapid development of new technologies and ensure that laws and regulations can be tested in a safe environment before being officially adopted. Member States should support local governments in the development of local policies, regulations and laws in line with national and international legal frameworks.

70. Member States should set clear requirements for AI system transparency and explainability so as to help ensure the trustworthiness of the full AI system life cycle. Such requirements should involve the design and implementation of impact mechanisms that take into consideration the nature of application domain, intended use, target audience and feasibility of each particular AI system.

POLICY AREA 3: DATA POLICY

71. Member States should work to develop data governance strategies that ensure the continual evaluation of the quality of training data for AI systems including the adequacy of the data collection and selection processes, proper data security and protection measures, as well as feedback mechanisms to learn from mistakes and share best practices among all AI actors.

72. Member States should put in place appropriate safeguards to protect the right to privacy in accordance with international law, including addressing concerns such as surveillance. Member States should, among others, adopt or enforce legislative frameworks that provide appropriate protection, compliant with international law. Member States should strongly encourage all AI actors, including business enterprises, to follow existing international standards and, in particular, to carry out adequate privacy impact assessments, as part of ethical impact assessments, which take into account the wider socio-economic impact of the intended data processing, and to apply privacy by design in their systems. Privacy should be respected, protected and promoted throughout the life cycle of AI systems.

73. Member States should ensure that individuals retain rights over their personal data and are protected by a framework, which notably foresees: transparency; appropriate safeguards for the processing of sensitive data; an appropriate level of data protection; effective and meaningful accountability schemes and mechanisms; the full enjoyment of the data subjects' rights and the ability to access and erase their personal data in AI systems, except for certain circumstances in compliance with international law; an appropriate level of protection in full compliance with data protection legislation where data are being used for commercial purposes such as enabling micro-targeted advertising, transferred crossborder; and an effective independent oversight as part of a data governance mechanism which keeps

individuals in control of their personal data and fosters the benefits of a free flow of information internationally, including access to data.

74. Member States should establish their data policies or equivalent frameworks, or reinforce existing ones, to ensure full security for personal data and sensitive data, which, if disclosed, may cause exceptional damage, injury or hardship to individuals. Examples include data relating to offences, criminal proceedings and convictions, and related security measures; biometric, genetic and health data; and -personal data such as that relating to race, colour, descent, gender, age, language, religion, political opinion, national origin, ethnic origin, social origin, economic or social condition of birth, or disability and any other characteristics.

75. Member States should promote open data. In this regard, Member States should consider reviewing their policies and regulatory frameworks, including on access to information and open government to reflect AI-specific requirements and promoting mechanisms, such as open repositories for publicly funded or publicly held data and source code and data trusts, to support the safe, fair, legal and ethical sharing of data, among others.

76. Member States should promote and facilitate the use of quality and robust datasets for training, development and use of AI systems, and exercise vigilance in overseeing their collection and use. This could, if possible and feasible, include investing in the creation of gold standard datasets, including open and trustworthy datasets, which are diverse, constructed on a valid legal basis, including consent of data subjects, when required by law. Standards for annotating datasets should be encouraged, including disaggregating data on gender and other bases, so it can easily be determined how a dataset is gathered and what properties it has.

77. Member States, as also suggested in the report of the United Nations Secretary-General's High-level Panel on Digital Cooperation, with the support of the United Nations and UNESCO, should adopt a digital commons approach to data where appropriate, increase interoperability of tools and datasets and interfaces of systems hosting data, and encourage private sector companies to share the data they collect with all stakeholders, as appropriate, for research, innovation or public benefits. They should also promote public and private efforts to create collaborative platforms to share quality data in trusted and secured data spaces.

POLICY AREA 4: DEVELOPMENT AND INTERNATIONAL COOPERATION

78. Member States and transnational corporations should prioritize AI ethics by including discussions of AI-related ethical issues into relevant international, intergovernmental and multi-stakeholder fora.

79. Member States should ensure that the use of AI in areas of development such as education, science, culture, communication and information, health care, agriculture and food supply, environment, natural resource and infrastructure management, economic planning and growth, among others, adheres to the values and principles set forth in this Recommendation.

80. Member States should work through international organizations to provide platforms for international cooperation on AI for development, including by contributing expertise, funding, data, domain knowledge, infrastructure, and facilitating multi-stakeholder collaboration to tackle challenging development problems, especially for LMICs, in particular LDCs, LLDCs and SIDS.

81. Member States should work to promote international collaboration on AI research and innovation, including research and innovation centres and networks that promote greater participation and leadership of researchers from LMICs and other countries, including LDCs, LLDCs and SIDS.

82. Member States should promote AI ethics research by engaging international organizations and research institutions, as well as transnational corporations, that can be a basis for the ethical use of AI systems by public and private entities, including research into the applicability of specific ethical frameworks in specific cultures and contexts, and the possibilities to develop technologically feasible solutions in line with these frameworks.

83. Member States should encourage international cooperation and collaboration in the field of AI to bridge geo-technological lines. Technological exchanges and consultations should take place between Member States and their populations, between the public and private sectors, and between and among the most and least technologically advanced countries in full respect of international law.

POLICY AREA 5: ENVIRONMENT AND ECOSYSTEMS

84. Member States and business enterprises should assess the direct and indirect environmental impact throughout the AI system life cycle, including, but not limited to, its carbon footprint, energy consumption and the environmental impact of raw material extraction for supporting the manufacturing of AI technologies, and reduce

the environmental impact of AI systems and data infrastructures. Member States should ensure compliance of all AI actors with environmental law, policies and practices.

85. Member States should introduce incentives, when needed and appropriate, to ensure the development and adoption of rights-based and ethical AI-powered solutions for disaster risk resilience; the monitoring, protection and regeneration of the environment and ecosystems; and the preservation of the planet. These AI systems should involve the participation of local and indigenous communities throughout the life cycle of AI systems and should support circular economy type approaches and sustainable consumption and production patterns. Some examples include using AI systems, when needed and appropriate, to:

(a) Support the protection, monitoring and management of natural resources.

(b) Support the prediction, prevention, control and mitigation of climate-related problems.

(c) Support a more efficient and sustainable food ecosystem.

(d) Support the acceleration of access to and mass adoption of sustainable energy.

(e) Enable and promote the mainstreaming of sustainable infrastructure, sustainable business models and sustainable finance for sustainable development.

(f) Detect pollutants or predict levels of pollution and thus help relevant stakeholders identify, plan and put in place targeted interventions to prevent and reduce pollution and exposure.

86. When choosing AI methods, given the potential data intensive or resource-intensive character of some of them and the respective impact on the environment, Member States should ensure that AI actors, in line with the principle of proportionality, favour data, energy and resource-efficient AI methods. Requirements should be developed to ensure that appropriate evidence is available to show that an AI application will have the intended effect, or that safeguards accompanying an AI application can support the justification for its use. If this cannot be done, the precautionary principle must be favoured, and in instances where there are disproportionate negative impacts on the environment, AI should not be used.

POLICY AREA 6: GENDER

87. Member States should ensure that the potential for digital technologies and artificial intelligence to contribute to achieving gender equality is fully maximized, and must ensure that the human rights and fundamental freedoms of girls and women, and their safety and integrity are not violated at any stage of the AI system life cycle. Moreover, Ethical Impact Assessment should include a transversal gender perspective.

88. Member States should have dedicated funds from their public budgets linked to financing gender-responsive schemes, ensure that national digital policies include a gender action plan, and develop relevant policies, for example, on labour education, targeted at supporting girls and women to make sure they are not left out of the digital economy powered by AI. Special investment in providing targeted programmes and genderspecific language, to increase the opportunities of girls' and women's participation in science, technology, engineering, and mathematics (STEM), including information and communication technologies (ICT) disciplines, preparedness, employability, equal career development and professional growth of girls and women, should be considered and implemented.

89. Member States should ensure that the potential of AI systems to advance the achievement of gender equality is realized. They should ensure that these technologies do not exacerbate the already wide gender gaps existing in several fields in the analogue world, and instead eliminate those gaps. These gaps include: the gender wage gap; the unequal representation in certain professions and activities; the lack of representation at top management positions, boards of directors, or research teams in the AI field; the education gap; the digital and AI access, adoption, usage and affordability gap; and the unequal distribution of unpaid work and of the caring responsibilities in our societies.

90. Member States should ensure that gender stereotyping and discriminatory biases are not translated into AI systems, and instead identify and proactively redress these. Efforts are necessary to avoid the compounding negative effect of technological divides in achieving gender equality and avoiding violence such as harassment, bullying or trafficking of girls and women and under-represented groups, including in the online domain.

91. Member States should encourage female entrepreneurship, participation and engagement in all stages of an AI system life cycle by offering and promoting economic, regulatory incentives, among other incentives and support schemes, as well as policies that aim at a balanced gender participation in AI research in academia, gender representation on digital and AI companies' top management

positions, boards of directors and research teams. Member States should ensure that public funds (for innovation, research and technologies) are channelled to inclusive programmes and companies, with clear gender representation, and that private funds are similarly encouraged through affirmative action principles. Policies on harassment-free environments should be developed and enforced, together with the encouragement of the transfer of best practices on how to promote diversity throughout the AI system life cycle.

92. Member States should promote gender diversity in AI research in academia and industry by offering incentives to girls and women to enter the field, putting in place mechanisms to fight gender stereotyping and harassment within the AI research community, and encouraging academic and private entities to share best practices on how to enhance gender diversity.

93. UNESCO can help form a repository of best practices for incentivizing the participation of girls, women and under-represented groups in all stages of the AI system life cycle.

POLICY AREA 7: CULTURE

94. Member States are encouraged to incorporate AI systems, where appropriate, in the preservation, enrichment, understanding, promotion, management and accessibility of tangible, documentary and intangible cultural heritage, including endangered languages as well as indigenous languages and knowledges, for example by introducing or updating educational programmes related to the application of AI systems in these areas, where appropriate, and by ensuring a participatory approach, targeted at institutions and the public.

95. Member States are encouraged to examine and address the cultural impact of AI systems, especially natural language processing (NLP) applications such as automated translation and voice assistants, on the nuances of human language and expression. Such assessments should provide input for the design and implementation of strategies that maximize the benefits from these systems by bridging cultural gaps and increasing human understanding, as well as addressing the negative implications such as the reduction of use, which could lead to the disappearance of endangered languages, local dialects, and tonal and cultural variations associated with human language and expression.

96. Member States should promote AI education and digital training for artists and creative professionals to assess the suitability of AI technologies for use in their profession, and contribute to the design and implementation of suitable AI technologies, as AI technologies are being used to create, produce, distribute,

broadcast and consume a variety of cultural goods and services, bearing in mind the importance of preserving cultural heritage, diversity and artistic freedom.

97. Member States should promote awareness and evaluation of AI tools among local cultural industries and small and medium enterprises working in the field of culture, to avoid the risk of concentration in the cultural market.

98. Member States should engage technology companies and other stakeholders to promote a diverse supply of and plural access to cultural expressions, and in particular to ensure that algorithmic recommendation enhances the visibility and discoverability of local content.

99. Member States should foster new research at the intersection between AI and intellectual property (IP), for example to determine whether or how to protect with IP rights the works created by means of AI technologies. Member States should also assess how AI technologies are affecting the rights or interests of IP owners, whose works are used to research, develop, train or implement AI applications.

100. Member States should encourage museums, galleries, libraries and archives at the national level to use AI systems to highlight their collections and enhance their libraries, databases and knowledge base, while also providing access to their users.

POLICY AREA 8: EDUCATION AND RESEARCH

101. Member States should work with international organizations, educational institutions and private and non-governmental entities to provide adequate AI literacy education to the public on all levels in all countries in order to empower people and reduce the digital divides and digital access inequalities resulting from the wide adoption of AI systems.

102. Member States should promote the acquisition of "prerequisite skills" for AI education, such as basic literacy, numeracy, coding and digital skills, and media and information literacy, as well as critical and creative thinking, teamwork, communication, socio-emotional and AI ethics skills, especially in countries and in regions or areas within countries where there are notable gaps in the education of these skills.

103. Member States should promote general awareness programmes about AI developments, including on data and the opportunities and challenges brought about by AI technologies, the impact of AI systems on human rights and their implications, including children's rights. These programmes should be accessible to nontechnical as well as technical groups.

104. Member States should encourage research initiatives on the responsible and ethical use of AI technologies in teaching, teacher training and e-learning, among other issues, to enhance opportunities and mitigate the challenges and risks involved in this area. The initiatives should be accompanied by an adequate assessment of the quality of education and impact on students and teachers of the use of AI technologies. Member States should also ensure that AI technologies empower students and teachers and enhance their experience, bearing in mind that relational and social aspects and the value of traditional forms of education are vital in teacher-student and student-student relationships and should be considered when discussing the adoption of AI technologies in education. AI systems used in learning should be subject to strict requirements when it comes to the monitoring, assessment of abilities, or prediction of the learners' behaviours. AI should support the learning process without reducing cognitive abilities and without extracting sensitive information, in compliance with relevant personal data protection standards. The data handed over to acquire knowledge collected during the learner's interactions with the AI system must not be subject to misuse, misappropriation or criminal exploitation, including for commercial purposes.

105. Member States should promote the participation and leadership of girls and women, diverse ethnicities and cultures, persons with disabilities, marginalized and vulnerable people or people in vulnerable situations, minorities and all persons not enjoying the full benefits of digital inclusion, in AI education programmes at all levels, as well as the monitoring and sharing of best practices in this regard with other Member States.

106. Member States should develop, in accordance with their national education programmes and traditions, AI ethics curricula for all levels, and promote cross-collaboration between AI technical skills education and humanistic, ethical and social aspects of AI education. Online courses and digital resources of AI ethics education should be developed in local languages, including indigenous languages, and take into account the diversity of environments, especially ensuring accessibility of formats for persons with disabilities.

107. Member States should promote and support AI research, notably AI ethics research, including for example through investing in such research or by creating incentives for the public and private sectors to invest in this area, recognizing that research contributes significantly to the further development and improvement of AI technologies with a view to promoting international law and the values and principles set forth in this Recommendation. Member States should also publicly promote the best practices of, and cooperation with, researchers and companies who develop AI in an ethical manner.

108. Member States should ensure that AI researchers are trained in research ethics and require them to include ethical considerations in their designs, products and publications, especially in the analyses of the datasets they use, how they are annotated, and the quality and scope of the results with possible applications.

109. Member States should encourage private sector companies to facilitate the access of the scientific community to their data for research, especially in LMICs, in particular LDCs, LLDCs and SIDS. This access should conform to relevant privacy and data protection standards.

110. To ensure a critical evaluation of AI research and proper monitoring of potential misuses or adverse effects, Member States should ensure that any future developments with regards to AI technologies should be based on rigorous and independent scientific research, and promote interdisciplinary AI research by including disciplines other than science, technology, engineering and mathematics (STEM), such as cultural studies, education, ethics, international relations, law, linguistics, philosophy, political science, sociology and psychology.

111. Recognizing that AI technologies present great opportunities to help advance scientific knowledge and practice, especially in traditionally model-driven disciplines, Member States should encourage scientific communities to be aware of the benefits, limits and risks of their use; this includes attempting to ensure that conclusions drawn from data-driven approaches, models and treatments are robust and sound. Furthermore, Member States should welcome and support the role of the scientific community in contributing to policy and in cultivating awareness of the strengths and weaknesses of AI technologies.

POLICY AREA 9: COMMUNICATION AND INFORMATION

112. Member States should use AI systems to improve access to information and knowledge. This can include support to researchers, academia, journalists, the general public and developers, to enhance freedom of expression, academic and scientific freedoms, access to information, and increased proactive disclosure of official data and information.

113. Member States should ensure that AI actors respect and promote freedom of expression as well as access to information with regard to automated content generation, moderation and curation. Appropriate frameworks, including regulation, should enable transparency of online communication and information operators and ensure users have access to a diversity of viewpoints, as well as processes for prompt notification to the users on the reasons for removal or other treatment of content, and appeal mechanisms that allow users to seek redress.

114. Member States should invest in and promote digital and media and information literacy skills to strengthen critical thinking and competencies needed to understand the use and implication of AI systems, in order to mitigate and counter disinformation, misinformation and hate speech. A better understanding and evaluation of both the positive and potentially harmful effects of recommender systems should be part of those efforts.

115. Member States should create enabling environments for media to have the rights and resources to effectively report on the benefits and harms of AI systems, and also encourage media to make ethical use of AI systems in their operations

POLICY AREA 10: ECONOMY AND LABOUR

116. Member States should assess and address the impact of AI systems on labour markets and its implications for education requirements, in all countries and with special emphasis on countries where the economy is labourintensive. This can include the introduction of a wider range of "core" and interdisciplinary skills at all education levels to provide current workers and new generations a fair chance of finding jobs in a rapidly changing market, and to ensure their awareness of the ethical aspects of AI systems. Skills such as "learning how to learn", communication, critical thinking, teamwork, empathy, and the ability to transfer one's knowledge across domains, should be taught alongside specialist, technical skills, as well as low-skilled tasks. Being transparent about what skills are in demand and updating curricula around these are key.

117. Member States should support collaboration agreements among governments, academic institutions, vocational education and training institutions, industry, workers' organizations and civil society to bridge the gap of skillset requirements to align training programmes and strategies with the implications of the future of work and the needs of industry, including small and medium enterprises. Project-based teaching and learning approaches for AI should be promoted, allowing for partnerships between public institutions, private sector companies, universities and research centres.

118. Member States should work with private sector companies, civil society organizations and other stakeholders, including workers and unions to ensure a fair transition for at-risk employees. This includes putting in place upskilling and reskilling programmes, finding effective mechanisms of retaining employees during those transition periods, and exploring "safety net" programmes for those who cannot be retrained. Member States should develop and implement programmes to research and address the challenges identified that could include upskilling and reskilling, enhanced social protection, proactive industry policies and interventions,

tax benefits, new taxation forms, among others. Member States should ensure that there is sufficient public funding to support these programmes. Relevant regulations, such as tax regimes, should be carefully examined and changed if needed to counteract the consequences of unemployment caused by AI-based automation.

119. Member States should encourage and support researchers to analyse the impact of AI systems on the local labour environment in order to anticipate future trends and challenges. These studies should have an interdisciplinary approach and investigate the impact of AI systems on economic, social and geographic sectors, as well as on human-robot interactions and human relationships, in order to advise on reskilling and redeployment best practices.

120. Member States should take appropriate steps to ensure competitive markets and consumer protection, considering possible measures and mechanisms at national, regional and international levels, to prevent abuse of dominant market positions, including by monopolies, in relation to AI systems throughout their life cycle, whether these are data, research, technology, or market. Member States should prevent the resulting inequalities, assess relevant markets and promote competitive markets. Due consideration should be given to LMICs, in particular LDCs, LLDCs and SIDS, which are more exposed and vulnerable to the possibility of abuses of market dominance as a result of a lack of infrastructure, human capacity and regulations, among other factors. AI actors developing AI systems in countries which have established or adopted ethical standards on AI should respect these standards when exporting these products, developing or applying their AI systems in countries where such standards may not exist, while respecting applicable international law and domestic legislation, standards and practices of these countries.

POLICY AREA 11: HEALTH AND SOCIAL WELL-BEING

121. Member States should endeavour to employ effective AI systems for improving human health and protecting the right to life, including mitigating disease outbreaks, while building and maintaining international solidarity to tackle global health risks and uncertainties, and ensure that their deployment of AI systems in health care be consistent with international law and their human rights law obligations. Member States should ensure that actors involved in health care AI systems take into consideration the importance of a patient's relationships with their family and with health care staff.

122. Member States should ensure that the development and deployment of AI systems related to health in general and mental health in particular, paying due attention to children and youth, is regulated to the effect that they are safe, effective,

efficient, scientifically and medically proven and enable evidence-based innovation and medical progress. Moreover, in the related area of digital health interventions, Member States are strongly encouraged to actively involve patients and their representatives in all relevant steps of the development of the system.

123. Member States should pay particular attention in regulating prediction, detection and treatment solutions for health care in AI applications by: (a) ensuring oversight to minimize and mitigate bias; (b) ensuring that the professional, the patient, caregiver or service user is included as a "domain expert" in the team in all relevant steps when developing the algorithms; (c) paying due attention to privacy because of the potential need for being medically monitored and ensuring that all relevant national and international data protection requirements are met; (d) ensuring effective mechanisms so that those whose personal data is being analysed are aware of and provide informed consent for the use and analysis of their data, without preventing access to health care; (e) ensuring the human care and final decision of diagnosis and treatment are taken always by humans while acknowledging that AI systems can also assist in their work; (f) ensuring, where necessary, the review of AI systems by an ethical research committee prior to clinical use.

124. Member States should establish research on the effects and regulation of potential harms to mental health related to AI systems, such as higher degrees of depression, anxiety, social isolation, developing addiction, trafficking, radicalization and misinformation, among others.

125. Member States should develop guidelines for human robot interactions and their impact on human-human relationships, based on research and directed at the future development of robots, and with special attention to the mental and physical health of human beings. Particular attention should be given to the use of robots in health care and the care for older persons and persons with disabilities, in education, and robots for use by children, toy robots, chatbots and companion robots for children and adults. Furthermore, assistance of AI technologies should be applied to increase the safety and ergonomic use of robots, including in a human-robot working environment. Special attention should be paid to the possibility of using AI to manipulate and abuse human cognitive biases.

126. Member States should ensure that human-robot interactions comply with the same values and principles that apply to any other AI systems, including human rights and fundamental freedoms, the promotion of diversity, and the protection of vulnerable people or people in vulnerable situations. Ethical questions related to AI-powered systems for neurotechnologies and brain computer interfaces should be considered in order to preserve human dignity and autonomy.

127. Member States should ensure that users can easily identify whether they are interacting with a living being, or with an AI system imitating human or animal characteristics, and can effectively refuse such interaction and request human intervention.

128. Member States should implement policies to raise awareness about the anthropomorphization of AI technologies and technologies that recognize and mimic human emotions, including in the language used to mention them, and assess the manifestations, ethical implications and possible limitations of such anthropomorphization, in particular in the context of robot-human interaction and especially when children are involved.

129. Member States should encourage and promote collaborative research into the effects of long-term interaction of people with AI systems, paying particular attention to the psychological and cognitive impact that these systems can have on children and young people. This should be done using multiple norms, principles, protocols, disciplinary approaches, and assessment of the modification of behaviours and habits, as well as careful evaluation of the downstream cultural and societal impacts. Furthermore, Member States should encourage research on the effect of AI technologies on health system performance and health outcomes.

130. Member States, as well as all stakeholders, should put in place mechanisms to meaningfully engage children and young people in conversations, debates and decisionmaking with regard to the impact of AI systems on their lives and futures.

Saudi Data & AI Authority, AI Ethics Principles
[excerpt]
(2023)

AI Ethics Principles and Controls

Principle 1 – Fairness

The fairness principle requires taking necessary actions to eliminate bias, discrimination or stigmatization of individuals, communities, or groups in the design, data, development, deployment and use of AI systems. Bias may occur due to data, representation or algorithms and could lead to discrimination against the historically disadvantaged groups.

When designing, selecting, and developing AI systems, it is essential to ensure just, fair, non-biased, non-discriminatory and objective standards that are inclusive, diverse, and representative of all or targeted segments of society. The functionality of an AI system should not be limited to a specific group based on gender, race, religion, disability, age, or sexual orientation. In addition, the potential risks, overall benefits, and purpose of utilizing sensitive personal data should be well-motivated and defined or articulated by the AI System Owner.

To ensure consistent AI systems that are based on fairness and inclusiveness, AI systems should be trained on data that are cleansed from bias and is representative of affected minority groups. AI algorithms should be built and developed in a manner that makes their composition free from bias and correlation fallacy.

Plan and Design:

1- At the initial stages of setting out the purpose of the AI system, the design team shall collaborate to pinpoint the objectives and how to reach them in an efficient and optimized manner. Planning the design of the AI system is an essential stage to translate the system's intended goals and outcomes. During this phase, it is important to implement a fairness-aware design that takes appropriate precautions across the AI system algorithm, processes, and mechanisms to prevent biases from having a discriminatory effect or lead to skewed and unwanted results or outcomes.

2- Fairness-aware design should start at the beginning of the AI System Lifecycle with a collaborative effort from technical and non-technical members to identify potential harm and benefits, affected individuals and vulnerable groups and evaluate how they are impacted by the results and whether the impact is justifiable given the general purpose of the AI system.

3- A fairness assessment of the AI system is crucial, and the metrics should be selected at this stage of the AI System Lifecycle. The metrics should be chosen based on the algorithm type (rule-based, classification, regression, etc.), the effect of the decision (punitive, selective, etc.), and the harm and benefit on correctly and incorrectly predicted samples.

4- Sensitive personal data attributes relating to persons or groups which are systematically or historically disadvantaged should be identified and defined at this stage. The allowed threshold which makes the assessment fair or unfair should be defined. The fairness assessment metrics to be applied to sensitive features should be measured during future steps.

Prepare Input Data:

1- Following the best practice of responsible data acquisition, handling, classification, and management must be a priority to ensure that results and outcomes align with the AI system's set goals and objectives. Effective data quality soundness and procurement begin by ensuring the integrity of the data source and data accuracy in representing all observations to avoid the systematic disadvantaging of under-represented or advantaging over-represented groups. The quantity and quality of the data sets should be sufficient and accurate to serve the purpose of the system. The sample size of the data collected or procured has a significant impact on the accuracy and fairness of the outputs of a trained model.

2- Sensitive personal data attributes which are defined in the plan and design phase should not be included in the model data not to feed the existing bias on them. Also, the proxies of the sensitive features should be analyzed and not included in the input data. In some cases, this may not be possible due to the accuracy or objective of the AI system. In this case, the justification of the usage of the sensitive personal data attributes or their proxies should be provided.

Build and Validate:

1- At the build and validate stage of the AI System Lifecycle, it is essential to take into consideration implementation fairness as a common theme when building, testing, and implementing the AI system. Model building and feature selection will require engineers and designers to be aware that the choices made about grouping or separating and including or excluding features as well as more general judgments about the reliability and security of the total set of features may have significant consequences for vulnerable or protected groups.

2- During the selection of the champion model, the fairness metric assessment should be considered. The champion model fairness metrics should be within the defined threshold for the sensitive features. The optimization approach of fairness and performance metrics should be clearly set throughout this phase. The fairness assessment should be justified if the champion model does not pass the assessment.

3- Causality-based feature selection should be ensured. Selected features should be verified with business owners and non-technical teams.

4- Automated decision-support technologies present major risks of bias and unwanted application at the deployment phase, so it is critical to set out mechanisms to prevent harmful and discriminatory results at this phase.

Deploy and Monitor:

1- Well-defined mechanisms and protocols should be set in place when deploying the AI system to measure the fairness and performance of the outcomes and how it impacts individuals and communities. When analyzing the outcomes of the predictive model, it should be assessed if represented groups in the data sample receive benefits in equal or similar portions and if the AI system disproportionately harms specific members based on demographic differences to ensure outcome fairness.

2- The predefined fairness metrics should be monitored in production. If there is any deviation from the allowed threshold, it should be investigated whether there is a need to renew the model.

3- The overall harm and benefit of the system should be quantified and materialized on the sensitive groups.

Principle 2 – Privacy & Security

The privacy and security principle represents overarching values that require AI systems; throughout the AI System Lifecycle; to be built in a safe way that respects the privacy of the data collected as well as upholds the highest levels of data security processes and procedures to keep the data confidential preventing data and system breaches which could lead to reputational, psychological, financial, professional, or other types of harm. AI systems should be designed with mechanisms and controls that provide the possibility to govern and monitor their outcomes and progress throughout their lifecycle to ensure continuous monitoring within the privacy and security principles and protocols set in place.

Plan and Design:

1- The planning and design of the AI system and its associated algorithm must be configured and modelled in a manner such that there is respect for the protection of the privacy of individuals, personal data is not misused and exploited, and the decision criteria of the automated technology is not based on personally identifying characteristics or information.

2- The use of personal information should be limited only to that which is necessary for the proper functioning of the system. The design of AI systems resulting in the profiling of individuals or communities may only occur if approved by Chief Compliance and Ethics Officer, Compliance Officer or in compliance with a code of ethics and conduct developed by a national regulatory authority for the specific sector or industry.

3- The security and protection blueprint of the AI system, including the data to be processed and the algorithm to be used, should be aligned to best practices to be able to withstand cyberattacks and data breach attempts.

4- Privacy and security legal frameworks and standards should be followed and customized for the particular use case or organization.

5- An important aspect of privacy and security is data architecture; consequently, data classification and profiling should be planned to define the levels of protection and usage of personal data.

6- Security mechanisms for de-identification should be planned for the sensitive or personal data in the system. Furthermore, read/write/update actions should be authorized for the relevant groups.

Prepare Input Data:

1- The exercise of data procurement, management, and organization should uphold the legal frameworks and standards of data privacy. Data privacy and security protect information from a wide range of threats.

2- The confidentiality of data ensures that information is accessible only to those who are authorized to access the information and that there are specific controls that manage the delegation of authority.

3- Designers and engineers of the AI system must exhibit the appropriate levels of integrity to safeguard the accuracy and completeness of information and processing methods to ensure that the privacy and security legal framework and standards are followed. They should also ensure that the availability and storage of data are protected through suitable security database systems.

4- All processed data should be classified to ensure that it receives the appropriate level of protection in accordance with its sensitivity or security classification and that AI system developers and owners are aware of the classification or sensitivity of the information they are handling and the associated requirements to keep it secure. All data shall be classified in terms of business requirements, criticality, and sensitivity in order to prevent unauthorized disclosure or modification. Data classification should be conducted in a contextual manner that does not result in the inference of personal information. Furthermore, de-identification mechanisms should be employed based on data classification as well as requirements relating to data protection laws.

5- Data backups and archiving actions should be taken in this stage to align with business continuity, disaster recovery and risk mitigation policies.

Build and Validate:

1- Privacy and security by design should be implemented while building the AI system. The security mechanisms should include the protection of various architectural dimensions of an AI model from malicious attacks. The structure and modules of the AI system should be protected from unauthorized modification or damage to any of its components.

2- The AI system should be secure to ensure and maintain the integrity of the information it processes. This ensures that the system remains continuously functional and accessible to authorized users. It is crucial that the system safeguards confidential and private information, even under hostile or adversarial conditions. Furthermore, appropriate measures should be in place to ensure that AI systems with automated decision-making capabilities uphold the necessary data privacy and security standards.

3- The AI System should be tested to ensure that the combination of available data does not reveal the sensitive data or break the anonymity of the observation.

Deploy and Monitor:

1- After the deployment of the AI system, when its outcomes are realized, there must be continuous monitoring to ensure that the AI system is privacy-preserving, safe and secure. The privacy impact assessment and risk management assessment should be continuously revisited to ensure that societal and ethical considerations are regularly evaluated.

2- AI System Owners should be accountable for the design and implementation of AI systems in such a way as to ensure that personal information is protected throughout the life cycle of the AI system. The components of the AI system should be updated based on continuous monitoring and privacy impact assessment.

Principle 3 – Humanity

The humanity principle highlights that AI systems should be built using an ethical methodology to be just and ethically permissible, based on intrinsic and fundamental human rights and cultural values to generate a beneficial impact on individual stakeholders and communities, in both the long and short-term goals and objectives to be used for the good of humanity. Predictive models should not be designed to deceive, manipulate, or condition behavior that is not meant to empower, aid, or augment human skills but should adopt a more human-centric design approach that allows for human choice and determination.

Plan and Design:

1- It is essential to design and build a model that is based on the fundamental human rights and cultural values and principles that are applied within and on the AI system's decisions, processes, and functionalities.

2- The designers of the AI model should define how the AI system will align with fundamental human rights and KSA's cultural values while designing, building, and testing the technology; as well as how the AI system and its outcomes will strive to achieve and positively contribute to augment and complement human skills and capabilities.

Prepare Input Data:

1- To ensure that AI models embody a human-centric build and design that requires adhering to practices of responsible and ethical data management frameworks and processes to be followed according to best practices and data regulations within KSA.

2- Data must be properly acquired, classified, processed, and accessible to ensure respect for human rights, and KSA's cultural values and preferences.

Build and Validate:

1- When constructing AI systems, designers and engineers should prioritize building AI systems and algorithms that allow and facilitate decision-making with an outlook of aligning with human rights and KSA's cultural values. The automated decisions that result from AI systems should not act in a partial and standalone manner without considering broader human rights and cultural values in their final outcomes and results.

2- Designers and Engineers should enable AI systems with the appropriate parameters and algorithm training to attain outcomes that advance humanity.

Deploy and Monitor:

1- Periodic assessments of the deployed AI system should be conducted to ensure that its results are aligned with human rights and cultural values, accuracy key performance indicators (KPIs), and impact on individuals or communities to ensure the continuous improvement of the technology.

2- Designers of AI models should establish mechanisms of assessing AI systems against fundamental human rights and cultural values to mitigate any negative and harmful outcomes resulting from the use of the AI system. If any negative and harmful outcomes are found, the owner of the AI system should identify the areas that need to be addressed and apply corrective measures to recursively improve the functioning and outcomes of the AI system.

Principle 4 – Social & Environmental Benefits

The social and environmental benefit principle embraces the beneficial and positive impact of social and environmental priorities that should benefit individuals and the wider community that focus on sustainable goals and objectives. AI systems should neither cause nor accelerate harm or otherwise adversely affect human beings but rather contribute to empowering and complementing social and environmental progress while addressing associated social and environmental ills. This entails the protection of social good as well as environmental sustainability.

Plan and Design:

1- AI systems have a significant impact on communities and the ecosystems that they live in; hence AI System Owners should have a high sense of awareness that these technologies may have disruptive and transformative effects on society and the environment. The design of AI systems should be approached in an ethical and sensitive manner in line with the values of prevention of harm to both human beings and the environment.

2- When planning and designing AI systems, due consideration should be given to preventing and helping address social and environmental issues in a way that will ensure sustainable social and ecological responsibility.

Prepare Input Data:

1- The processes and policies that govern data management should be followed when preparing the categorization and structuring of data that will feed into the AI system.

2- The data pertaining to the social and environmental topics should be accessible to the public data infrastructure and must clearly articulate the social benefit of the data presented.

Build and Validate:

1- The models and algorithms must have, as their ultimate goal, a result linked to a socially recognized end, with the ability to demonstrate how the expected results relate to that social or environmental purpose through transformative and impactful benefits where applicable.

2- It is best practice to measure and maintain acceptable levels of resource usage and energy consumption during this phase setting the tone that AI systems not only strive to foster AI solutions that address global concerns relating to social and environmental issues but also practice sustainable and ecological responsibilities.

Deploy and Monitor:

1- After the deployment of the AI system, the AI System Owner should ensure that continuous assessment of the human, social, cultural, economic and environmental impact of AI technologies are carried out with full cognizance of the implications of the AI system for sustainability as a set of constantly evolving goals across a range of dimensions against the priority objectives that were set at the Plan and Design phase.

2- The AI System Owner should also foster and encourage the power of AI solutions in addressing areas of global concern aligning with sustainable development goals.

Principle 5 – Reliability & Safety

The reliability and safety principle ensures that the AI system adheres to the set specifications and that the AI system behaves exactly as its designers intended and anticipated. Reliability is a measure of consistency and provides confidence in how robust a system is. It is a measure of dependability with which it operationally conforms to its intended functionality and the outcomes it produces. On the other hand, safety is a measure of how the AI system does not pose a risk of harm or danger to society and individuals. As an illustration, AI systems such as autonomous vehicles can pose a risk to people's lives if living organisms are not properly recognized, certain scenarios are not trained for or if the system malfunctions. A reliable working system should be safe by not posing a danger to society and should have built-in mechanisms to prevent harm. The risk mitigation framework is closely related to this principle. Potential risks and unintended harms should be minimized in this aspect. The predictive model should be monitored and controlled in a periodic and continuous manner to check if its operations and functionality are aligned with the designed structure and frameworks in place. The AI system should be technically sound, robust, and developed to prevent malicious usage to exploit its data and outcomes to harm entities, individuals or communities. A continuous implementation/continuous development approach is essential to ensure reliability.

Plan and Design:

1- Designing and developing an AI system that can withstand the uncertainty, instability, and volatility that it might encounter is crucial.

2- Planning to set out a robust and reliable AI system that works with different sets of inputs and situations is essential to prevent unintended harm and mitigate risks of system failures when positioned against unknown and unforeseen events.

3- Establishing a set of standards and protocols for assessing the reliability of an AI system is necessary to secure the safety of the system's algorithm and data output. It is essential to keep a sustainable technical outlay and outcomes generated from the system to maintain the public's trust and confidence in the AI system.

4- The documentation standards are essential to track the evolution of the system, foresee possible risks and fix vulnerabilities.

5- All critical decision points in the system design should be subject to sign-off by relevant stakeholders to minimize risks and make stakeholders accountable for the decisions.

Prepare Input Data:

1- Adequate steps and actions should be taken to measure the data sample's quality, accuracy, suitability, and credibility when dealing with the data sets of an AI model. This is essential to ensure the accuracy of data interpretation by the AI system, the consistency of avoiding misleading measurements, as well as ensuring the relevance of the AI system's outcomes to the purpose of the model.

2- It is crucial for the build and validate step to test how the system behaves under outlier events, extreme parameters, etc. In this step, stress test data should be prepared for extreme scenarios.

Build and Validate:

1- To develop a sound and functional AI system that is both reliable and safe, the AI system's technical construct should be accompanied by a comprehensive methodology to test the quality of the predictive data-based systems and models according to standard policies and protocols.

2- To ensure the technical robustness of an AI system rigorous testing, validation, and re-assessment as well as the integration of adequate mechanisms of oversight and controls into its development is required. System integration test sign-off should be done with relevant stakeholders to minimize risks and liability.

3- Automated AI systems involving scenarios where decisions are understood to have an impact that is irreversible or difficult to reverse or may involve life-and-death decisions should trigger human oversight and final determination. Furthermore, AI systems should not be used for social scoring or mass surveillance purposes.

Deploy and Monitor:

1- Monitoring the robustness of the AI system should be adopted and undertaken in a periodic and continuous manner to measure and assess any risks related to the technicalities of the AI system (an inward perspective) as well as the magnitude of the risk posed by the system and its capabilities (an outward perspective).

2- The model must also be monitored in a periodic and continuous manner to verify whether its operations and functions are compatible with the designed structure and frameworks. The AI system must also be safe to prevent destructive use to exploit its data and results to harm entities, individuals, or groups. It is necessary to continuously work on implementation and development to ensure system reliability.

Principle 6 – Transparency & Explainability

The transparency and explainability principle is crucial for building and maintaining trust in AI systems and technologies. AI systems must be built with a high level of clarity and explainability as well as features to track the stages of automated decision-making, particularly those that may lead to detrimental effects on data subjects. It follows that data, algorithms, capabilities, processes, and purpose of the AI system need to be transparent and communicated as well as explainable to those who are directly and indirectly affected. The degree to which the system is traceable, auditable, transparent, and explainable is dependent on the context and purpose of the AI system and the severity of the outcomes that may result from the technology. AI systems and their designers should be able to justify how the rationale behind their design, practices, processes, algorithms, and decisions or behaviors are ethically permissible, nondiscriminatory, and nonharmful to the public.

Plan and Design:

1- When designing a transparent and trusted AI system, it is vital to ensure that stakeholders affected by AI systems are fully aware and informed of how outcomes are processed. They should further be given access to and an explanation of the rationale for decisions made by the AI technology in an understandable and contextual manner. Decisions should be traceable. AI system owners must define the level of transparency for different stakeholders on the technology based on data privacy, sensitivity, and authorization of the stakeholders.

2- The AI system should be designed to include an information section in the platform to give an overview of the AI model decisions as part of the overall transparency application of the technology. Information sharing as a sub-principle should be adhered to with end-users and stakeholders of the AI system upon request or open to the public, depending on the nature of the AI system and target market. The model should establish a process mechanism to log and address issues and complaints that arise to be able to resolve them in a transparent and explainable manner.

Prepare Input Data:

1- The data sets and the processes that yield the AI system's decision should be documented to the best possible standard to allow for traceability and an increase in transparency.

2- The data sets should be assessed in the context of their accuracy, suitability, validity, and source. This has a direct effect on the training and implementation of these systems since the criteria for the data's organization, and structuring must be transparent and explainable in their acquisition and collection adhering to data privacy regulations and intellectual property standards and controls.

Build and Validate:

1- Transparency in AI is thought about from two perspectives, the first is the process behind it (the design and implementation practices that lead to an algorithmically supported outcome) and the second is in terms of its product (the content and justification of that outcome). Algorithms should be developed in a transparent way to ensure that input transparency is evident and explainable to the end-users of the AI system to be able to provide evidence and information on the data used to process the decisions that have been processed.

2- Transparent and explainable algorithms ensure that stakeholders affected by AI systems, both individuals and communities, are fully informed when an outcome is processed by the AI system by providing the opportunity to request explanatory information from the AI system owner. This enables the identification of the AI decision and its respective analysis which facilitates its auditability as well as its explainability.

3- If the AI system is built by a third party, AI system owners should make sure that an AI Ethics due diligence is carried out and all the documentation are accessible and traceable before procurement or sign-off.

Deploy and Monitor:

1- Upon deployment of the AI system, performance metrics relating the AI system's output, accuracy and alignment to priorities and objectives, as well as its measured impact on individuals and communities should be documented, available and accessible to stakeholders of the AI technology.

2- Information on any system failures, data breaches, system breakdowns, etc. should be logged and stakeholders should be informed about these instances keeping the performance and execution of the AI system transparent. Periodic UI and UX testing should be conducted to avoid the risk of confusion, confirmation of biases, or cognitive fatigue of the AI system.

Principle 7 – Accountability & Responsibility

The accountability and responsibility principle holds designers, vendors, procurers, developers, owners and assessors of AI systems and the technology itself ethically responsible and liable for the decisions and actions that may result in potential risk and negative effects on individuals and communities. Human oversight, governance, and proper management should be demonstrated across the entire AI System Lifecycle to ensure that proper mechanisms are in place to avoid harm and misuse of this technology. AI systems should never lead to people being deceived or unjustifiably impaired in their freedom of choice. The designers, developers, and people who implement the AI system should be identifiable and assume responsibility and accountability for any potential damage the technology has on individuals or communities, even if the adverse impact is unintended. The liable parties should take necessary preventive actions as well as set risk assessment and mitigation strategy to minimize the harm due to the AI system. The accountability and responsibility principle is closely related to the fairness principle. The parties responsible for the AI system should ensure that the fairness of the system is maintained and sustained through control mechanisms. All parties involved in the AI System Lifecycle should consider and action these values in their decisions and execution.

Plan and Design:

1- This step is crucial to design or procure an AI System in an accountable and responsible manner. The ethical responsibility and liability for the outcomes of the AI system should be attributable to stakeholders who are responsible for certain actions in the AI System Lifecycle. It is essential to set a robust governance structure that defines the authorization and responsibility areas of the internal and external stakeholders without leaving any areas of uncertainty to achieve this principle. The design approach of the AI system should respect human rights, and fundamental freedoms as well as the national laws and cultural values of the kingdom.

2- Organizations can put in place additional instruments such as impact assessments, risk mitigation frameworks, audit and due diligence mechanisms, redress, and disaster recovery plans.

3- It is essential to build and design a human-controlled AI system where decisions on the processes and functionality of the technology are monitored and executed, and are susceptible to intervention from authorized users. Human governance and oversight establish the necessary control and levels of autonomy through set mechanisms.

Prepare Input Data:

1- An important aspect of the Accountability and Responsibility principle during Prepare Input Data step in the AI System Lifecycle is data quality as it affects the outcome of the AI model and decisions accordingly. It is, therefore, important to do necessary data quality checks, clean data and ensure the integrity of the data in order to get accurate results and capture intended behavior in supervised and unsupervised models.

2- Data sets should be approved and signed-off before commencing with developing the AI model. Furthermore, the data should be cleansed from societal biases. In parallel with the fairness principle, the sensitive features should not be included in the model data. In the event that sensitive features need to be included, the rationale or trade-off behind the decision for such inclusion should be clearly explained. The data preparation process and data quality checks should be documented and validated by responsible parties.

3- The documentation of the process is necessary for auditing and risk mitigation. Data must be properly acquired, classified, processed, and accessible to ease human intervention and control at later stages when needed.

Build and Validate:

1- Model development of the AI system and algorithm should consist of the selection of features, hyperparameter tuning and performance metric selection. To achieve this, the technical stakeholders who build and validate models should be responsible for these decisions.

2- Assigning the appropriate ownership and communicating responsibilities will set the tone for accountability that would aid in steering the development of the AI system on good reasons, solid interference, and will allow the intervention of human critical judgement and expertise.

3- The decisions should be supported with quantitative (performance measures on train/test datasets, consistency of the performance on different sensitive groups, performance comparison for each set of hyperparameters, etc.) and qualitative indicators (decisions to mitigate and correct unintended risks from inaccurate predictions).

4- The appropriate stakeholders and owners of the AI technology should review and sign off the model after successful testing and validation of user acceptance testing rounds have been conducted and completed before the AI models can be productionized.

Deploy and Monitor:

1- The responsibility and associated liability in the Deploy and Monitor step should be set clearly. The outcomes and decisions set in the build and validate step should be monitored continuously and should result in periodic performance reports.

2- Predefined triggers/alerts should be defined for this step on the data and performance metrics. Setting these triggers is a rigorous process and each trigger should be assigned to the appropriate stakeholder. These triggers/alerts can be defined as part of the risk mitigation or disaster recovery procedure and may need human oversight.

Hiroshima Process International Guiding Principles for All AI Actors (2023)

1. We emphasize the responsibilities of all AI actors in promoting, as relevant and appropriate, safe, secure and trustworthy AI. We recognize that actors across the lifecycle will have different responsibilities and different needs with regard to the safety, security, and trustworthiness of AI. We encourage all AI actors to read and understand the "Hiroshima Process International Guiding Principles for Organizatioons Developing Advanced AI Systems (October 30, 2023)" with due consideration to their capacity and their role within the lifecycle.

2. The following 11 principles of the "Hiroshima Process International Guiding Principles for Organizations Developing Advanced AI Systems" should be applied to all AI actors when and as relevant and appropriate, in appropriate forms, to cover the design, development, deployment, provision and use of advanced AI systems, recognizing that some elements are only possible to apply to organizations developing advanced AI systems.

I. Take appropriate measures throughout the development of advanced AI systems, including prior to and throughout their deployment and placement on the market, to identify, evaluate, and mitigate risks across the AI lifecycle.

II. Identify and mitigate vulnerabilities, and, where appropriate, incidents and patterns of misuse, after deployment including placement on the market.

III. Publicly report advanced AI systems' capabilities, limitations and domains of appropriate and inappropriate use, to support ensuring sufficient transparency, thereby contributing to increase accountability.

IV. Work towards responsible information sharing and reporting of incidents among organizations developing advanced AI systems including with industry, governments, civil society, and academia.

V. Develop, implement and disclose AI governance and risk management policies, grounded in a risk-based approach – including privacy policies, and mitigation measures, in particular for organizations developing advanced AI systems.

VI. Invest in and implement robust security controls, including physical security, cybersecurity and insider threat safeguards across the AI lifecycle.

VII. Develop and deploy reliable content authentication and provenance mechanisms, where technically feasible, such as watermarking or other techniques to enable users to identify AI-generated content.

VIII. Prioritize research to mitigate societal, safety and security risks and prioritize investment in effective mitigation measures.

IX. Prioritize the development of advanced AI systems to address the world's greatest challenges, notably but not limited to the climate crisis, global health and education.

X. Advance the development of and, where appropriate, adoption of international technical standards.

XI. Implement appropriate data input measures and protections for personal data and intellectual property.

3. In addition, AI actors should follow the 12th principle.

XII. Promote and contribute to trustworthy and responsible use of advanced AI systems.

AI actors should seek opportunities to improve their own and, where appropriate, others' digital literacy, training and awareness, including on issues such as how advanced AI systems may exacerbate certain risks (e.g. with regard to the spread of disinformation) and/or create new ones.

All relevant AI actors are encouraged to cooperate and share information, as appropriate, to identify and address emerging risks and vulnerabilities of advanced AI systems.

The Bletchley Declaration by Countries Attending the AI Safety Summit
(2023)

Artificial Intelligence (AI) presents enormous global opportunities: it has the potential to transform and enhance human wellbeing, peace and prosperity. To realise this, we affirm that, for the good of all, AI should be designed, developed, deployed, and used, in a manner that is safe, in such a way as to be human-centric, trustworthy and responsible. We welcome the international community's efforts so far to cooperate on AI to promote inclusive economic growth, sustainable development and innovation, to protect human rights and fundamental freedoms, and to foster public trust and confidence in AI systems to fully realise their potential.

AI systems are already deployed across many domains of daily life including housing, employment, transport, education, health, accessibility, and justice, and their use is likely to increase. We recognise that this is therefore a unique moment to act and affirm the need for the safe development of AI and for the transformative opportunities of AI to be used for good and for all, in an inclusive manner in our countries and globally. This includes for public services such as health and education, food security, in science, clean energy, biodiversity, and climate, to realise the enjoyment of human rights, and to strengthen efforts towards the achievement of the United Nations Sustainable Development Goals.

Alongside these opportunities, AI also poses significant risks, including in those domains of daily life. To that end, we welcome relevant international efforts to examine and address the potential impact of AI systems in existing fora and other relevant initiatives, and the recognition that the protection of human rights, transparency and explainability, fairness, accountability, regulation, safety, appropriate human oversight, ethics, bias mitigation, privacy and data protection needs to be addressed. We also note the potential for unforeseen risks stemming from the capability to manipulate content or generate deceptive content. All of these issues are critically important and we affirm the necessity and urgency of addressing them.

Particular safety risks arise at the 'frontier' of AI, understood as being those highly capable general-purpose AI models, including foundation models, that could perform a wide variety of tasks - as well as relevant specific narrow AI that could exhibit capabilities that cause harm - which match or exceed the capabilities present in today's most advanced models. Substantial risks may arise from potential intentional misuse or unintended issues of control relating to alignment with human intent. These issues are in part because those capabilities are not fully understood and are therefore hard to predict. We are especially concerned by such risks in domains such as cybersecurity and biotechnology, as well as where frontier AI systems may amplify risks such as disinformation. There is potential for serious, even catastrophic, harm, either deliberate or unintentional, stemming from the most significant capabilities of these AI models. Given the rapid and uncertain rate of change of AI, and in the context of the acceleration of investment in technology, we affirm that deepening our understanding of these potential risks and of actions to address them is especially urgent.

Many risks arising from AI are inherently international in nature, and so are best addressed through international cooperation. We resolve to work together in an inclusive manner to ensure human-centric, trustworthy and responsible AI that is safe, and supports the good of all through existing international fora and other relevant initiatives, to promote cooperation to address the broad range of risks posed by AI. In doing so, we recognise that countries should consider the importance of a pro-innovation and proportionate governance and regulatory approach that maximises the benefits and takes into account the risks associated with AI. This could include making, where appropriate, classifications and categorisations of risk based on national circumstances and applicable legal frameworks. We also note the relevance of cooperation, where appropriate, on approaches such as common principles and codes of conduct. With regard to the specific risks most likely found in relation to frontier AI, we resolve to intensify and sustain our cooperation, and broaden it with further countries, to identify, understand and as appropriate act, through existing international fora and other relevant initiatives, including future international AI Safety Summits.

All actors have a role to play in ensuring the safety of AI: nations, international fora and other initiatives, companies, civil society and academia will need to work together. Noting the importance of inclusive AI and bridging the digital divide, we reaffirm that international collaboration should endeavour to engage and involve a broad range of partners as appropriate, and welcome development-orientated approaches and policies that could help developing countries strengthen AI capacity building and leverage the enabling role of AI to support sustainable growth and address the development gap.

We affirm that, whilst safety must be considered across the AI lifecycle, actors developing frontier AI capabilities, in particular those AI systems which are unusually powerful and potentially harmful, have a particularly strong responsibility for ensuring the safety of these AI systems, including through systems for safety testing, through evaluations, and by other appropriate measures. We encourage all relevant actors to provide context-appropriate transparency and accountability on their plans to measure, monitor and mitigate potentially harmful capabilities and the associated effects that may emerge, in particular to prevent misuse and issues of control, and the amplification of other risks.

In the context of our cooperation, and to inform action at the national and international levels, our agenda for addressing frontier AI risk will focus on:

- identifying AI safety risks of shared concern, building a shared scientific and evidence-based understanding of these risks, and sustaining that understanding as capabilities continue to increase, in the context of a wider global approach to understanding the impact of AI in our societies.

- building respective risk-based policies across our countries to ensure safety in light of such risks, collaborating as appropriate while recognising our approaches may differ based on national circumstances and applicable legal frameworks. This includes, alongside increased transparency by private actors developing frontier AI capabilities, appropriate evaluation metrics, tools for safety testing, and developing relevant public sector capability and scientific research.

In furtherance of this agenda, we resolve to support an internationally inclusive network of scientific research on frontier AI safety that encompasses and complements existing and new multilateral, plurilateral and bilateral collaboration, including through existing international fora and other relevant initiatives, to facilitate the provision of the best science available for policy making and the public good.

In recognition of the transformative positive potential of AI, and as part of ensuring wider international cooperation on AI, we resolve to sustain an inclusive global dialogue that engages existing international fora and other relevant initiatives and contributes in an open manner to broader international discussions, and to continue research on frontier AI safety to ensure that the benefits of the technology can be harnessed responsibly for good and for all. We look forward to meeting again in 2024.

The countries represented were:

- Australia

- Brazil

- Canada

- Chile

- China

- European Union

- France

- Germany

- India

- Indonesia

- Ireland

- Israel

- Italy

- Japan

- Kenya

- Kingdom of Saudi Arabia

- Netherlands

- Nigeria

- The Philippines

- Republic of Korea

- Rwanda

- Singapore

- Spain

- Switzerland

- Türkiye

- Ukraine

- United Arab Emirates

- United Kingdom of Great Britain and Northern Ireland

- United States of America

References to 'governments' and 'countries' include international organisations acting in accordance with their legislative or executive competences.

Santiago Declaration 2024, Extraordinary Meeting of Ministers of Education of Latin America and the Caribbean: Towards a Regional Reference Framework for Educational Reactivation, Recovery and Transformation
(2024)

Preamble

1. The Ministers of Education and high-ranking authorities representing the countries of Latin America and the Caribbean, gathered in Santiago, Chile, on January 25 and 26, 2024, express our gratitude to the Ministry of Education of Chile, UNESCO, the Development Bank of Latin America and the Caribbean (CAF), the World Bank, the Economic Commission for Latin America and the Caribbean (ECLAC), UNICEF, and collaborating institutions, such as the Inter-American Development Bank, Santa María Foundation, among others for organizing this meeting, which marks a milestone in cooperation for the reactivation, recovery, and educational transformation of the region.

2. We recognize, renew, and reaffirm our commitments to the agreements established in the I Regional Meeting of Ministers of Education of Latin America and the Caribbean (Buenos Aires, 2017), which outlined the regional educational vision for 2030 and reiterated our commitment to SDG 4. Likewise, we reaffirm the decisions made in the II Regional Meeting (Cochabamba, 2018), where educational quality, equity, inclusion, teaching staff, and lifelong learning were prioritized. Furthermore, we confirm the commitments made in the III Regional Meeting (Buenos Aires, 2022), aiming to deepen our efforts to drive educational reactivation, recovery, and transformation as accelerators for achieving the goals of SDG 4.

3. We congratulate the approval of the Roadmap for the Regional Coordination Mechanism for SDG 4, as well as its Work Plan for the period 2022-2025 by the Regional Steering Committee (RSC) of SDG 4.

4. We appreciate the efforts of the Ministry of Education of the Republic of Colombia, the Ministry of Education of the Republic of Argentina, and the Organization of Ibero-American States for Education, Science, and Culture (OEI) during the period in which they represented the Latin American and Caribbean region in the High-Level Steering Committee of SDG 4-Education 2030.

5. We celebrate the election of the Federative Republic of Brazil and the Republic of Chile, as well as the re-election of the OEI, to represent the Latin American and

Caribbean region in the High-Level Steering Committee of SDG 4-Education 2030 for the biennium 2024-2025. Additionally, we highlight that the President of the Republic of Chile, S.E. Gabriel Boric Font, has assumed the cochairmanship of the Committee for the mentioned biennium.

We acknowledge:

6. That the region has faced enormous challenges in achieving SDG 4, which have been further exacerbated by the impact of emergencies and prolonged crises, including the COVID-19 pandemic. This pandemic heightened educational inequalities and restricted the right to education, particularly for vulnerable and historically marginalized individuals. Additionally, indicators for SDG 4, which had shown significant progress in recent decades, especially those related to access, coverage, and completion, are now exhibiting signs of deceleration.

7. That the achievement of SDG 4 targets is more urgent than ever to ensure the exercise of the human right to education, making schools inclusive, safe, and healthy, ensuring fundamental learning and lifelong learning, promoting education as a necessary condition for sustainable development, active participation of all educational stakeholders, planned and monitored digital transformation addressing all kinds of gaps, including gender disparities, and prioritizing public investment in the education sector.

8. That the global education community at the Transforming Education Summit (TES), convened by the Secretary-General of the United Nations in September 2022, emphasized the need to move 'from commitments to action' and leverage the social mobilization it generated to take urgent actions that transform the fundamental elements of education: its purpose, content, and provision, elevating education on regional and national agendas. This call was reiterated at the regional level during the High-Level Meeting 'Commitment to Action on Basic Learning and its Recovery' (Bogotá, March 2023).

9. That the importance of implementing the agreements of the Commitment to Action for Basic Learning, signed by various countries in the region, emphasizing the need to reduce learning poverty.

We highlight:

10. That reactivation, recovery, transformation, and financing are the priority axes for accelerating SDG 4 targets, with education financing being one of the essential enabling elements. Achieving SDG 4 requires a local and regional, comprehensive, collaborative, and coordinated response to ensure improved education systems in the aftermath of the COVID-19 pandemic.

11. The importance of expanding the scope, financing, and deepening of actions that effectively address the reactivation, recovery, and transformation of educational systems, as well as the relevance of ongoing exchange of best practices, research, and lessons learned based on countries' experiences in these areas.

12. That education is a fundamental right, and its substantive and harmonious exercise with other rights, such as the right to a healthy environment, a life free from violence and discrimination, mobility, and citizen security, among others, allows for the reduction of inequalities and the strengthening of educational communities in the region.

13. That the learning crisis of the last decade was exacerbated by the closure of schools during the pandemic, impacting the well-being of girls, boys, adolescents, young people, and adults, despite the tremendous efforts of various educational stakeholders to prioritize the need for recovery and socio-emotional stability of educational communities.

14. That reactivation and recovery require strengthening educational management practices, pedagogical approaches, and enabling conditions for the application of shared experiences and methods among the countries of the region to enhance learning. This is particularly crucial for the acquisition of knowledge, skills, abilities, attitudes, and values recognized as essential for navigating a world undergoing multiple contextual changes. Examples include digital skills and education for sustainable development, enabling individuals to effectively, securely, freely, and opportunistically navigate an increasingly digital and interconnected world to address local and global challenges.

15. That educational transformation must recognize and strengthen the key role of teachers, prioritizing their training and professional development, promoting innovation, and embracing evidence-based pedagogical approaches where having information and assessments are relevant for comprehensive education and well-being, recognizing diverse purposes and modalities. These efforts should focus not only on enhancing cognitive skills but also on fostering socio-emotional and civic competencies in girls, boys, adolescents, young people, and adults.

16. That transformation entails rethinking what is taught and how it is taught, recognizing the need to implement pedagogies rooted in cooperation, solidarity, critical thinking, and dialogue. It involves developing and consolidating research and assessment of learning at the regional level, as well as designing curricula that draw from the richness of intercultural, interdisciplinary knowledge to advance in reducing all kinds of gaps, including gender disparities, and protecting educational trajectories.

17. That these actions are only possible if they are contextualized, placing the needs of our countries with their specific challenges at the center. This is to strengthen the resilience of educational systems in changing contexts, connecting responses to the educational emergency, sustainable development, and peacebuilding.

18. That young people be part of the educational discussions in our region, as reflected in the priorities and proposals presented in the 'Open Letter from Youth to Ministers of Education of Latin America and the Caribbean.' Likewise, we reaffirm the importance of having a permanent representative in the RSC and that in future meetings of Ministers, they continue to be part of the discussion on education and the future of youth in our region.

We reaffirm:

19. Our commitment to promoting both equity and inclusion in our educational systems, fostering a culture that recognizes, values, and builds upon diversity to enrich the quality of the educational process. In this regard, the positive outcomes achieved during the International Meeting on Educational Quality, organized by the Bolivarian Republic of Venezuela in November 2023, and reflected in the Caracas Declaration, as part of the celebration of the 30th anniversary of UNESCO's Latin American Laboratory for Assessment of the Quality of Education (LLECE), are highly appreciated.

20. That it is essential to ensure enabling conditions for educational reactivation, recovery, and transformation, safeguarding the interests of girls, boys, adolescents, young people, adults, teachers, and families, promoting their active participation in various processes. Digital technologies should be used to support innovative educational experiences based on human interaction, leaving no one behind, especially indigenous populations, Afro-descendants, people with disabilities, and those in mobility situations.

21. The need to prioritize education on the public agenda of our nations, ensuring adequate and sustainable financing to advance actions for educational reactivation, recovery, and transformation. This involves recognizing the necessity of finding optimal budgetary opportunities and practices to promote investment and efficiency in the use of these resources.

Considering the above, we have decided:

22. Expressing that the comprehensive response to educational reactivation, recovery, and transformation must consider that the educational process promotes not only fundamental cognitive learning and competencies but also socio-emotional and civic skills, including digital literacy, education for sustainable development,

and a culture of peace and non-violence. Additionally, attention should be given to the attendance and reintegration of girls, boys, adolescents, young people, and adults into educational systems to safeguard educational trajectories.

23. Strengthening a curricular policy within the framework of flexibility to ensure the development and evaluation of both basic learning and well-being, coexistence, mental and emotional health. This includes teaching and learning processes that acknowledge indigenous knowledge, different contexts, and promote memory, reconciliation, critical and creative thinking, arts, and sports.

24. To boost high-quality initial and ongoing training for our teachers, as well as for the supervision and pedagogical support systems in the classroom, which are essential to strengthen the capacities of teachers, pedagogical innovation, and the promotion of learning for girls, boys, adolescents, young people, and adults.

25. To promote education and literacy policies that foster equity and inclusion, implementing strategies to eliminate access barriers and fostering environments for learning that are free from gender stereotypes, among others, based on the recognition and appreciation of diversity.

26. To urge the states of the region, civil society, United Nations organizations, and cooperation entities to develop comprehensive actions, with national ownership and leadership, to protect the right to education and every educational space from any form of violence, and to provide effective support to families.

27. To strengthen the response to emergencies and prolonged crises, ensuring the continuity of existing initiatives and coordination mechanisms. Additionally, address the specific needs of those most affected by the consequences of pandemics, social and political crises, natural disasters, climate change, and forced displacement.

28. To reiterate that States are called upon to ensure the human right to education, coordinating and encouraging all societal actors who should be part of a comprehensive, plural, and diverse response for the reactivation, recovery, and transformation of education. Driven by a call to action, they should generate education plans with a regional perspective and long-term outlook.

29. To promote initiatives that foster regional cooperation in the design and implementation of policies aimed at ensuring the right to education and facilitating the regional integration of educational systems.

30. To maintain, seek, and expand financial sustainability alternatives for educational systems in Latin America and the Caribbean, keeping an ongoing

dialogue with fiscal authorities and other national and subnational stakeholders involved.

31. That in this context, to generate new approaches for the financing of the education sector, policies and programs will be promoted for an efficient distribution and effective, equitable, and inclusive use of education resources. This involves seeking the integration of financial sustainability of the education sector with the guarantee of access, quality, and relevance of education. To achieve this goal, the generation and installation of capacities that enable better resource management at all levels of the education system will be encouraged. For the development of this component, the support of specialized multilateral organizations and mechanisms may be sought.

32. To promote a conception of financial sustainability in education that, alongside the continuity and increase of educational investment, considers a logic of evaluation and support to enhance efficiency, participation, learning, and educational justice.

33. To promote that educational financing does not regress, but rather achieves and surpasses the financing goals already established at the international level to progress towards the targets of SDG 4.

34. To urge international organizations and multilateral banks collaborating in the educational field to facilitate lines of work and financing for the consolidation of the regional agenda defined in meetings of Ministers of Education.

35. To reiterate support for the continuity of the work carried out by cooperation agencies to conduct a comprehensive monitoring of all SDG 4 targets. Likewise, declare the intention to strengthen national and regional evaluation and monitoring mechanisms for decision-making for the Education 2030 Agenda, and national and regional data and information production, with special attention to the necessary disaggregation to visualize the educational situation of the entire population.

36. To emphasize the importance of considering the achievement of SDG 4 as a cross-cutting element in the discussion of various points on the agenda of the United Nations Summit of the Future (September 2024) and as a substantive element of the United Nations World Social Summit to be held in 2025.

37. To reiterate our commitment to ensuring the right to education for people in mobility situation. We are determined to advance collective, innovative, and intergovernmental actions such as the Regional Framework for Monitoring Students in Displacement, coordinated by UNESCO and the Regional Education Group.

38. To entrust the RSC, the conveners, and co-organizers of this meeting with the approval and establishment of an ad-hoc Working Group and a preliminary proposal for the creation of a Regional Reference Framework on public policies for educational reactivation, recovery, and transformation. This framework should serve as a practical reference to move 'from commitment to action' and accelerate the achievement of SDG 4 goals, deepening the dialogue on education financing as an essential enabling condition for these matters.

39. To entrust the RSC with establishing an ad-hoc Working Group to develop a preliminary proposal on inclusive and effective financing within the Regional Reference Framework on public policies for educational reactivation, recovery, and transformation. This Working Group should be composed of conveners, co-organizers, collaborating organizations, and strategic partners of this ministerial meeting. Additionally, request UNESCO to provide technical support to the mentioned Working Group for a period of two (2) years, ensuring the prevention of duplication with other entities to enhance efficiency.

40. To promote the institutionalization of educational coordination mechanisms, for which we request the Executive Secretariat of the RSC to analyze and present scenarios for the establishment of this Ministers' Assembly as a platform for intergovernmental coordination.

41. With the aim of enriching the dialogue within the RSC, henceforth, it will be composed of all countries in Latin America and the Caribbean, led by the countries representing us in the HighLevel Steering Committee of SDG 4.

42. To express gratitude to the Government of Chile for its outstanding work in the preparation of this ministerial meeting and for its support, along with Brazil, for SDG 4 at the regional and global levels to ensure inclusive, equitable, and quality education, promoting lifelong learning opportunities for all.

43. To recognize UNESCO, CAF, World Bank, ECLAC, UNICEF, and collaborating institutions, such as the Inter-American Development Bank, Santa María Foundation, among others for their technical and financial support for the realization of this meeting and their contribution to fostering discussions and the exchange of best practices and experiences on education. This constitutes a valuable exercise in knowledge building and mobilization from and for the region.

UN General Assembly, Resolution A/78/L.49, Seizing the Opportunities of Safe, Secure and Trustworthy Artificial Intelligence Systems for Sustainable Development (2024)

The General Assembly,

Reaffirming international law, in particular the Charter of the United Nations, and recalling the Universal Declaration of Human Rights,

Reaffirming also its resolution 70/1 of 25 September 2015, entitled "Transforming our world: the 2030 Agenda for Sustainable Development", its resolution 69/313 of 27 July 2015 on the Addis Ababa Action Agenda of the Third International Conference on Financing for Development, and the political declaration adopted at the high-level political forum on sustainable development convened under the auspices of the General Assembly contained in the annex to its resolution 78/1 of 29 September 2023,

Recalling its resolutions 77/320 of 25 July 2023 on the impact of rapid technological change on the achievement of the Sustainable Development Goals and targets, 78/132 of 19 December 2023 on information and communications technologies for sustainable development, 78/160 of 19 December 2023 on science, technology and innovation for sustainable development, 78/213 of 19 December 2023 on the promotion and protection of human rights in the context of digital technologies, 77/211 of 15 December 2022 on the right to privacy in the digital age, 70/125 of 16 December 2015 on the overall review of the implementation of the outcomes of the World Summit on the Information Society, all the outcomes of the World Summit on the Information Society, including the Geneva Declaration of Principles, the Geneva Plan of Action, the Tunis Commitment4 and the Tunis Agenda for the Information Society, and the declaration on the commemoration of the seventy-fifth anniversary of the United Nations contained in its resolution 75/1 of 21 September 2020,

Taking note of the efforts of the International Telecommunication Union, in partnership with 40 United Nations bodies, to convene the Artificial Intelligence for Good platform, including its annual summit and the launch of the International Telecommunication Union's Artificial Intelligence Repository to identify responsible and practical applications of artificial intelligence to advance the Sustainable Development Goals; and the adoption by the General Conference of the United Nations Educational, Scientific and Cultural Organization of its

Recommendation on the Ethics of Artificial Intelligence of 23 November 2021, its implementation plan, including the Readiness Assessment Methodology and the Ethical Impact Assessment and the Global Forum on the Ethics of Artificial Intelligence; as well as taking note of the Guiding Principles on Business and Human Rights: Implementing the United Nations "Protect, Respect and Remedy" Framework, as endorsed by the Human Rights Council in its resolution 17/4 of 16 June 2011; and the work of the Office of the United Nations High Commissioner for Human Rights regarding artificial intelligence,

Taking note also of the report of the Secretary-General of the United Nations entitled "Road map for digital cooperation", as well as the establishment of the Office of the Secretary-General's Envoy on Technology to coordinate its implementation and the establishment by the Secretary-General of a multistakeholder High-level Advisory Body on Artificial Intelligence and of its interim report issued on 21 December 2023, and looking forward to its final report,

Recognizing that safe, secure and trustworthy artificial intelligence systems – which, for the purpose of this resolution, refers to artificial intelligence systems in the non-military domain, whose life cycle includes the stages: pre-design, design, development, evaluation, testing, deployment, use, sale, procurement, operation and decommissioning, are such that they are human-centric, reliable, explainable, ethical, inclusive, in full respect, promotion and protection of human rights and international law, privacy preserving, sustainable development oriented, and responsible – have the potential to accelerate and enable progress towards the achievement of all 17 Sustainable Development Goals and sustainable development in its three dimensions – economic, social and environmental – in a balanced and integrated manner; promote digital transformation; promote peace; overcome digital divides between and within countries; and promote and protect the enjoyment of human rights and fundamental freedoms for all, while keeping the human person at the centre,

Recognizing also that the improper or malicious design, development, deployment and use of artificial intelligence systems, such as without adequate safeguards or in a manner inconsistent with international law, pose risks that could hinder progress towards the achievement of the 2030 Agenda for Sustainable Development and its Sustainable Development Goals and undermine sustainable development in its three dimensions – economic, social and environmental; widen digital divides between and within countries; reinforce structural inequalities and biases; lead to discrimination; undermine information integrity and access to information; undercut the protection, promotion and enjoyment of human rights and fundamental freedoms, including the right not to be subject to unlawful or arbitrary interference with one's privacy; and increase the potential risk for accidents and compound threats from malicious actors,

Recognizing further the rapid acceleration of the design, development, deployment and use of artificial intelligence systems and rapid technological change, and their potential impact in accelerating the achievement of the Sustainable Development Goals, therefore stressing the urgency of achieving global consensus on safe, secure and trustworthy artificial intelligence systems; facilitating inclusive international cooperation to formulate and use effective, internationally interoperable safeguards, practices and standards that promote innovation and prevent the fragmentation of the governance of safe, secure and trustworthy artificial intelligence systems; and recognizing also existing artificial intelligence and other digital divides, and the varying levels of technological development between and within countries, that developing countries face unique challenges in keeping pace with this rapid acceleration, which cause obstacles to sustainable development, the need to narrow the existing disparities between developed and developing countries in terms of conditions, possibilities and capacities, therefore stressing also the urgency of strengthening capacity building and technical and financial assistance to developing countries to close digital divides between and within countries and support developing countries' effective, equitable and meaningful participation and representation in international processes and forums on the governance of artificial intelligence systems,

Recognizing also that the governance of artificial intelligence systems is an evolving area and the need for continued discussions on possible governance approaches that are appropriate, based on international law, interoperable, agile, adaptable, inclusive, responsive to the different needs and capacities of developed and developing countries alike and for the benefit of all, as the technology and our understanding of it develops,

1. *Resolves* to bridge the artificial intelligence and other digital divides between and within countries;

2. *Resolves* to promote safe, secure and trustworthy artificial intelligence systems to accelerate progress towards the full realization of the 2030 Agenda for Sustainable Development, further bridging the artificial intelligence and other digital divides between and within countries; and stresses the need for the standard of safe, secure and trustworthy artificial intelligence systems to promote, not hinder, digital transformation and equitable access to their benefits in order to achieve all Sustainable Development Goals and sustainable development in its three dimensions – economic, social and environmental – and address other shared global challenges, particularly for developing countries;

3. *Encourages* Member States and invites multi-stakeholders from all regions and countries, within their respective roles and responsibilities, including from the private sector, international and regional organizations, civil society, the media, academia and research institutions and technical communities and individuals, to develop and support regulatory and governance approaches and frameworks related to safe, secure and trustworthy artificial intelligence systems that create an enabling ecosystem at all levels, including for innovation, entrepreneurship and the dissemination of knowledge and technologies on mutually agreed terms, recognizing that effective partnership and cooperation between Governments and multi-stakeholders is necessary in developing such approaches and frameworks;

4. *Calls upon* Member States and invites other stakeholders to take action to cooperate with and provide assistance to developing countries towards inclusive and equitable access to the benefits of digital transformation and safe, secure and trustworthy artificial intelligence systems, including by:

(a) Expanding participation of all countries, in particular developing countries, in digital transformation to harness the benefits and effectively participate in the development, deployment and use of safe, secure and trustworthy artificial intelligence systems, including by capacity building relating to artificial intelligence systems, recognizing that promoting knowledge sharing activities and the transfer of technology on mutually agreed terms is an important aspect of building capacity, stressing the need to close the artificial intelligence and other digital divides; and increase digital literacy;

(b) Enhancing digital infrastructure connectivity and access to technological innovations through stronger partnerships to help developing countries effectively participate throughout the life cycle of artificial intelligence systems and accelerate the inclusive and positive contribution of artificial intelligence systems to society, including towards the full realization of the 2030 Agenda and its Sustainable Development Goals, while ensuring that artificial intelligence systems around the world are safe, secure and trustworthy throughout their life cycle;

(c) Enhancing the ability of developing countries, in particular the least developed countries, to address major structural impediments and lift obstacles to accessing the benefits of new and emerging technologies and artificial intelligence innovation to achieve all 17 Sustainable Development Goals, including through scaling up the use of scientific sources, affordable technology, research and development, including through strengthened partnerships;

(d) Aiming to increase funding for Sustainable Development Goals related research and innovation related to digital technologies and safe, secure and trustworthy artificial intelligence systems and build capacity in all regions and countries to contribute to and benefit from this research;

(e) Enabling international innovation-based environments to enhance the ability of developing countries to develop technical expertise and capacities, harness data and compute resources, and national regulatory and governance approaches, frameworks and procurement capacity, and create an inclusive enabling environment at all levels for safe, secure and trustworthy artificial intelligence systems-based solutions;

(f) Urgently mobilizing means of implementation such as technology transfer on mutually agreed terms, capacity building to close the artificial intelligence and other digital divides, technical assistance and financing to developing countries related to artificial intelligence systems in accordance with developing countries' national needs, policies and priorities;

(g) Promoting the access to and design, development, deployment and use of safe, secure and trustworthy artificial intelligence systems to achieve sustainable development in its three dimensions – economic, social and environmental;

5. *Emphasizes* that human rights and fundamental freedoms must be respected, protected and promoted throughout the life cycle of artificial intelligence systems, calls upon all Member States and, where applicable, other stakeholders to refrain from or cease the use of artificial intelligence systems that are impossible to operate in compliance with international human rights law or that pose undue risks to the enjoyment of human rights, especially of those who are in vulnerable situations, and reaffirms that the same rights that people have offline must also be protected online, including throughout the life cycle of artificial intelligence systems;

6. *Encourages* all Member States, where appropriate, in line with their national priorities and circumstances and while implementing their distinct national regulatory and governance approaches and frameworks, and, where applicable, other stakeholders to promote safe, secure and trustworthy artificial intelligence systems in an inclusive and equitable manner, and for the benefit of all, and foster an enabling environment for such systems to address the world's greatest challenges, including achieving sustainable development in its three dimensions – economic, social and environmental – with specific consideration of developing countries and leaving no one behind by:

(a) Promoting the development and implementation of domestic regulatory and governance approaches and frameworks, in line with their respective national, and where applicable subnational, policies and priorities and obligations under international law, to support responsible and inclusive artificial intelligence innovation and investment for sustainable development, while simultaneously promoting safe, secure and trustworthy artificial intelligence systems;

(b) Encouraging effective measures, that promote innovation for the internationally interoperable identification, classification, evaluation, testing, prevention and mitigation of vulnerabilities and risks during the design and development and prior to the deployment and use of artificial intelligence systems;

(c) Encouraging the incorporation of feedback mechanisms to allow evidence based discovery and reporting by end-users and third parties of technical vulnerabilities and, as appropriate, misuses of artificial intelligence systems and artificial intelligence incidents following their development, testing and deployment to address them;

(d) Raising public awareness and understanding of the core functions, capabilities, limitations and domains of appropriate civil use of artificial intelligence systems;

(e) Fostering the development, implementation and disclosure of mechanisms of risk monitoring and management, mechanisms for securing data, including personal data protection and privacy policies, as well as impact assessments as appropriate, across the life cycle of artificial intelligence systems;

(f) Strengthening investment in developing and implementing effective safeguards, including physical security, artificial intelligence systems security, and risk management across the life cycle of artificial intelligence systems;

(g) Encouraging the development and deployment of effective, accessible, adaptable, internationally interoperable technical tools, standards or practices, including reliable content authentication and provenance mechanisms – such as watermarking or labelling, where technically feasible and appropriate, that enable users to identify information manipulation, distinguish or determine the origins of authentic digital content and artificial intelligence-generated or manipulated digital content – and increasing media and information literacy; 24-04670 5/8 A/78/L.49

(h) Facilitating the development and implementation of effective, internationally interoperable frameworks, practices and standards for training and testing artificial intelligence systems to enhance policymaking and to help protect individuals from all forms of discrimination, bias, misuse or other harm, and avoid reinforcing or perpetuating discriminatory or biased applications and outcomes throughout the life cycle of artificial intelligence systems, including, for example, by analysing and mitigating bias encoded in datasets and otherwise combating algorithmic discrimination and bias, while not inadvertently or disproportionally impacting the positive development, access and uses of other users and beneficiaries;

(i) Encouraging, where appropriate and relevant, the implementation of appropriate safeguards to respect intellectual property rights, including copyright protected content, while promoting innovation;

(j) Safeguarding privacy and the protection of personal data when testing and evaluating systems, and for transparency and reporting requirements in compliance with applicable international, national and subnational legal frameworks, including on the use of personal data throughout the life cycle of artificial intelligence systems;

(k) Promoting transparency, predictability, reliability and understandability throughout the life cycle of artificial intelligence systems that make or support decisions impacting end-users, including providing notice and explanation, and promoting human oversight, such as, for example, through review of automated decisions and related processes or, where appropriate and relevant, human decisionmaking alternatives or effective redress and accountability for those adversely impacted by automated decisions of artificial intelligence systems;

(l) Strengthening investment in developing and implementing effective safeguards, including risk and impact assessments, throughout the life cycle of artificial intelligence systems to protect the exercise of and mitigate against the potential impact on the full and effective enjoyment of human rights and fundamental freedoms;

(m) Promoting artificial intelligence systems that advance, protect and preserve linguistic and cultural diversity, taking into account multilingualism in their training data and throughout the life cycle of the artificial intelligence system, particularly for the large language models;

(n) Intensifying information-sharing on mutually agreed terms among entities with roles across the life cycle of artificial intelligence systems to identify, understand and act using scientific and evidence-based best practices, policies and approaches to artificial intelligence systems to maximize the benefits while mitigating the potential risks across the life cycle of artificial intelligence systems, including advanced artificial intelligence systems;

(o) Encouraging research and international cooperation to understand, balance and address the potential benefits and risks related to the role of artificial intelligence systems in bridging digital divides and achieving all 17 Sustainable Development Goals, including the role of scaling up of digital solutions such as open-source artificial intelligence systems;

(p) Calling upon Member States to adopt specific measures to close the gender digital divide and to ensure that particular attention is paid to access, affordability, digital literacy, privacy and online safety, to enhance the use of digital technologies, including artificial intelligence systems, and to mainstream a disability, gender and racial equality perspective in policy decisions and the frameworks that guide them;

(q) Encouraging research and international cooperation to develop measures for the identification and assessment of the impacts of the deployment of artificial 24-04670 6/8 A/78/L.49 intelligence systems on labour markets, and providing support for the mitigation of potential negative consequences for workforces, especially in developing countries, in particular the least developed countries, and fostering programmes aimed at digital training, capacity building, supporting innovation and enhancing access to benefits of artificial intelligence systems;

7. *Recognizes also* that data is fundamental to the development and operation of artificial intelligence systems; emphasizes that the fair, inclusive, responsible and effective data governance, improving data generation, accessibility and infrastructure, and the use of digital public goods are essential to harnessing the potential of safe, secure and trustworthy artificial intelligence systems for sustainable development, and urges Member States to share best practices on data governance and to promote international cooperation, collaboration and assistance on data governance for greater consistency and interoperability, where feasible, of approaches for advancing trusted cross-border data flows for safe, secure and trustworthy artificial intelligence systems, and make its development more inclusive, equitable, effective and beneficial to all;

8. *Acknowledges* the importance of continuing the discussion on developments in the area of artificial intelligence governance so that international approaches keep pace with the evolution of artificial intelligence systems and their uses; and encourages continued efforts by the international community to promote inclusive research, mapping and analysis that benefit all parties on the potential impacts and applications that artificial intelligence systems and rapid technological change can have in the development of existing and new and emerging technologies and on accelerating the achievement of all 17 Sustainable Development Goals, and to inform how to develop, promote and implement effective, internationally interoperable safeguards, practices, standards and tools for artificial intelligence designers, developers, evaluators, deployers, users and other stakeholders for safe, secure and trustworthy artificial intelligence systems; as well as stresses the need for Governments, the private sector, civil society, international and regional organizations, academia and research institutions and technical communities and all other stakeholders to continue to work together, as appropriate; as well as acknowledges the need for more cohesive, effective, coordinated and inclusive

engagement and participation of all communities, particularly from developing countries, in the inclusive governance of safe, secure and trustworthy artificial intelligence systems;

9. *Encourages* the private sector to adhere to applicable international and domestic laws and act in line with the United Nations Guiding Principles on Business and Human Rights: Implementing the United Nations "Protect, Respect and Remedy" Framework; acknowledges the importance of more inclusive and equitable access to the benefits of safe, secure and trustworthy artificial intelligence systems; and recognizes the need for increased collaboration, including between and within the public and private sectors and civil society, academia and research institutions and technical communities, to provide and promote fair, open, inclusive and non-discriminatory business environment, economic and commercial activities, competitive ecosystems and marketplaces across the life cycle of safe, secure and trustworthy artificial intelligence; as well as encourages Member States to develop policies and regulations to promote competition in safe, secure and trustworthy artificial intelligence systems and related technologies, including by supporting and enabling new opportunities for small businesses and entrepreneurs and technical talent, and enabling fair competition in the artificial intelligence marketplace, through critical investment, especially for developing countries;

10. *Calls upon* specialized agencies, funds, programmes, other entities, bodies and offices, and related organizations of the United Nations system, within their respective mandates and resources, to continue to assess and enhance their response 24-04670 7/8 A/78/L.49 to leverage the opportunities and address the challenges posed by artificial intelligence systems in a collaborative, coordinated and inclusive manner, through appropriate inter-agency mechanisms, including by conducting research, mapping and analysis that benefit all parties on the potential impacts and applications; reporting on progress and challenges in addressing issues; and cooperating with and assisting developing countries in capacity building, access and sharing the benefits of safe, secure and trustworthy artificial intelligence systems in achieving all 17 Sustainable Development Goals and sustainable development in its three dimensions – economic, social and environmental; stressing the need to close artificial intelligence and other digital divides between and within countries;

11. *Recalls* its resolution 76/307 of 8 September 2022 and its decision 77/568 of 1 September 2023 on the modalities and the scope of the Summit of the Future and, in this regard, looks forward to the development of a global digital compact;

12. *Looks forward* also to the overall review by the General Assembly, in 2025, of the progress made since the World Summit on the Information Society;

13. *Acknowledges* that the United Nations system, consistent with its mandate, uniquely contributes to reaching global consensus on safe, secure and trustworthy artificial intelligence systems, that is consistent with international law, in particular the Charter of the United Nations; the Universal Declaration of Human Rights; and the 2030 Agenda for Sustainable Development, including by promoting inclusive international cooperation and facilitating the inclusion, participation and representation of developing countries in deliberations.

UN General Assembly, Resolution A/RES/78/311, Enhancing International Cooperation on Capacity-Building of Artificial Intelligence (2024)

The General Assembly,

Reaffirming international law, in particular the Charter of the United Nations, and recalling the Universal Declaration of Human Rights,

Reaffirming also its resolution 70/1 of 25 September 2015, entitled "Transforming our world: the 2030 Agenda for Sustainable Development", its resolution 69/313 of 27 July 2015 on the Addis Ababa Action Agenda of the Third International Conference on Financing for Development, the political declaration adopted at the high-level political forum on sustainable development convened under the auspices of the General Assembly contained in the annex to its resolution 78/1 of 29 September 2023, and its resolution 70/125 of 16 December 2015, entitled "Outcome document of the high-level meeting of the General Assembly on the overall review of the implementation of the outcomes of the World Summit on the Information Society",

Reaffirming its resolutions 78/265 of 21 March 2024, entitled "Seizing the opportunities of safe, secure and trustworthy artificial intelligence systems for sustainable development", 77/320 of 25 July 2023, entitled "Impact of rapid technological change on the achievement of the Sustainable Development Goals and targets", 78/132 of 19 December 2023, entitled "Information and communications technologies for sustainable development", and 78/213 of 19 December 2023, entitled "Promotion and protection of human rights in the context of digital technologies",

Recognizing that rapid technological change, including the rapid advancement of artificial intelligence has the potential to bring new opportunities for socioeconomic development and accelerate the progress and achievement of the Sustainable Development Goals and sustainable development in its three dimensions – economic, social and environmental – in a balanced and integrated manner, while recognizing also that the improper or malicious design, development, deployment and use of artificial intelligence systems, such as without adequate safeguards or in a manner inconsistent with international law, could pose potential risks and challenges,

Recognizing that eradicating poverty in all its forms and dimensions, including extreme poverty, is the greatest global challenge for sustainable development and an indispensable requirement for sustainable development,

Stressing that artificial intelligence systems should be safe, secure and trustworthy – whose life cycle includes the stages: pre-design, design, development, evaluation, testing, deployment, use, sale, procurement, operation and decommissioning, are such that they are human-centric, reliable, explainable, ethical, inclusive, in full respect, promotion and protection of human rights and international law – in line with the principle of artificial intelligence for good for all, with the vision of a people-centred, inclusive and development-oriented information society, bearing in mind that this resolution focuses on international cooperation on artificial intelligence capacity-building in the non-military domain and does not touch the development or use of artificial intelligence for military purposes,

Emphasizing that Member States should enjoy equal opportunities in the design, development, deployment, decommissioning and use of artificial intelligence, while respecting intellectual property rights and promoting innovation,

Noting with concern that artificial intelligence and other digital divides between and within countries continue to widen, and developing countries face unique challenges in keeping pace with the rapid acceleration of artificial intelligence development, in particular in terms of the design, development, deployment and use of artificial intelligence, thus stressing the need and urgency to narrow the disparities and assist developing countries in artificial intelligence capacity-building so that they will not be further left behind,

Noting also with concern that actions inconsistent with international law and the Charter of the United Nations could hinder the development and innovation of artificial intelligence and global cooperation,

Noting both the positive benefits of open-source software, open models and open data, among other methods and business models, in spreading the benefits of artificial intelligence, as well as the potential risks, and recognizing the importance of the participation of developing countries in relevant global cooperation,

Recognizing that the lack of digital infrastructure connectivity and skills, including education, expertise and human capacity, remains a fundamental challenge in many developing countries, in particular the least developed countries, which can contribute to artificial intelligence and other digital divides, including different levels of readiness to make use of and benefit from artificial intelligence,

Stressing the need and urgency to bridge artificial intelligence and other digital divides between and within countries, and emphasizing the quality of access to the benefits of artificial intelligence, therefore stressing the importance of providing adequate public and private financing, mobilizing action-oriented means of implementation, such as knowledge sharing activities and the transfer of technology on mutually agreed terms, and capacity-building assistance for developing countries in line with their national needs, policies and priorities,

Noting the important role that international, regional and subregional organizations and international financial institutions, the private sector, companies, civil society, the media, academia and research institutions and technical communities and individuals, and other relevant stakeholders could play in enhancing artificial intelligence capacity-building and promoting safe, secure and trustworthy artificial intelligence systems for sustainable development,

Stressing the importance of enhancing coordination and cooperation among and within countries and increasing investment in artificial intelligence capacity-building, with specific consideration of the needs, priorities and conditions of developing countries, and encouraging developed countries and developing countries in a position to do so, as well as international organizations and all relevant stakeholders, to take active measures,

Recognizing that the United Nations system has an important role to play in artificial intelligence capacity-building, and emphasizing that it is important to enhance international cooperation and multi-stakeholder collaboration on capacity building and support developing countries' effective, equitable and meaningful participation and representation in international processes and forums on the governance of artificial intelligence systems,

1. *Resolves* to bridge the artificial intelligence and other digital divides between and within countries, and to enhance international cooperation on capacity building in developing countries, including through North-South, South-South and triangular cooperation, with full consideration of the needs, policies and priorities of developing countries, with the aim of harnessing the benefits of artificial intelligence, minimizing its risks, and accelerating innovation and progress toward the achievement of all 17 Sustainable Development Goals;

2. *Encourages* Member States, where appropriate and in line with their national circumstances and priorities, to incorporate capacity-building and the design, development, deployment and use of artificial intelligence into their national development plans and strategies, in full respect, promotion and protection of human rights and international law, and make necessary investments in this regard, and also

encourages the international community, as well as relevant stakeholders, to increase financing and technical assistance to developing countries in the field of capacity building, including by drawing on voluntary cooperative initiatives;

3. *Calls upon* the international community to foster an enabling environment for international cooperation on artificial intelligence capacity-building, and emphasizes that international law, in particular the Charter of the United Nations, as well as human rights and fundamental freedoms, must be respected;

4. *Calls upon* Member States to strengthen cooperation and partnerships that bring benefits for all to address major structural impediments and lift obstacles, including through expansion of access to infrastructure, knowledge and skills, with a view to promoting universal and meaningful digital connectivity, which can help lay the foundation for digital transformation and equitable and inclusive access to the benefits of digital and artificial intelligence development and innovations;

5. *Encourages* Member States, in particular developed countries and those developing countries in a position to do so, to increase capacity-building cooperation, including policy exchanges, knowledge sharing activities and the transfer of technology on mutually agreed terms, technical assistance, lifelong learning, personnel training, skilling of workforce, international research cooperation, voluntary joint international research laboratories and artificial intelligence capacity building centres, with full consideration of developing countries' national needs, policies and priorities, and to hold training courses, seminars and workshops, among others for sharing experiences and best practices;

6. *Calls upon* the international community to provide and promote a fair, open, inclusive and non-discriminatory business environment across the life cycle of safe, secure and trustworthy artificial intelligence systems;

7. *Encourages* Member States to consider the benefits and risks when scaling up the responsible use of digital solutions, such as open-source artificial intelligence and digital public infrastructure, among other methods and business models; to promote, protect and preserve linguistic and cultural diversity, taking into account multilingualism in their training data, particularly for the large language models; to adopt proactive measures to counteract racism, discrimination and other forms of algorithmic bias; and to enable fair competition in the artificial intelligence market place, promote an innovation environment and the use of digital public goods to harness the potential of safe, secure and trustworthy artificial intelligence systems;

8. *Encourages* Governments, the private sector, international, regional and subregional organizations, civil society, the media, academia and research institutions and technical communities and individuals, to strengthen international

cooperation on artificial intelligence capacity-building and take necessary and positive steps, as appropriate, to eliminate barriers faced by all people, women and girls, persons with disabilities, Indigenous Peoples, local communities, children and youth, those living in poverty and in rural and remote areas, and those in vulnerable situations, and to ensure the full enjoyment of artificial intelligence benefits for all, improve digital inclusion and realize gender equality and the empowerment of women and girls, through investment, education, training, technological innovation, use and application;

9. Calls upon specialized agencies, funds, programmes, other entities, bodies and offices, and related organizations of the United Nations system, within their respective mandates and resources, to leverage the opportunities and address the challenges in enhancing international cooperation on artificial intelligence capacity building, including through using appropriate inter-agency mechanisms, conducting research, mapping and analysis, reporting on progress and challenges in this field, and leveraging their resources and expertise to provide tailored assistance;

10. *Calls upon* other international, regional and subregional organizations and international financial institutions and all relevant stakeholders, within their respective mandates and resources, to strengthen artificial intelligence capacity building in developing countries, and to carry out relevant cooperation and coordination, fully utilizing the existing international and regional mechanisms and platforms;

11. *Encourages* Member States, in particular developing countries, to make use of capacity-building resources relating to artificial intelligence in the United Nations system and other international, regional and subregional organizations, to enhance capacity-building, including support for the preparation of national digital and artificial intelligence strategies in line with their national needs, policies and priorities, ensuring that those investments lead to long-term and sustainable results;

12. *Supports* the United Nations in playing a central and coordinating role in international development cooperation;

13. *Calls upon* the United Nations system to enhance action-oriented international cooperation on artificial intelligence capacity-building, including through voluntary and transparent cooperation frameworks or initiatives, which also promotes the meaningful participation of developing countries in artificial intelligence processes;

14. *Looks forward* to the final report of the High-level Advisory Body on Artificial Intelligence, and the discussions on capacity-building related to artificial intelligence in the context of the Global Digital Compact, the Pact for the Future,

and relevant follow-up processes in the United Nations as appropriate; 15. Requests the Secretary-General to report to the General Assembly at its eightieth session, within existing resources, processes and documents, on unique challenges faced by developing countries in artificial intelligence capacity-building, with recommendations that address those challenges.

Seoul Declaration for Safe, Innovative and Inclusive AI by Participants Attending the Leaders' Session of the AI Seoul Summit
(2024)

1. We, world leaders representing Australia, Canada, the European Union, France, Germany, Italy, Japan, the Republic of Korea, the Republic of Singapore, the United Kingdom, and the United States of America, gathered at the AI Seoul Summit on 21st May 2024, affirm our common dedication to fostering international cooperation and dialogue on artificial intelligence (AI) in the face of its unprecedented advancements and the impact on our economies and societies.

2. Building on the work laid out at the AI Safety Summit held at Bletchley Park in the United Kingdom on November 2023, we recognize that AI safety, innovation, and inclusivity are inter-related goals and that it is important to encompass these priorities in international discussions on AI governance to address the broad spectrum of opportunities and challenges that the design, development, deployment, and use of AI presents and may present.

3. We recognize the importance of interoperability between AI governance frameworks in line with a risk-based approach to maximize the benefits and address the broad range of risks from AI, to ensure the safe, secure, and trustworthy design, development, deployment, and use of AI. We continue to focus on supporting the operationalisation of the Hiroshima Process International Code of Conduct for Organizations Developing Advanced AI Systems. We recognize the particular responsibility of organizations developing and deploying frontier AI, and, in this regard, note the Frontier AI Safety Commitments.

4. We support existing and ongoing efforts of the participants to this Declaration to create or expand AI safety institutes, research programmes and/or other relevant institutions including supervisory bodies, and we strive to promote cooperation on safety research and to share best practices by nurturing networks between these organizations. In this regard, we welcome the Seoul Statement of Intent toward International Cooperation on AI Safety Science, which is annexed to this Declaration.

5. We call for enhanced international cooperation to advance AI safety, innovation and inclusivity to harness human-centric AI to address the world's greatest challenges, to protect and promote democratic values, the rule of law and human rights, fundamental freedoms and privacy, to bridge AI and digital divides between

and within countries, thereby contributing to the advancement of human well-being, and to support practical applications of Al including to advance the UN Sustainable Development Goals.

6. We advocate for policy and governance frameworks, including risk-based approaches, that foster safe, innovative and inclusive Al ecosystems. Frameworks should facilitate a virtuous cycle between human creativity and the development and use of Al, promote socio-cultural, linguistic, and gender diversity, and promote environmentally sustainable development and use of technology and infrastructure throughout the life-cycle of commercially and publicly available AI systems.

7. We affirm the importance of active multi-stakeholder collaboration, including governments, the private sector, academia, and civil society to cultivate safe, innovative and inclusive Al ecosystems, and the importance of cross-border and cross-disciplinary collaboration. Recognizing that all states will be affected by the benefits and risks of AI, we will actively include a wide range of international stakeholders in conversations around AI governance.

8. We aim to strengthen international cooperation on Al governance through engagement with other international initiatives at the UN and its bodies, G7, G20, the Organization for Economic Co-operation and Development (OECD), the Council of Europe, and the Global Partnership on AI (GPAI). In this light, we acknowledge the Hiroshima AI Process Friends Group, welcome the recently updated OECD AI principles, and the recent adoption by consensus of the United Nations General Assembly resolution "Seizing the opportunities of safe, secure and trustworthy artificial intelligence systems for sustainable development" that solidified the global understanding on the need for safeguards for Al systems and the imperative to develop, deploy, and use Al for good, and welcome discussions on the Global Digital Compact in advance of the Summit of the Future in September 2024 and look forward to the final report of the UN Secretary-General's High-level Advisory Body on AI (HLAB).

9. Acknowledging the value of Al Summit dialogues as a high-level forum to advance discussion on Al governance which facilitates Al safety, innovation and inclusivity, we look forward to our third gathering at the upcoming AI Action Summit to be held in France.

Cartagena de Indias Declaration for Governance, the Construction of Artificial Intelligence (AI) Ecosystems and the Promotion of AI Education in an Ethical and Responsible Manner in Latin America and the Caribbean (2024)

1. In Cartagena de Indias, Colombia, on August 8 and 9, 2024, the Ministers and High Authorities attending this Summit representing the following countries: Argentina, Brazil, Chile, Colombia, Costa Rica, Cuba, Curacao, Dominican Republic, Ecuador, Guatemala, Guyana, Honduras, Panama, Paraguay, Peru, Suriname and Uruguay.

2. Highlighted the importance of the regional process initiated with the first Ministerial and High Authorities Summit on the Ethics of Artificial Intelligence (AI) in Latin America and the Caribbean, hosted by the Government of Chile, in Santiago in 2023; and the corresponding AI Working Group for the region, with a view to holding the second summit in Montevideo in 2024, organized by the Government of Uruguay, and the development of a regional Roadmap to address AI priorities.

3. They recognized the importance of strengthening coordination and cooperation in the digital sphere, especially in the context of the Global Digital Compact to be adopted in the framework of the UN Summit of the Future on "Multilateral Solutions for a Better Tomorrow". The Latin American and Caribbean Ministerial Summit held in Colombia seeks to make a substantive contribution to the debate on the construction of ecosystems, education and governance of AI, to address the challenge of channeling technological development with full respect for human rights and dignity, sustainability and gender equity, avoiding any form of bias and discrimination, and taking advantage of its potential for the strength of our democratic systems.

4. They highlighted the support and commitment of the United Nations system through organizations such as the United Nations Educational, Scientific and Cultural Organization (UNESCO), and also acknowledge the support of other international and multilateral organizations such as the Development Bank of Latin America and the Caribbean (CAF), the European Union (EU), the Inter-American Development Bank (IDB), the Organization of American States (OAS), and the Economic Commission for Latin America and the Caribbean (ECLAC), in the development of this Latin American Summit on IA and the promotion of spaces for discussion regarding ecosystems, education and governance in the area of IA1.

5. In this context, the signatory countries, bearing in mind the progress achieved in other regional groups and processes, among them the Digital Agenda for Latin America and the Caribbean (eLAC), and the importance of generating synergies and not duplicating efforts, express the following:

Development of Ecosystems for the deployment of AI.

5.1 Recognize that cooperation among the countries of Latin America and the Caribbean is important for the construction of ecosystems that allow the development and deployment of AI in an ethical, safe, inclusive, efficient and dynamic manner, so that this technology becomes the catalyst for local innovation, sustainable development and growth of Latin American and Caribbean economies with the objective of reducing economic, social and digital gaps in the region.

5.2 Reaffirm the commitment to promote exchanges of knowledge, information, best practices and other resources that can be made available in accordance with the laws of each country, in order to develop favorable ecosystems for the deployment of AI in the region. In this line, exchanges and collaborations will be promoted that, among others, may occur in relation to infrastructure, computational capacity, scientific and technical advances for the development and sustainable implementation of AI. In addition, an exchange of knowledge on digital solutions will be promoted, as well as on AI applications and projects through collaboration among experts.

5.3 They recognize the importance of having standardized diagnostics to share information, lessons and good practices, such as the UNESCO Readiness Assessment Methodology (RAM) or the AI Readiness Assessment (AIRA) of the United Nations Development Programme (UNDP), among others. In addition, the countries recognize the importance of other national and international instruments or initiatives that are highly relevant for Latin American and Caribbean countries in the development of AI ecosystems.

5.4 In order to strengthen AI ecosystems in Latin America and the Caribbean, they will promote the development and deployment of AI-based solutions to achieve the United Nations Sustainable Development Goals.

5.5 Promote the exchange of best practices for the construction and development of general purpose and advanced models in Spanish and Portuguese, with data that take into account the cultural, social and linguistic characteristics of the countries of Latin America and the Caribbean.

5.6 Foster the exchange of knowledge and best practices regarding the use of AI in the public sector. Likewise, the countries will exchange experiences in relation to the development of solutions that improve the interaction between people and public entities.

Education and use of AI

5.7 Promote education, training and capacity building on digital issues, as well as the exchange of best practices regarding the use of AI in the educational system, in order to train people in digital capabilities and skills, so that the human resources of each country are prepared for the development and evolution of AI in the workplace, taking as a reference other instruments or initiatives developed by international and multilateral organizations with an impact on the region.

5 .8 Promote the exchange of best practices regarding the development of AI training and certification programs for professionals and students that address basic concepts to advanced skills, encouraging the workforce to be trained and equipped to work in an AI-driven environment.

5.9 Promote the appropriation and democratization of access, use and development of AI, so that the population in Latin America and the Caribbean contributes to the efficient, inclusive, ethical, safe and responsible development and use of this technology. Likewise, the signatory countries will promote the exchange of best practices in the implementation of AI centers, AI laboratories and educational programs, which will be developed to promote scientific research in AI and promote the appropriation of this technology.

5.10 They reaffirm the importance of promoting the development of adaptive educational systems that foster inclusive digital literacy and emphasize, within the framework of media and information literacy (MIL), as well as critical and humanistic perspectives, recognizing the important societal benefits of a safe, inclusive, ethical and responsible implementation of AI. This approach underscores the importance of AI education, as well as its application at different levels of education and its incorporation into job training.

Governance of Artificial Intelligence (AI)

5.11 Reaffirm the commitment to promote AI Governance frameworks for the development and use of this technology in a safe, inclusive and ethical and responsible manner, respecting human rights - including in particular the right to the protection of personal data - and fostering innovation and sustainability.

5.12 Recognize the importance of the regional articulation work that began with the Ministerial and High Authorities Summit on the Ethics of Artificial Intelligence hosted by the Government of Chile and the Santiago Declaration of 2023, and the important contribution of the Latin American and Caribbean Ministerial Summit held in Colombia in 2024, particularly with respect to the Global Digital Pact to be adopted in September 2024 at the United Nations Summit on the Future. Recognizing that these efforts will be consolidated at the Montevideo Summit on October 3 and 4, 2024, as well as the importance of instances or initiatives to be developed by international and multilateral organizations.

5.13 Promote active dialogue in the aforementioned instances in order to agree on coordinated positions in international scenarios and forums related to AI, taking into account mechanisms for multi-stakeholder participation, to promote and defend the interests of the region with respect to issues related to AI governance. This dialogue will seek to develop, implement and use this technology in an ethical, safe and responsible manner, in accordance with the efforts that have been made in the region.

5.14 Highlight the importance of any international framework for AI governance to integrate the regional particularities of Latin America and the Caribbean, promoting the participation of countries in the region in its design and implementation, paying particular attention to closing gaps in access, talent, data, infrastructure and regulatory capacity.

5.15 Urge to explore and recommend taking into consideration recommendations, guidelines and instruments developed by Latin American and Caribbean countries, international and multilateral organizations in the region, in harmony with their applicable legal frameworks and in accordance with their local circumstances and priorities, in pursuit of the objectives set forth in this Declaration. Likewise, the signatory countries urge [sic] to take UNESCO's Ethics of Artificial Intelligence methods, as well as the tools that emanate from it (Readiness Assessment Methodology and Ethics Impact Methodology), drawing on the contributions of its network of experts as well as the recommendations stemming from the AI for Good of the Telecommunications Union Sumit.

The EU AI ACT
Regulation (EU) 2024/1689 of the European Parliament and of the Council of 13 June 2024 laying down harmonised rules on artificial intelligence and amending Regulations (EC) No 300/2008, (EU) No 167/2013, (EU) No 168/2013, (EU) 2018/858, (EU) 2018/1139 and (EU) 2019/2144 and Directives 2014/90/EU, (EU) 2016/797 and (EU) 2020/1828
(EU 2024)

CHAPTER I

GENERAL PROVISIONS

Article 1

Subject matter

1. The purpose of this Regulation is to improve the functioning of the internal market and promote the uptake of human-centric and trustworthy artificial intelligence (AI), while ensuring a high level of protection of health, safety, fundamental rights enshrined in the Charter, including democracy, the rule of law and environmental protection, against the harmful effects of AI systems in the Union and supporting innovation.

2. This Regulation lays down:

(a) harmonised rules for the placing on the market, theprohibitions of certain putting into service, and the use of AI systems in the Union;AI practices;

(b) prohibitions on certain practices;

(c) specific requirements for high-risk AI systems and obligations for operators of such systems;

(d) harmonised transparency rules for certain AI systems;

(e) harmonised rules for the placing on the market of general-purpose AI models;

(f) rules on market monitoring, market surveillance, governance and enforcement;

(g)measures to support innovation, with a particular focus on SMEs, including start-ups.

Article 2

Scope

1. This Regulation applies to:

(a) providers placing on the market or putting into service AI systems or placing on the market general-purpose AI models in the Union, irrespective of whether those providers are established or located within the Union or in a third country;

(b) deployers of AI systems that have their place of establishment or are located within the Union;

(c) providers and deployers of AI systems that have their place of establishment or are located in a third country, where the output produced by the AI system is used in the Union;

(d) importers and distributors of AI systems;

(e) product manufacturers placing on the market or putting into service an AI system together with their product and under their own name or trademark;

(f) authorised representatives of providers, which are not established in the Union;

(g) affected persons that are located in the Union.

2. For AI systems classified as high-risk AI systems in accordance with Article 6(1) related to products covered by the Union harmonisation legislation listed in Section B of Annex I, only Article 6(1), Articles 102 to 109 and Article 112 apply. Article 57 applies only in so far as the requirements for high-risk AI systems under this Regulation have been integrated in that Union harmonisation legislation.

3. This Regulation does not apply to areas outside the scope of Union law, and shall not, in any event, affect the competences of the Member States concerning national security, regardless of the type of entity entrusted by the Member States with carrying out tasks in relation to those competences.

This Regulation does not apply to AI systems where and in so far they are placed on the market, put into service, or used with or without modification exclusively for military, defence or national security purposes, regardless of the type of entity carrying out those activities.

This Regulation does not apply to AI systems which are not placed on the market or put into service in the Union, where the output is used in the Union exclusively for military, defence or national security purposes, regardless of the type of entity carrying out those activities.

4. This Regulation applies neither to public authorities in a third country nor to international organisations falling within the scope of this Regulation pursuant to paragraph 1, where those authorities or organisations use AI systems in the framework of international cooperation or agreements for law enforcement and judicial cooperation with the Union or with one or more Member States, provided that such a third country or international organisation provides adequate safeguards with respect to the protection of fundamental rights and freedoms of individuals.

5. This Regulation shall not affect the application of the provisions on the liability of providers of intermediary services as set out in Chapter II of Regulation (EU) 2022/2065.

6. This Regulation does not apply to AI systems or AI models, including their output, specifically developed and put into service for the sole purpose of scientific research and development.

7. Union law on the protection of personal data, privacy and the confidentiality of communications applies to personal data processed in connection with the rights and obligations laid down in this Regulation. This Regulation shall not affect Regulation (EU) 2016/679 or (EU) 2018/1725, or Directive 2002/58/EC or (EU) 2016/680, without prejudice to Article 10(5) and Article 59 of this Regulation.

8. This Regulation does not apply to any research, testing or development activity regarding AI systems or AI models prior to their being placed on the market or put into service. Such activities shall be conducted in accordance with applicable Union law. Testing in real world conditions shall not be covered by that exclusion.

9. This Regulation is without prejudice to the rules laid down by other Union legal acts related to consumer protection and product safety.

10. This Regulation does not apply to obligations of deployers who are natural persons using AI systems in the course of a purely personal non-professional activity.

11. This Regulation does not preclude the Union or Member States from maintaining or introducing laws, regulations or administrative provisions which are more favourable to workers in terms of protecting their rights in respect of the use

of AI systems by employers, or from encouraging or allowing the application of collective agreements which are more favourable to workers.

12. This Regulation does not apply to AI systems released under free and open-source licences, unless they are placed on the market or put into service as high-risk AI systems or as an AI system that falls under Article 5 or 50.

Article 3

Definitions

For the purposes of this Regulation, the following definitions apply:

(1) 'AI system' means a machine-based system that is designed to operate with varying levels of autonomy and that may exhibit adaptiveness after deployment, and that, for explicit or implicit objectives, infers, from the input it receives, how to generate outputs such as predictions, content, recommendations, or decisions that can influence physical or virtual environments;

(2) 'risk' means the combination of the probability of an occurrence of harm and the severity of that harm;

(3) 'provider' means a natural or legal person, public authority, agency or other body that develops an AI system or a general-purpose AI model or that has an AI system or a general-purpose AI model developed and places it on the market or puts the AI system into service under its own name or trademark, whether for payment or free of charge;

(4) 'deployer' means a natural or legal person, public authority, agency or other body using an AI system under its authority except where the AI system is used in the course of a personal non-professional activity;

(5) 'authorised representative' means a natural or legal person located or established in the Union who has received and accepted a written mandate from a provider of an AI system or a general-purpose AI model to, respectively, perform and carry out on its behalf the obligations and procedures established by this Regulation;

(6) 'importer' means a natural or legal person located or established in the Union that places on the market an AI system that bears the name or trademark of a natural or legal person established in a third country;

(7) 'distributor' means a natural or legal person in the supply chain, other than the provider or the importer, that makes an AI system available on the Union market;

(8) 'operator' means a provider, product manufacturer, deployer, authorised representative, importer or distributor;

(9) 'placing on the market' means the first making available of an AI system or a general-purpose AI model on the Union market;

(10) 'making available on the market' means the supply of an AI system or a general-purpose AI model for distribution or use on the Union market in the course of a commercial activity, whether in return for payment or free of charge;

(11) 'putting into service' means the supply of an AI system for first use directly to the deployer or for own use in the Union for its intended purpose;

(12) 'intended purpose' means the use for which an AI system is intended by the provider, including the specific context and conditions of use, as specified in the information supplied by the provider in the instructions for use, promotional or sales materials and statements, as well as in the technical documentation;

(13) 'reasonably foreseeable misuse' means the use of an AI system in a way that is not in accordance with its intended purpose, but which may result from reasonably foreseeable human behaviour or interaction with other systems, including other AI systems;

(14) 'safety component' means a component of a product or of an AI system which fulfils a safety function for that product or AI system, or the failure or malfunctioning of which endangers the health and safety of persons or property;

(15) 'instructions for use' means the information provided by the provider to inform the deployer of, in particular, an AI system's intended purpose and proper use;

(16) 'recall of an AI system' means any measure aiming to achieve the return to the provider or taking out of service or disabling the use of an AI system made available to deployers;

(17) 'withdrawal of an AI system' means any measure aiming to prevent an AI system in the supply chain being made available on the market;

(18) 'performance of an AI system' means the ability of an AI system to achieve its intended purpose;

(19) 'notifying authority' means the national authority responsible for setting up and carrying out the necessary procedures for the assessment, designation and notification of conformity assessment bodies and for their monitoring;

(20) 'conformity assessment' means the process of demonstrating whether the requirements set out in Chapter III, Section 2 relating to a high-risk AI system have been fulfilled;

(21) 'conformity assessment body' means a body that performs third-party conformity assessment activities, including testing, certification and inspection;

(22) 'notified body' means a conformity assessment body notified in accordance with this Regulation and other relevant Union harmonisation legislation;

(23) 'substantial modification' means a change to an AI system after its placing on the market or putting into service which is not foreseen or planned in the initial conformity assessment carried out by the provider and as a result of which the compliance of the AI system with the requirements set out in Chapter III, Section 2 is affected or results in a modification to the intended purpose for which the AI system has been assessed;

(24) 'CE marking' means a marking by which a provider indicates that an AI system is in conformity with the requirements set out in Chapter III, Section 2 and other applicable Union harmonisation legislation providing for its affixing;

(25) 'post-market monitoring system' means all activities carried out by providers of AI systems to collect and review experience gained from the use of AI systems they place on the market or put into service for the purpose of identifying any need to immediately apply any necessary corrective or preventive actions;

(26) 'market surveillance authority' means the national authority carrying out the activities and taking the measures pursuant to Regulation (EU) 2019/1020;

(27) 'harmonised standard' means a harmonised standard as defined in Article 2(1), point (c), of Regulation (EU) No 1025/2012;

(28) 'common specification' means a set of technical specifications as defined in Article 2, point (4) of Regulation (EU) No 1025/2012, providing means to comply with certain requirements established under this Regulation;

(29) 'training data' means data used for training an AI system through fitting its learnable parameters;

(30) 'validation data' means data used for providing an evaluation of the trained AI system and for tuning its non-learnable parameters and its learning process in order, inter alia, to prevent underfitting or overfitting;

(31) 'validation data set' means a separate data set or part of the training data set, either as a fixed or variable split;

(32) 'testing data' means data used for providing an independent evaluation of the AI system in order to confirm the expected performance of that system before its placing on the market or putting into service;

(33) 'input data' means data provided to or directly acquired by an AI system on the basis of which the system produces an output;

(34) 'biometric data' means personal data resulting from specific technical processing relating to the physical, physiological or behavioural characteristics of a natural person, such as facial images or dactyloscopic data;

(35) 'biometric identification' means the automated recognition of physical, physiological, behavioural, or psychological human features for the purpose of establishing the identity of a natural person by comparing biometric data of that individual to biometric data of individuals stored in a database;

(36) 'biometric verification' means the automated, one-to-one verification, including authentication, of the identity of natural persons by comparing their biometric data to previously provided biometric data;

(37) 'special categories of personal data' means the categories of personal data referred to in Article 9(1) of Regulation (EU) 2016/679, Article 10 of Directive (EU) 2016/680 and Article 10(1) of Regulation (EU) 2018/1725;

(38) 'sensitive operational data' means operational data related to activities of prevention, detection, investigation or prosecution of criminal offences, the disclosure of which could jeopardise the integrity of criminal proceedings;

(39) 'emotion recognition system' means an AI system for the purpose of identifying or inferring emotions or intentions of natural persons on the basis of their biometric data;

(40) 'biometric categorisation system' means an AI system for the purpose of assigning natural persons to specific categories on the basis of their biometric data, unless it is ancillary to another commercial service and strictly necessary for objective technical reasons;

(41) 'remote biometric identification system' means an AI system for the purpose of identifying natural persons, without their active involvement, typically at a distance through the comparison of a person's biometric data with the biometric data contained in a reference database;

(42) 'real-time remote biometric identification system' means a remote biometric identification system, whereby the capturing of biometric data, the comparison and the identification all occur without a significant delay, comprising not only instant identification, but also limited short delays in order to avoid circumvention;

(43) 'post-remote biometric identification system' means a remote biometric identification system other than a real-time remote biometric identification system;

(44) 'publicly accessible space' means any publicly or privately owned physical place accessible to an undetermined number of natural persons, regardless of whether certain conditions for access may apply, and regardless of the potential capacity restrictions;

(45) 'law enforcement authority' means:

(a) any public authority competent for the prevention, investigation, detection or prosecution of criminal offences or the execution of criminal penalties, including the safeguarding against and the prevention of threats to public security; or

(b) any other body or entity entrusted by Member State law to exercise public authority and public powers for the purposes of the prevention, investigation, detection or prosecution of criminal offences or the execution of criminal penalties, including the safeguarding against and the prevention of threats to public security;

(46) 'law enforcement' means activities carried out by law enforcement authorities or on their behalf for the prevention, investigation, detection or prosecution of criminal offences or the execution of criminal penalties, including safeguarding against and preventing threats to public security;

(47) 'AI Office' means the Commission's function of contributing to the implementation, monitoring and supervision of AI systems and general-purpose AI models, and AI governance, provided for in Commission Decision of 24 January 2024; references in this Regulation to the AI Office shall be construed as references to the Commission;

(48) 'national competent authority' means a notifying authority or a market surveillance authority; as regards AI systems put into service or used by Union institutions, agencies, offices and bodies, references to national competent authorities or market surveillance authorities in this Regulation shall be construed as references to the European Data Protection Supervisor;

(49) 'serious incident' means an incident or malfunctioning of an AI system that directly or indirectly leads to any of the following:

(a) the death of a person, or serious harm to a person's health;

(b) a serious and irreversible disruption of the management or operation of critical infrastructure;

(c) the infringement of obligations under Union law intended to protect fundamental rights;

(d) serious harm to property or the environment;

(50) 'personal data' means personal data as defined in Article 4, point (1), of Regulation (EU) 2016/679;

(51) 'non-personal data' means data other than personal data as defined in Article 4, point (1), of Regulation (EU) 2016/679;

(52) 'profiling' means profiling as defined in Article 4, point (4), of Regulation (EU) 2016/679;

(53) 'real-world testing plan' means a document that describes the objectives, methodology, geographical, population and temporal scope, monitoring, organisation and conduct of testing in real-world conditions;

(54) 'sandbox plan' means a document agreed between the participating provider and the competent authority describing the objectives, conditions, timeframe, methodology and requirements for the activities carried out within the sandbox;

(55) 'AI regulatory sandbox' means a controlled framework set up by a competent authority which offers providers or prospective providers of AI systems the possibility to develop, train, validate and test, where appropriate in real-world conditions, an innovative AI system, pursuant to a sandbox plan for a limited time under regulatory supervision;

(56) 'AI literacy' means skills, knowledge and understanding that allow providers, deployers and affected persons, taking into account their respective rights and obligations in the context of this Regulation, to make an informed deployment of AI systems, as well as to gain awareness about the opportunities and risks of AI and possible harm it can cause;

(57) 'testing in real-world conditions' means the temporary testing of an AI system for its intended purpose in real-world conditions outside a laboratory or otherwise simulated environment, with a view to gathering reliable and robust data and to assessing and verifying the conformity of the AI system with the requirements of this Regulation and it does not qualify as placing the AI system on the market or putting it into service within the meaning of this Regulation, provided that all the conditions laid down in Article 57 or 60 are fulfilled;

(58) 'subject', for the purpose of real-world testing, means a natural person who participates in testing in real-world conditions;

(59) 'informed consent' means a subject's freely given, specific, unambiguous and voluntary expression of his or her willingness to participate in a particular testing in real-world conditions, after having been informed of all aspects of the testing that are relevant to the subject's decision to participate;

(60) 'deep fake' means AI-generated or manipulated image, audio or video content that resembles existing persons, objects, places, entities or events and would falsely appear to a person to be authentic or truthful;

(61) 'widespread infringement' means any act or omission contrary to Union law protecting the interest of individuals, which:

(a) has harmed or is likely to harm the collective interests of individuals residing in at least two Member States other than the Member State in which:

(i) the act or omission originated or took place;

(ii) the provider concerned, or, where applicable, its authorised representative is located or established; or

(iii) the deployer is established, when the infringement is committed by the deployer;

(b) has caused, causes or is likely to cause harm to the collective interests of individuals and has common features, including the same unlawful practice or the same interest being infringed, and is occurring concurrently, committed by the same operator, in at least three Member States;

(62) 'critical infrastructure' means critical infrastructure as defined in Article 2, point (4), of Directive (EU) 2022/2557;

(63) 'general-purpose AI model' means an AI model, including where such an AI model is trained with a large amount of data using self-supervision at scale, that displays significant generality and is capable of competently performing a wide range of distinct tasks regardless of the way the model is placed on the market and that can be integrated into a variety of downstream systems or applications, except AI models that are used for research, development or prototyping activities before they are placed on the market;

(64) 'high-impact capabilities' means capabilities that match or exceed the capabilities recorded in the most advanced general-purpose AI models;

(65) 'systemic risk' means a risk that is specific to the high-impact capabilities of general-purpose AI models, having a significant impact on the Union market due to their reach, or due to actual or reasonably foreseeable negative effects on public health, safety, public security, fundamental rights, or the society as a whole, that can be propagated at scale across the value chain;

(66) 'general-purpose AI system' means an AI system which is based on a general-purpose AI model and which has the capability to serve a variety of purposes, both for direct use as well as for integration in other AI systems;

(67) 'floating-point operation' means any mathematical operation or assignment involving floating-point numbers, which are a subset of the real numbers typically represented on computers by an integer of fixed precision scaled by an integer exponent of a fixed base;

(68) 'downstream provider' means a provider of an AI system, including a general-purpose AI system, which integrates an AI model, regardless of whether the AI model is provided by themselves and vertically integrated or provided by another entity based on contractual relations.

Article 4

AI literacy

Providers and deployers of AI systems shall take measures to ensure, to their best extent, a sufficient level of AI literacy of their staff and other persons dealing with the operation and use of AI systems on their behalf, taking into account their technical knowledge, experience, education and training and the context the AI systems are to be used in, and considering the persons or groups of persons on whom the AI systems are to be used.

CHAPTER II

PROHIBITED AI PRACTICES

Article 5

Prohibited AI practices

1. The following AI practices shall be prohibited:

(a) the placing on the market, the putting into service or the use of an AI system that deploys subliminal techniques beyond a person's consciousness or purposefully manipulative or deceptive techniques, with the objective, or the effect of materially distorting the behaviour of a person or a group of persons by appreciably impairing their ability to make an informed decision, thereby causing them to take a decision that they would not have otherwise taken in a manner that causes or is reasonably likely to cause that person, another person or group of persons significant harm;

(b) the placing on the market, the putting into service or the use of an AI system that exploits any of the vulnerabilities of a natural person or a specific group of persons due to their age, disability or a specific social or economic situation, with the objective, or the effect, of materially distorting the behaviour of that person or a person belonging to that group in a manner that causes or is reasonably likely to cause that person or another person significant harm;

(c) the placing on the market, the putting into service or the use of AI systems for the evaluation or classification of natural persons or groups of persons over a certain period of time based on their social behaviour or known, inferred or predicted personal or personality characteristics, with the social score leading to either or both of the following:

(i) detrimental or unfavourable treatment of certain natural persons or groups of persons in social contexts that are unrelated to the contexts in which the data was originally generated or collected;

(ii) detrimental or unfavourable treatment of certain natural persons or groups of persons that is unjustified or disproportionate to their social behaviour or its gravity;

(d) the placing on the market, the putting into service for this specific purpose, or the use of an AI system for making risk assessments of natural persons in order to assess or predict the risk of a natural person committing a criminal offence, based solely on the profiling of a natural person or on assessing their personality traits and characteristics; this prohibition shall not apply to AI systems used to support the human assessment of the involvement of a person in a criminal activity, which is already based on objective and verifiable facts directly linked to a criminal activity;

(e) the placing on the market, the putting into service for this specific purpose, or the use of AI systems that create or expand facial recognition databases through the untargeted scraping of facial images from the internet or CCTV footage;

(f) the placing on the market, the putting into service for this specific purpose, or the use of AI systems to infer emotions of a natural person in the areas of workplace and education institutions, except where the use of the AI system is intended to be put in place or into the market for medical or safety reasons;

(g) the placing on the market, the putting into service for this specific purpose, or the use of biometric categorisation systems that categorise individually natural persons based on their biometric data to deduce or infer their race, political opinions, trade union membership, religious or philosophical beliefs, sex life or sexual orientation; this prohibition does not cover any labelling or filtering of lawfully acquired biometric datasets, such as images, based on biometric data or categorizing of biometric data in the area of law enforcement;

(h) The use of 'real-time' remote biometric identification systems in publicly accessible spaces for the purposes of law enforcement, unless and in so far as such use is strictly necessary for one of the following objectives:

(i) the targeted search for specific victims of abduction, trafficking in human beings or sexual exploitation of human beings, as well as the search for missing persons;

(ii) the prevention of a specific, substantial and imminent threat to the life or physical safety of natural persons or a genuine and present or genuine and foreseeable threat of a terrorist attack;

(iii) the localisation or identification of a person suspected of having committed a criminal offence, for the purpose of conducting a criminal investigation or prosecution or executing a criminal penalty for offences referred to in Annex II and punishable in the Member State concerned by a custodial sentence or a detention order for a maximum period of at least four years.

Point (h) of the first subparagraph is without prejudice to Article 9 of Regulation (EU) 2016/679 for the processing of biometric data for purposes other than law enforcement.

2. The use of 'real-time' remote biometric identification systems in publicly accessible spaces for the purposes of law enforcement for any of the objectives referred to in paragraph 1, first subparagraph, point (h), shall be deployed for the purposes set out in that point only to confirm the identity of the specifically targeted individual, and it shall take into account the following elements:

(a) the nature of the situation giving rise to the possible use, in particular the seriousness, probability and scale of the harm that would be caused if the system were not used;

(b) the consequences of the use of the system for the rights and freedoms of all persons concerned, in particular the seriousness, probability and scale of those consequences.

In addition, the use of 'real-time' remote biometric identification systems in publicly accessible spaces for the purposes of law enforcement for any of the objectives referred to in paragraph 1, first subparagraph, point (h), of this Article shall comply with necessary and proportionate safeguards and conditions in relation to the use in accordance with the national law authorising the use thereof, in particular as regards the temporal, geographic and personal limitations. The use of the 'real-time' remote biometric identification system in publicly accessible spaces shall be authorised only if the law enforcement authority has completed a fundamental rights impact assessment as provided for in Article 27 and has registered the system in the EU database according to Article 49. However, in duly justified cases of urgency, the use of such systems may be commenced without the registration in the EU database, provided that such registration is completed without undue delay.

3. For the purposes of paragraph 1, first subparagraph, point (h) and paragraph 2, each use for the purposes of law enforcement of a 'real-time' remote biometric

identification system in publicly accessible spaces shall be subject to a prior authorisation granted by a judicial authority or an independent administrative authority whose decision is binding of the Member State in which the use is to take place, issued upon a reasoned request and in accordance with the detailed rules of national law referred to in paragraph 5. However, in a duly justified situation of urgency, the use of such system may be commenced without an authorisation provided that such authorisation is requested without undue delay, at the latest within 24 hours. If such authorisation is rejected, the use shall be stopped with immediate effect and all the data, as well as the results and outputs of that use shall be immediately discarded and deleted.

The competent judicial authority or an independent administrative authority whose decision is binding shall grant the authorisation only where it is satisfied, on the basis of objective evidence or clear indications presented to it, that the use of the 'real-time' remote biometric identification system concerned is necessary for, and proportionate to, achieving one of the objectives specified in paragraph 1, first subparagraph, point (h), as identified in the request and, in particular, remains limited to what is strictly necessary concerning the period of time as well as the geographic and personal scope. In deciding on the request, that authority shall take into account the elements referred to in paragraph 2. No decision that produces an adverse legal effect on a person may be taken based solely on the output of the 'real-time' remote biometric identification system.

4. Without prejudice to paragraph 3, each use of a 'real-time' remote biometric identification system in publicly accessible spaces for law enforcement purposes shall be notified to the relevant market surveillance authority and the national data protection authority in accordance with the national rules referred to in paragraph 5. The notification shall, as a minimum, contain the information specified under paragraph 6 and shall not include sensitive operational data.

5. A Member State may decide to provide for the possibility to fully or partially authorise the use of 'real-time' remote biometric identification systems in publicly accessible spaces for the purposes of law enforcement within the limits and under the conditions listed in paragraph 1, first subparagraph, point (h), and paragraphs 2 and 3. Member States concerned shall lay down in their national law the necessary detailed rules for the request, issuance and exercise of, as well as supervision and reporting relating to, the authorisations referred to in paragraph 3. Those rules shall also specify in respect of which of the objectives listed in paragraph 1, first subparagraph, point (h), including which of the criminal offences referred to in point (h)(iii) thereof, the competent authorities may be authorised to use those systems for the purposes of law enforcement. Member States shall notify those rules to the Commission at the latest 30 days following the adoption thereof. Member States

may introduce, in accordance with Union law, more restrictive laws on the use of remote biometric identification systems.

6. National market surveillance authorities and the national data protection authorities of Member States that have been notified of the use of 'real-time' remote biometric identification systems in publicly accessible spaces for law enforcement purposes pursuant to paragraph 4 shall submit to the Commission annual reports on such use. For that purpose, the Commission shall provide Member States and national market surveillance and data protection authorities with a template, including information on the number of the decisions taken by competent judicial authorities or an independent administrative authority whose decision is binding upon requests for authorisations in accordance with paragraph 3 and their result.

7. The Commission shall publish annual reports on the use of real-time remote biometric identification systems in publicly accessible spaces for law enforcement purposes, based on aggregated data in Member States on the basis of the annual reports referred to in paragraph 6. Those annual reports shall not include sensitive operational data of the related law enforcement activities.

8. This Article shall not affect the prohibitions that apply where an AI practice infringes other Union law.

CHAPTER III

HIGH-RISK AI SYSTEMS

SECTION 1

Classification of AI systems as high-risk

Article 6

Classification rules for high-risk AI systems

1. Irrespective of whether an AI system is placed on the market or put into service independently of the products referred to in points (a) and (b), that AI system shall be considered to be high-risk where both of the following conditions are fulfilled:

(a) the AI system is intended to be used as a safety component of a product, or the AI system is itself a product, covered by the Union harmonisation legislation listed in Annex I;

(b) product whose safety component pursuant to point (a) is the AI system, or the AI system itself as a product, is required to undergo a third-party conformity assessment, with a view to the placing on the market or the putting into service of that product pursuant to the Union harmonisation legislation listed in Annex I.

2.   In addition to the high-risk AI systems referred to in paragraph 1, AI systems referred to in Annex III shall be considered to be high-risk.

3.   By derogation from paragraph 2, an AI system referred to in Annex III shall not be considered to be high-risk where it does not pose a significant risk of harm to the health, safety or fundamental rights of natural persons, including by not materially influencing the outcome of decision making.

The first subparagraph shall apply where any of the following conditions is fulfilled:

(a) the AI system is intended to perform a narrow procedural task;

(b) the AI system is intended to improve the result of a previously completed human activity;

(c) the AI system is intended to detect decision-making patterns or deviations from prior decision-making patterns and is not meant to replace or influence the previously completed human assessment, without proper human review; or

(d) the AI system is intended to perform a preparatory task to an assessment relevant for the purposes of the use cases listed in Annex III.

Notwithstanding the first subparagraph, an AI system referred to in Annex III shall always be considered to be high-risk where the AI system performs profiling of natural persons.

4.   A provider who considers that an AI system referred to in Annex III is not high-risk shall document its assessment before that system is placed on the market or put into service. Such provider shall be subject to the registration obligation set out in Article 49(2). Upon request of national competent authorities, the provider shall provide the documentation of the assessment.

5.   The Commission shall, after consulting the European Artificial Intelligence Board (the 'Board'), and no later than 2 February 2026, provide guidelines specifying the practical implementation of this Article in line with Article 96 together with a comprehensive list of practical examples of use cases of AI systems that are high-risk and not high-risk.

6. The Commission is empowered to adopt delegated acts in accordance with Article 97 in order to amend paragraph 3, second subparagraph, of this Article by adding new conditions to those laid down therein, or by modifying them, where there is concrete and reliable evidence of the existence of AI systems that fall under the scope of Annex III, but do not pose a significant risk of harm to the health, safety or fundamental rights of natural persons.

7. The Commission shall adopt delegated acts in accordance with Article 97 in order to amend paragraph 3, second subparagraph, of this Article by deleting any of the conditions laid down therein, where there is concrete and reliable evidence that this is necessary to maintain the level of protection of health, safety and fundamental rights provided for by this Regulation.

8. Any amendment to the conditions laid down in paragraph 3, second subparagraph, adopted in accordance with paragraphs 6 and 7 of this Article shall not decrease the overall level of protection of health, safety and fundamental rights provided for by this Regulation and shall ensure consistency with the delegated acts adopted pursuant to Article 7(1), and take account of market and technological developments.

Article 7

Amendments to Annex III

1. The Commission is empowered to adopt delegated acts in accordance with Article 97 to amend Annex III by adding or modifying use-cases of high-risk AI systems where both of the following conditions are fulfilled:

(a) the AI systems are intended to be used in any of the areas listed in Annex III;

(b) the AI systems pose a risk of harm to health and safety, or an adverse impact on fundamental rights, and that risk is equivalent to, or greater than, the risk of harm or of adverse impact posed by the high-risk AI systems already referred to in Annex III.

2. When assessing the condition under paragraph 1, point (b), the Commission shall take into account the following criteria:

(a) the intended purpose of the AI system;

(b) the extent to which an AI system has been used or is likely to be used;

(c) the nature and amount of the data processed and used by the AI system, in particular whether special categories of personal data are processed;

(d) the extent to which the AI system acts autonomously and the possibility for a human to override a decision or recommendations that may lead to potential harm;

(e) the extent to which the use of an AI system has already caused harm to health and safety, has had an adverse impact on fundamental rights or has given rise to significant concerns in relation to the likelihood of such harm or adverse impact, as demonstrated, for example, by reports or documented allegations submitted to national competent authorities or by other reports, as appropriate;

(f) the potential extent of such harm or such adverse impact, in particular in terms of its intensity and its ability to affect multiple persons or to disproportionately affect a particular group of persons;

(g) the extent to which persons who are potentially harmed or suffer an adverse impact are dependent on the outcome produced with an AI system, in particular because for practical or legal reasons it is not reasonably possible to opt-out from that outcome;

(h) the extent to which there is an imbalance of power, or the persons who are potentially harmed or suffer an adverse impact are in a vulnerable position in relation to the deployer of an AI system, in particular due to status, authority, knowledge, economic or social circumstances, or age;

(i) the extent to which the outcome produced involving an AI system is easily corrigible or reversible, taking into account the technical solutions available to correct or reverse it, whereby outcomes having an adverse impact on health, safety or fundamental rights, shall not be considered to be easily corrigible or reversible;

(j) the magnitude and likelihood of benefit of the deployment of the AI system for individuals, groups, or society at large, including possible improvements in product safety;

(k) the extent to which existing Union law provides for:

(i) effective measures of redress in relation to the risks posed by an AI system, with the exclusion of claims for damages;

(ii) effective measures to prevent or substantially minimise those risks.

3. The Commission is empowered to adopt delegated acts in accordance with Article 97 to amend the list in Annex III by removing high-risk AI systems where both of the following conditions are fulfilled:

(a) the high-risk AI system concerned no longer poses any significant risks to fundamental rights, health or safety, taking into account the criteria listed in paragraph 2;

(b) the deletion does not decrease the overall level of protection of health, safety and fundamental rights under Union law.

SECTION 2

Requirements for high-risk AI systems

Article 8

Compliance with the requirements

1. High-risk AI systems shall comply with the requirements laid down in this Section, taking into account their intended purpose as well as the generally acknowledged state of the art on AI and AI-related technologies. The risk management system referred to in Article 9 shall be taken into account when ensuring compliance with those requirements.

2. Where a product contains an AI system, to which the requirements of this Regulation as well as requirements of the Union harmonisation legislation listed in Section A of Annex I apply, providers shall be responsible for ensuring that their product is fully compliant with all applicable requirements under applicable Union harmonisation legislation. In ensuring the compliance of high-risk AI systems referred to in paragraph 1 with the requirements set out in this Section, and in order to ensure consistency, avoid duplication and minimise additional burdens, providers shall have a choice of integrating, as appropriate, the necessary testing and reporting processes, information and documentation they provide with regard to their product into documentation and procedures that already exist and are required under the Union harmonisation legislation listed in Section A of Annex I.

Article 9

Risk management system

1. A risk management system shall be established, implemented, documented and maintained in relation to high-risk AI systems.

———

140

2. The risk management system shall be understood as a continuous iterative process planned and run throughout the entire lifecycle of a high-risk AI system, requiring regular systematic review and updating. It shall comprise the following steps:

(a) the identification and analysis of the known and the reasonably foreseeable risks that the high-risk AI system can pose to health, safety or fundamental rights when the high-risk AI system is used in accordance with its intended purpose;

(b) the estimation and evaluation of the risks that may emerge when the high-risk AI system is used in accordance with its intended purpose, and under conditions of reasonably foreseeable misuse;

(c) the evaluation of other risks possibly arising, based on the analysis of data gathered from the post-market monitoring system referred to in Article 72;

(d) the adoption of appropriate and targeted risk management measures designed to address the risks identified pursuant to point (a).

3. The risks referred to in this Article shall concern only those which may be reasonably mitigated or eliminated through the development or design of the high-risk AI system, or the provision of adequate technical information.

4. The risk management measures referred to in paragraph 2, point (d), shall give due consideration to the effects and possible interaction resulting from the combined application of the requirements set out in this Section, with a view to minimising risks more effectively while achieving an appropriate balance in implementing the measures to fulfil those requirements.

5. The risk management measures referred to in paragraph 2, point (d), shall be such that the relevant residual risk associated with each hazard, as well as the overall residual risk of the high-risk AI systems is judged to be acceptable.

In identifying the most appropriate risk management measures, the following shall be ensured:

(a) elimination or reduction of risks identified and evaluated pursuant to paragraph 2 in as far as technically feasible through adequate design and development of the high-risk AI system;

(b) where appropriate, implementation of adequate mitigation and control measures addressing risks that cannot be eliminated;

(c) provision of information required pursuant to Article 13 and, where appropriate, training to deployers.

With a view to eliminating or reducing risks related to the use of the high-risk AI system, due consideration shall be given to the technical knowledge, experience, education, the training to be expected by the deployer, and the presumable context in which the system is intended to be used.

6. High-risk AI systems shall be tested for the purpose of identifying the most appropriate and targeted risk management measures. Testing shall ensure that high-risk AI systems perform consistently for their intended purpose and that they are in compliance with the requirements set out in this Section.

7. Testing procedures may include testing in real-world conditions in accordance with Article 60.

8. The testing of high-risk AI systems shall be performed, as appropriate, at any time throughout the development process, and, in any event, prior to their being placed on the market or put into service. Testing shall be carried out against prior defined metrics and probabilistic thresholds that are appropriate to the intended purpose of the high-risk AI system.

9. When implementing the risk management system as provided for in paragraphs 1 to 7, providers shall give consideration to whether in view of its intended purpose the high-risk AI system is likely to have an adverse impact on persons under the age of 18 and, as appropriate, other vulnerable groups.

10. For providers of high-risk AI systems that are subject to requirements regarding internal risk management processes under other relevant provisions of Union law, the aspects provided in paragraphs 1 to 9 may be part of, or combined with, the risk management procedures established pursuant to that law.

Article 10

Data and data governance

1. High-risk AI systems which make use of techniques involving the training of AI models with data shall be developed on the basis of training, validation and testing data sets that meet the quality criteria referred to in paragraphs 2 to 5 whenever such data sets are used.

2.   Training, validation and testing data sets shall be subject to data governance and management practices appropriate for the intended purpose of the high-risk AI system. Those practices shall concern in particular:

(a) the relevant design choices;
(b) data collection processes and the origin of data, and in the case of personal data, the original purpose of the data collection;

(c) relevant data-preparation processing operations, such as annotation, labelling, cleaning, updating, enrichment and aggregation;

(d) the formulation of assumptions, in particular with respect to the information that the data are supposed to measure and represent;

(e) an assessment of the availability, quantity and suitability of the data sets that are needed;

(f) examination in view of possible biases that are likely to affect the health and safety of persons, have a negative impact on fundamental rights or lead to discrimination prohibited under Union law, especially where data outputs influence inputs for future operations;

(g) appropriate measures to detect, prevent and mitigate possible biases identified according to point (f);

(h) the identification of relevant data gaps or shortcomings that prevent compliance with this Regulation, and how those gaps and shortcomings can be addressed.

3. Training, validation and testing data sets shall be relevant, sufficiently representative, and to the best extent possible, free of errors and complete in view of the intended purpose. They shall have the appropriate statistical properties, including, where applicable, as regards the persons or groups of persons in relation to whom the high-risk AI system is intended to be used. Those characteristics of the data sets may be met at the level of individual data sets or at the level of a combination thereof.

4.   Data sets shall take into account, to the extent required by the intended purpose, the characteristics or elements that are particular to the specific geographical, contextual, behavioural or functional setting within which the high-risk AI system is intended to be used.

5.   To the extent that it is strictly necessary for the purpose of ensuring bias detection and correction in relation to the high-risk AI systems in accordance with

paragraph (2), points (f) and (g) of this Article, the providers of such systems may exceptionally process special categories of personal data, subject to appropriate safeguards for the fundamental rights and freedoms of natural persons. In addition to the provisions set out in Regulations (EU) 2016/679 and (EU) 2018/1725 and Directive (EU) 2016/680, all the following conditions must be met in order for such processing to occur:

(a) The bias detection and correction cannot be effectively fulfilled by processing other data, including synthetic or anonymised data;

(b) the special categories of personal data are subject to technical limitations on the re-use of the personal data, and state-of-the-art security and privacy-preserving measures, including pseudonymisation;

(c) the special categories of personal data are subject to measures to ensure that the personal data processed are secured, protected, subject to suitable safeguards, including strict controls and documentation of the access, to avoid misuse and ensure that only authorised persons have access to those personal data with appropriate confidentiality obligations;

(d) the special categories of personal data are not to be transmitted, transferred or otherwise accessed by other parties;

(e) the special categories of personal data are deleted once the bias has been corrected or the personal data has reached the end of its retention period, whichever comes first;

(f) the records of processing activities pursuant to Regulations (EU) 2016/679 and (EU) 2018/1725 and Directive (EU) 2016/680 include the reasons why the processing of special categories of personal data was strictly necessary to detect and correct biases, and why that objective could not be achieved by processing other data.

6. For the development of high-risk AI systems not using techniques involving the training of AI models, paragraphs 2 to 5 apply only to the testing data sets.

Article 11

Technical documentation

1. The technical documentation of a high-risk AI system shall be drawn up before that system is placed on the market or put into service and shall be kept up-to date.

The technical documentation shall be drawn up in such a way as to demonstrate that the high-risk AI system complies with the requirements set out in this Section and to provide national competent authorities and notified bodies with the necessary information in a clear and comprehensive form to assess the compliance of the AI system with those requirements. It shall contain, at a minimum, the elements set out in Annex IV. SMEs, including start-ups, may provide the elements of the technical documentation specified in Annex IV in a simplified manner. To that end, the Commission shall establish a simplified technical documentation form targeted at the needs of small and microenterprises. Where an SME, including a start-up, opts to provide the information required in Annex IV in a simplified manner, it shall use the form referred to in this paragraph. Notified bodies shall accept the form for the purposes of the conformity assessment.

2. Where a high-risk AI system related to a product covered by the Union harmonisation legislation listed in Section A of Annex I is placed on the market or put into service, a single set of technical documentation shall be drawn up containing all the information set out in paragraph 1, as well as the information required under those legal acts.

3. The Commission is empowered to adopt delegated acts in accordance with Article 97 in order to amend Annex IV, where necessary, to ensure that, in light of technical progress, the technical documentation provides all the information necessary to assess the compliance of the system with the requirements set out in this Section.

Article 12

Record-keeping

1. High-risk AI systems shall technically allow for the automatic recording of events (logs) over the lifetime of the system.

2. In order to ensure a level of traceability of the functioning of a high-risk AI system that is appropriate to the intended purpose of the system, logging capabilities shall enable the recording of events relevant for:

(a) identifying situations that may result in the high-risk AI system presenting a risk within the meaning of Article 79(1) or in a substantial modification;

(b) facilitating the post-market monitoring referred to in Article 72; and

(c) monitoring the operation of high-risk AI systems referred to in Article 26(5).

---

3. For high-risk AI systems referred to in point 1 (a), of Annex III, the logging capabilities shall provide, at a minimum:

(a) recording of the period of each use of the system (start date and time and end date and time of each use);

(b) the reference database against which input data has been checked by the system;

(c) the input data for which the search has led to a match;

(d) the identification of the natural persons involved in the verification of the results, as referred to in Article 14(5).

Article 13

Transparency and provision of information to deployers

1. High-risk AI systems shall be designed and developed in such a way as to ensure that their operation is sufficiently transparent to enable deployers to interpret a system's output and use it appropriately. An appropriate type and degree of transparency shall be ensured with a view to achieving compliance with the relevant obligations of the provider and deployer set out in Section 3.

2. High-risk AI systems shall be accompanied by instructions for use in an appropriate digital format or otherwise that include concise, complete, correct and clear information that is relevant, accessible and comprehensible to deployers.

3. The instructions for use shall contain at least the following information:

(a) the identity and the contact details of the provider and, where applicable, of its authorised representative;

(b) the characteristics, capabilities and limitations of performance of the high-risk AI system, including:

  (i) its intended purpose;

  (ii) the level of accuracy, including its metrics, robustness and cybersecurity referred to in Article 15 against which the high-risk AI system has been tested and validated and which can be expected, and any known and foreseeable circumstances that may have an impact on that expected level of accuracy, robustness and cybersecurity;

(iii) any known or foreseeable circumstance, related to the use of the high-risk AI system in accordance with its intended purpose or under conditions of reasonably foreseeable misuse, which may lead to risks to the health and safety or fundamental rights referred to in Article 9(2);

(iv) where applicable, the technical capabilities and characteristics of the high-risk AI system to provide information that is relevant to explain its output;

(v) when appropriate, its performance regarding specific persons or groups of persons on which the system is intended to be used;

(vi) when appropriate, specifications for the input data, or any other relevant information in terms of the training, validation and testing data sets used, taking into account the intended purpose of the high-risk AI system;

(vii) where applicable, information to enable deployers to interpret the output of the high-risk AI system and use it appropriately;

(c) the changes to the high-risk AI system and its performance which have been pre-determined by the provider at the moment of the initial conformity assessment, if any;

(d) the human oversight measures referred to in Article 14, including the technical measures put in place to facilitate the interpretation of the outputs of the high-risk AI systems by the deployers;

(e) the computational and hardware resources needed, the expected lifetime of the high-risk AI system and any necessary maintenance and care measures, including their frequency, to ensure the proper functioning of that AI system, including as regards software updates;

(f) where relevant, a description of the mechanisms included within the high-risk AI system that allows deployers to properly collect, store and interpret the logs in accordance with Article 12.

Article 14

Human oversight

1. High-risk AI systems shall be designed and developed in such a way, including with appropriate human-machine interface tools, that they can be effectively overseen by natural persons during the period in which they are in use.

2.   Human oversight shall aim to prevent or minimise the risks to health, safety or fundamental rights that may emerge when a high-risk AI system is used in accordance with its intended purpose or under conditions of reasonably foreseeable misuse, in particular where such risks persist despite the application of other requirements set out in this Section.

3.   The oversight measures shall be commensurate with the risks, level of autonomy and context of use of the high-risk AI system, and shall be ensured through either one or both of the following types of measures:

(a) measures identified and built, when technically feasible, into the high-risk AI system by the provider before it is placed on the market or put into service;

(b) measures identified by the provider before placing the high-risk AI system on the market or putting it into service and that are appropriate to be implemented by the deployer.

4.   For the purpose of implementing paragraphs 1, 2 and 3, the high-risk AI system shall be provided to the deployer in such a way that natural persons to whom human oversight is assigned are enabled, as appropriate and proportionate:

(a) To properly understand the relevant capacities and limitations of the high-risk AI system and be able to duly monitor its operation, including in view of detecting and addressing anomalies, dysfunctions and unexpected performance;

(b) to remain aware of the possible tendency of automatically relying or over-relying on the output produced by a high-risk AI system (automation bias), in particular for high-risk AI systems used to provide information or recommendations for decisions to be taken by natural persons;

(c) to correctly interpret the high-risk AI system's output, taking into account, for example, the interpretation tools and methods available;

(d) To decide, in any particular situation, not to use the high-risk AI system or to otherwise disregard, override or reverse the output of the high-risk AI system;

(e) to intervene in the operation of the high-risk AI system or interrupt the system through a 'stop' button or a similar procedure that allows the system to come to a halt in a safe state.

5.   For high-risk AI systems referred to in point 1(a) of Annex III, the measures referred to in paragraph 3 of this Article shall be such as to ensure that, in addition, no action or decision is taken by the deployer on the basis of the identification

resulting from the system unless that identification has been separately verified and confirmed by at least two natural persons with the necessary competence, training and authority.

The requirement for a separate verification by at least two natural persons shall not apply to high-risk AI systems used for the purposes of law enforcement, migration, border control or asylum, where Union or national law considers the application of this requirement to be disproportionate.

Article 15

Accuracy, robustness and cybersecurity

1. High-risk AI systems shall be designed and developed in such a way that they achieve an appropriate level of accuracy, robustness, and cybersecurity, and that they perform consistently in those respects throughout their lifecycle.

2. To address the technical aspects of how to measure the appropriate levels of accuracy and robustness set out in paragraph 1 and any other relevant performance metrics, the Commission shall, in cooperation with relevant stakeholders and organisations such as metrology and benchmarking authorities, encourage, as appropriate, the development of benchmarks and measurement methodologies.

3. The levels of accuracy and the relevant accuracy metrics of high-risk AI systems shall be declared in the accompanying instructions of use.

4. High-risk AI systems shall be as resilient as possible regarding errors, faults or inconsistencies that may occur within the system or the environment in which the system operates, in particular due to their interaction with natural persons or other systems. Technical and organisational measures shall be taken in this regard.

The robustness of high-risk AI systems may be achieved through technical redundancy solutions, which may include backup or fail-safe plans.

High-risk AI systems that continue to learn after being placed on the market or put into service shall be developed in such a way as to eliminate or reduce as far as possible the risk of possibly biased outputs influencing input for future operations (feedback loops), and as to ensure that any such feedback loops are duly addressed with appropriate mitigation measures.

5. High-risk AI systems shall be resilient against attempts by unauthorised third parties to alter their use, outputs or performance by exploiting system vulnerabilities.

The technical solutions aiming to ensure the cybersecurity of high-risk AI systems shall be appropriate to the relevant circumstances and the risks.

The technical solutions to address AI specific vulnerabilities shall include, where appropriate, measures to prevent, detect, respond to, resolve and control for attacks trying to manipulate the training data set (data poisoning), or pre-trained components used in training (model poisoning), inputs designed to cause the AI model to make a mistake (adversarial examples or model evasion), confidentiality attacks or model flaws.

SECTION 3

Obligations of providers and deployers of high-risk AI systems and other parties

Article 16

Obligations of providers of high-risk AI systems

Providers of high-risk AI systems shall:

(a) ensure that their high-risk AI systems are compliant with the requirements set out in Section 2;

(b) indicate on the high-risk AI system or, where that is not possible, on its packaging or its accompanying documentation, as applicable, their name, registered trade name or registered trade mark, the address at which they can be contacted;

(c) have a quality management system in place which complies with Article 17;

(d) keep the documentation referred to in Article 18;

(e) when under their control, keep the logs automatically generated by their high-risk AI systems as referred to in Article 19;

(f) ensure that the high-risk AI system undergoes the relevant conformity assessment procedure as referred to in Article 43, prior to its being placed on the market or put into service;

(g) draw up an EU declaration of conformity in accordance with Article 47;

(h) affix the CE marking to the high-risk AI system or, where that is not possible, on its packaging or its accompanying documentation, to indicate conformity with this Regulation, in accordance with Article 48;

(i) comply with the registration obligations referred to in Article 49(1);

(j) take the necessary corrective actions and provide information as required in Article 20;

(k) upon a reasoned request of a national competent authority, demonstrate the conformity of the high-risk AI system with the requirements set out in Section 2;

(l) ensure that the high-risk AI system complies with accessibility requirements in accordance with Directives (EU) 2016/2102 and (EU) 2019/882.

Article 17

Quality management system

1. Providers of high-risk AI systems shall put a quality management system in place that ensures compliance with this Regulation. That system shall be documented in a systematic and orderly manner in the form of written policies, procedures and instructions, and shall include at least the following aspects:

(a) a strategy for regulatory compliance, including compliance with conformity assessment procedures and procedures for the management of modifications to the high-risk AI system;

(b) techniques, procedures and systematic actions to be used for the design, design control and design verification of the high-risk AI system;

(c) techniques, procedures and systematic actions to be used for the development, quality control and quality assurance of the high-risk AI system;

(d) examination, test and validation procedures to be carried out before, during and after the development of the high-risk AI system, and the frequency with which they have to be carried out;

(e) Technical specifications, including standards, to be applied and, where the relevant harmonised standards are not applied in full or do not cover all of the relevant requirements set out in Section 2, the means to be used to ensure that the high-risk AI system complies with those requirements;

(f) systems and procedures for data management, including data acquisition, data collection, data analysis, data labelling, data storage, data filtration, data mining, data aggregation, data retention and any other operation regarding the data that is performed before and for the purpose of the placing on the market or the putting into service of high-risk AI systems;

(g) the risk management system referred to in Article 9;

(h) the setting-up, implementation and maintenance of a post-market monitoring system, in accordance with Article 72;

(i) procedures related to the reporting of a serious incident in accordance with Article 73;

(j) the handling of communication with national competent authorities, other relevant authorities, including those providing or supporting the access to data, notified bodies, other operators, customers or other interested parties;

(k) systems and procedures for record-keeping of all relevant documentation and information;

(l) resource management, including security-of-supply related measures;

(m) an accountability framework setting out the responsibilities of the management and other staff with regard to all the aspects listed in this paragraph.

2. The implementation of the aspects referred to in paragraph 1 shall be proportionate to the size of the provider's organisation. Providers shall, in any event, respect the degree of rigour and the level of protection required to ensure the compliance of their high-risk AI systems with this Regulation.

3. Providers of high-risk AI systems that are subject to obligations regarding quality management systems or an equivalent function under relevant sectoral Union law may include the aspects listed in paragraph 1 as part of the quality management systems pursuant to that law.

4. For providers that are financial institutions subject to requirements regarding their internal governance, arrangements or processes under Union financial services law, the obligation to put in place a quality management system, with the exception of paragraph 1, points (g), (h) and (i) of this Article, shall be deemed to be fulfilled by complying with the rules on internal governance arrangements or processes pursuant to the relevant Union financial services law. To that end, any harmonised standards referred to in Article 40 shall be taken into account.

Article 18

Documentation keeping

1. The provider shall, for a period ending 10 years after the high-risk AI system has been placed on the market or put into service, keep at the disposal of the national competent authorities:

(a) the technical documentation referred to in Article 11;

(b) the documentation concerning the quality management system referred to in Article 17;

(c) the documentation concerning the changes approved by notified bodies, where applicable;

(d) the decisions and other documents issued by the notified bodies, where applicable;

(e) the EU declaration of conformity referred to in Article 47.

2. Each Member State shall determine conditions under which the documentation referred to in paragraph 1 remains at the disposal of the national competent authorities for the period indicated in that paragraph for the cases when a provider or its authorised representative established on its territory goes bankrupt or ceases its activity prior to the end of that period.

3. Providers that are financial institutions subject to requirements regarding their internal governance, arrangements or processes under Union financial services law shall maintain the technical documentation as part of the documentation kept under the relevant Union financial services law.

Article 19

Automatically generated logs

1. Providers of high-risk AI systems shall keep the logs referred to in Article 12(1), automatically generated by their high-risk AI systems, to the extent such logs are under their control. Without prejudice to applicable Union or national law, the logs shall be kept for a period appropriate to the intended purpose of the high-risk AI system, of at least six months, unless provided otherwise in the applicable Union or national law, in particular in Union law on the protection of personal data.

2. Providers that are financial institutions subject to requirements regarding their internal governance, arrangements or processes under Union financial services law shall maintain the logs automatically generated by their high-risk AI systems as part of the documentation kept under the relevant financial services law.

Article 20

Corrective actions and duty of information

1. Providers of high-risk AI systems which consider or have reason to consider that a high-risk AI system that they have placed on the market or put into service is not in conformity with this Regulation shall immediately take the necessary corrective actions to bring that system into conformity, to withdraw it, to disable it, or to recall it, as appropriate. They shall inform the distributors of the high-risk AI system concerned and, where applicable, the deployers, the authorised representative and importers accordingly.

2. Where the high-risk AI system presents a risk within the meaning of Article 79(1) and the provider becomes aware of that risk, it shall immediately investigate the causes, in collaboration with the reporting deployer, where applicable, and inform the market surveillance authorities competent for the high-risk AI system concerned and, where applicable, the notified body that issued a certificate for that high-risk AI system in accordance with Article 44, in particular, of the nature of the non-compliance and of any relevant corrective action taken.

Article 21

Cooperation with competent authorities

1. Providers of high-risk AI systems shall, upon a reasoned request by a competent authority, provide that authority all the information and documentation necessary to demonstrate the conformity of the high-risk AI system with the requirements set out in Section 2, in a language which can be easily understood by the authority in one of the official languages of the institutions of the Union as indicated by the Member State concerned.

2. Upon a reasoned request by a competent authority, providers shall also give the requesting competent authority, as applicable, access to the automatically generated logs of the high-risk AI system referred to in Article 12(1), to the extent such logs are under their control.

3. Any information obtained by a competent authority pursuant to this Article shall be treated in accordance with the confidentiality obligations sct out in Article 78.

Article 22

Authorised representatives of providers of high-risk AI systems

1. Prior to making their high-risk AI systems available on the Union market, providers established in third countries shall, by written mandate, appoint an authorised representative which is established in the Union.

2. The provider shall enable its authorised representative to perform the tasks specified in the mandate received from the provider.

3. The authorised representative shall perform the tasks specified in the mandate received from the provider. It shall provide a copy of the mandate to the market surveillance authorities upon request, in one of the official languages of the institutions of the Union, as indicated by the competent authority. For the purposes of this Regulation, the mandate shall empower the authorised representative to carry out the following tasks:

(a) verify that the EU declaration of conformity referred to in Article 47 and the technical documentation referred to in Article 11 have been drawn up and that an appropriate conformity assessment procedure has been carried out by the provider;

(b) keep at the disposal of the competent authorities and national authorities or bodies referred to in Article 74(10), for a period of 10 years after the high-risk AI system has been placed on the market or put into service, the contact details of the provider that appointed the authorised representative, a copy of the EU declaration of conformity referred to in Article 47, the technical documentation and, if applicable, the certificate issued by the notified body;

(c) provide a competent authority, upon a reasoned request, with all the information and documentation, including that referred to in point (b) of this subparagraph, necessary to demonstrate the conformity of a high-risk AI system with the requirements set out in Section 2, including access to the logs, as referred to in Article 12(1), automatically generated by the high-risk AI system, to the extent such logs are under the control of the provider;

(d) cooperate with competent authorities, upon a reasoned request, in any action the latter take in relation to the high-risk AI system, in particular to reduce and mitigate the risks posed by the high-risk AI system;

(e) where applicable, comply with the registration obligations referred to in Article 49(1), or, if the registration is carried out by the provider itself, ensure that the information referred to in point 3 of Section A of Annex VIII is correct.

The mandate shall empower the authorised representative to be addressed, in addition to or instead of the provider, by the competent authorities, on all issues related to ensuring compliance with this Regulation.

4.   The authorised representative shall terminate the mandate if it considers or has reason to consider the provider to be acting contrary to its obligations pursuant to this Regulation. In such a case, it shall immediately inform the relevant market surveillance authority, as well as, where applicable, the relevant notified body, about the termination of the mandate and the reasons therefor.

Article 23

Obligations of importers

1.   Before placing a high-risk AI system on the market, importers shall ensure that the system is in conformity with this Regulation by verifying that:

(a) the relevant conformity assessment procedure referred to in Article 43 has been carried out by the provider of the high-risk AI system;

(b) the provider has drawn up the technical documentation in accordance with Article 11 and Annex IV;

(c) the system bears the required CE marking and is accompanied by the EU declaration of conformity referred to in Article 47 and instructions for use;

(d) the provider has appointed an authorised representative in accordance with Article 22(1).

2.   Where an importer has sufficient reason to consider that a high-risk AI system is not in conformity with this Regulation, or is falsified, or accompanied by falsified documentation, it shall not place the system on the market until it has been brought into conformity. Where the high-risk AI system presents a risk within the meaning of Article 79(1), the importer shall inform the provider of the system, the authorised representative and the market surveillance authorities to that effect.

3.   Importers shall indicate their name, registered trade name or registered trade mark, and the address at which they can be contacted on the high-risk AI system and on its packaging or its accompanying documentation, where applicable.

4. Importers shall ensure that, while a high-risk AI system is under their responsibility, storage or transport conditions, where applicable, do not jeopardise its compliance with the requirements set out in Section 2.

5. Importers shall keep, for a period of 10 years after the high-risk AI system has been placed on the market or put into service, a copy of the certificate issued by the notified body, where applicable, of the instructions for use, and of the EU declaration of conformity referred to in Article 47.

6. Importers shall provide the relevant competent authorities, upon a reasoned request, with all the necessary information and documentation, including that referred to in paragraph 5, to demonstrate the conformity of a high-risk AI system with the requirements set out in Section 2 in a language which can be easily understood by them. For this purpose, they shall also ensure that the technical documentation can be made available to those authorities.

7. Importers shall cooperate with the relevant competent authorities in any action those authorities take in relation to a high-risk AI system placed on the market by the importers, in particular to reduce and mitigate the risks posed by it.

Article 24

Obligations of distributors

1. Before making a high-risk AI system available on the market, distributors shall verify that it bears the required CE marking, that it is accompanied by a copy of the EU declaration of conformity referred to in Article 47 and instructions for use, and that the provider and the importer of that system, as applicable, have complied with their respective obligations as laid down in Article 16, points (b) and (c) and Article 23(3).

2. Where a distributor considers or has reason to consider, on the basis of the information in its possession, that a high-risk AI system is not in conformity with the requirements set out in Section 2, it shall not make the high-risk AI system available on the market until the system has been brought into conformity with those requirements. Furthermore, where the high-risk AI system presents a risk within the meaning of Article 79(1), the distributor shall inform the provider or the importer of the system, as applicable, to that effect.

3. Distributors shall ensure that, while a high-risk AI system is under their responsibility, storage or transport conditions, where applicable, do not jeopardise the compliance of the system with the requirements set out in Section 2.

4. A distributor that considers or has reason to consider, on the basis of the information in its possession, a high-risk AI system which it has made available on the market not to be in conformity with the requirements set out in Section 2, shall take the corrective actions necessary to bring that system into conformity with those requirements, to withdraw it or recall it, or shall ensure that the provider, the importer or any relevant operator, as appropriate, takes those corrective actions. Where the high-risk AI system presents a risk within the meaning of Article 79(1), the distributor shall immediately inform the provider or importer of the system and the authorities competent for the high-risk AI system concerned, giving details, in particular, of the non-compliance and of any corrective actions taken.

5. Upon a reasoned request from a relevant competent authority, distributors of a high-risk AI system shall provide that authority with all the information and documentation regarding their actions pursuant to paragraphs 1 to 4 necessary to demonstrate the conformity of that system with the requirements set out in Section 2.

6. Distributors shall cooperate with the relevant competent authorities in any action those authorities take in relation to a high-risk AI system made available on the market by the distributors, in particular to reduce or mitigate the risk posed by it.

Article 25

Responsibilities along the AI value chain

1. Any distributor, importer, deployer or other third-party shall be considered to be a provider of a high-risk AI system for the purposes of this Regulation and shall be subject to the obligations of the provider under Article 16, in any of the following circumstances:

(a) they put their name or trademark on a high-risk AI system already placed on the market or put into service, without prejudice to contractual arrangements stipulating that the obligations are otherwise allocated;

(b) they make a substantial modification to a high-risk AI system that has already been placed on the market or has already been put into service in such a way that it remains a high-risk AI system pursuant to Article 6;

(c) they modify the intended purpose of an AI system, including a general-purpose AI system, which has not been classified as high-risk and has already been placed on the market or put into service in such a way that the AI system concerned becomes a high-risk AI system in accordance with Article 6.

2. Where the circumstances referred to in paragraph 1 occur, the provider that initially placed the AI system on the market or put it into service shall no longer be considered to be a provider of that specific AI system for the purposes of this Regulation. That initial provider shall closely cooperate with new providers and shall make available the necessary information and provide the reasonably expected technical access and other assistance that are required for the fulfilment of the obligations set out in this Regulation, in particular regarding the compliance with the conformity assessment of high-risk AI systems. This paragraph shall not apply in cases where the initial provider has clearly specified that its AI system is not to be changed into a high-risk AI system and therefore does not fall under the obligation to hand over the documentation.

3. In the case of high-risk AI systems that are safety components of products covered by the Union harmonisation legislation listed in Section A of Annex I, the product manufacturer shall be considered to be the provider of the high-risk AI system, and shall be subject to the obligations under Article 16 under either of the following circumstances:

(a) the high-risk AI system is placed on the market together with the product under the name or trademark of the product manufacturer;

(b) the high-risk AI system is put into service under the name or trademark of the product manufacturer after the product has been placed on the market.

4. The provider of a high-risk AI system and the third party that supplies an AI system, tools, services, components, or processes that are used or integrated in a high-risk AI system shall, by written agreement, specify the necessary information, capabilities, technical access and other assistance based on the generally acknowledged state of the art, in order to enable the provider of the high-risk AI system to fully comply with the obligations set out in this Regulation. This paragraph shall not apply to third parties making accessible to the public tools, services, processes, or components, other than general-purpose AI models, under a free and open-source licence.

The AI Office may develop and recommend voluntary model terms for contracts between providers of high-risk AI systems and third parties that supply tools, services, components or processes that are used for or integrated into high-risk AI systems. When developing those voluntary model terms, the AI Office shall take into account possible contractual requirements applicable in specific sectors or business cases. The voluntary model terms shall be published and be available free of charge in an easily usable electronic format.

5. Paragraphs 2 and 3 are without prejudice to the need to observe and protect intellectual property rights, confidential business information and trade secrets in accordance with Union and national law.

Article 26

Obligations of deployers of high-risk AI systems

1. Deployers of high-risk AI systems shall take appropriate technical and organisational measures to ensure they use such systems in accordance with the instructions for use accompanying the systems, pursuant to paragraphs 3 and 6.

2. Deployers shall assign human oversight to natural persons who have the necessary competence, training and authority, as well as the necessary support.

3. The obligations set out in paragraphs 1 and 2, are without prejudice to other deployer obligations under Union or national law and to the deployer's freedom to organise its own resources and activities for the purpose of implementing the human oversight measures indicated by the provider.

4. Without prejudice to paragraphs 1 and 2, to the extent the deployer exercises control over the input data, that deployer shall ensure that input data is relevant and sufficiently representative in view of the intended purpose of the high-risk AI system.

5. Deployers shall monitor the operation of the high-risk AI system on the basis of the instructions for use and, where relevant, inform providers in accordance with Article 72. Where deployers have reason to consider that the use of the high-risk AI system in accordance with the instructions may result in that AI system presenting a risk within the meaning of Article 79(1), they shall, without undue delay, inform the provider or distributor and the relevant market surveillance authority, and shall suspend the use of that system. Where deployers have identified a serious incident, they shall also immediately inform first the provider, and then the importer or distributor and the relevant market surveillance authorities of that incident. If the deployer is not able to reach the provider, Article 73 shall apply *mutatis mutandis*. This obligation shall not cover sensitive operational data of deployers of AI systems which are law enforcement authorities.

For deployers that are financial institutions subject to requirements regarding their internal governance, arrangements or processes under Union financial services law, the monitoring obligation set out in the first subparagraph shall be deemed to be fulfilled by complying with the rules on internal governance arrangements, processes and mechanisms pursuant to the relevant financial service law.

6. Deployers of high-risk AI systems shall keep the logs automatically generated by that high-risk AI system to the extent such logs are under their control, for a period appropriate to the intended purpose of the high-risk AI system, of at least six months, unless provided otherwise in applicable Union or national law, in particular in Union law on the protection of personal data.

Deployers that are financial institutions subject to requirements regarding their internal governance, arrangements or processes under Union financial services law shall maintain the logs as part of the documentation kept pursuant to the relevant Union financial service law.

7. Before putting into service or using a high-risk AI system at the workplace, deployers who are employers shall inform workers' representatives and the affected workers that they will be subject to the use of the high-risk AI system. This information shall be provided, where applicable, in accordance with the rules and procedures laid down in Union and national law and practice on information of workers and their representatives.

8. Deployers of high-risk AI systems that are public authorities, or Union institutions, bodies, offices or agencies shall comply with the registration obligations referred to in Article 49. When such deployers find that the high-risk AI system that they envisage using has not been registered in the EU database referred to in Article 71, they shall not use that system and shall inform the provider or the distributor.

9. Where applicable, deployers of high-risk AI systems shall use the information provided under Article 13 of this Regulation to comply with their obligation to carry out a data protection impact assessment under Article 35 of Regulation (EU) 2016/679 or Article 27 of Directive (EU) 2016/680.

10. Without prejudice to Directive (EU) 2016/680, in the framework of an investigation for the targeted search of a person suspected or convicted of having committed a criminal offence, the deployer of a high-risk AI system for post-remote biometric identification shall request an authorisation, *ex ante*, or without undue delay and no later than 48 hours, by a judicial authority or an administrative authority whose decision is binding and subject to judicial review, for the use of that system, except when it is used for the initial identification of a potential suspect based on objective and verifiable facts directly linked to the offence. Each use shall be limited to what is strictly necessary for the investigation of a specific criminal offence.

If the authorisation requested pursuant to the first subparagraph is rejected, the use of the post-remote biometric identification system linked to that requested authorisation shall be stopped with immediate effect and the personal data linked to the use of the high-risk AI system for which the authorisation was requested shall be deleted.

In no case shall such high-risk AI system for post-remote biometric identification be used for law enforcement purposes in an untargeted way, without any link to a criminal offence, a criminal proceeding, a genuine and present or genuine and foreseeable threat of a criminal offence, or the search for a specific missing person. It shall be ensured that no decision that produces an adverse legal effect on a person may be taken by the law enforcement authorities based solely on the output of such post-remote biometric identification systems.

This paragraph is without prejudice to Article 9 of Regulation (EU) 2016/679 and Article 10 of Directive (EU) 2016/680 for the processing of biometric data.

Regardless of the purpose or deployer, each use of such high-risk AI systems shall be documented in the relevant police file and shall be made available to the relevant market surveillance authority and the national data protection authority upon request, excluding the disclosure of sensitive operational data related to law enforcement. This subparagraph shall be without prejudice to the powers conferred by Directive (EU) 2016/680 on supervisory authorities.

Deployers shall submit annual reports to the relevant market surveillance and national data protection authorities on their use of post-remote biometric identification systems, excluding the disclosure of sensitive operational data related to law enforcement. The reports may be aggregated to cover more than one deployment.

Member States may introduce, in accordance with Union law, more restrictive laws on the use of post-remote biometric identification systems.

11.   Without prejudice to Article 50 of this Regulation, deployers of high-risk AI systems referred to in Annex III that make decisions or assist in making decisions related to natural persons shall inform the natural persons that they are subject to the use of the high-risk AI system. For high-risk AI systems used for law enforcement purposes Article 13 of Directive (EU) 2016/680 shall apply.

12.   Deployers shall cooperate with the relevant competent authorities in any action those authorities take in relation to the high-risk AI system in order to implement this Regulation.

Article 27

Fundamental rights impact assessment for high-risk AI systems

1. Prior to deploying a high-risk AI system referred to in Article 6(2), with the exception of high-risk AI systems intended to be used in the area listed in point 2 of Annex III, deployers that are bodies governed by public law, or are private entities providing public services, and deployers of high-risk AI systems referred to in points 5 (b) and (c) of Annex III, shall perform an assessment of the impact on fundamental rights that the use of such system may produce. For that purpose, deployers shall perform an assessment consisting of:

(a) a description of the deployer's processes in which the high-risk AI system will be used in line with its intended purpose;

(b) a description of the period of time within which, and the frequency with which, each high-risk AI system is intended to be used;

(c) the categories of natural persons and groups likely to be affected by its use in the specific context;

(d) the specific risks of harm likely to have an impact on the categories of natural persons or groups of persons identified pursuant to point (c) of this paragraph, taking into account the information given by the provider pursuant to Article 13;

(e) a description of the implementation of human oversight measures, according to the instructions for use;

(f) the measures to be taken in the case of the materialisation of those risks, including the arrangements for internal governance and complaint mechanisms.

2. The obligation laid down in paragraph 1 applies to the first use of the high-risk AI system. The deployer may, in similar cases, rely on previously conducted fundamental rights impact assessments or existing impact assessments carried out by provider. If, during the use of the high-risk AI system, the deployer considers that any of the elements listed in paragraph 1 has changed or is no longer up to date, the deployer shall take the necessary steps to update the information.

3. Once the assessment referred to in paragraph 1 of this Article has been performed, the deployer shall notify the market surveillance authority of its results, submitting the filled-out template referred to in paragraph 5 of this Article as part of the notification. In the case referred to in Article 46(1), deployers may be exempt from that obligation to notify.

4.   If any of the obligations laid down in this Article is already met through the data protection impact assessment conducted pursuant to Article 35 of Regulation (EU) 2016/679 or Article 27 of Directive (EU) 2016/680, the fundamental rights impact assessment referred to in paragraph 1 of this Article shall complement that data protection impact assessment.

5.   The AI Office shall develop a template for a questionnaire, including through an automated tool, to facilitate deployers in complying with their obligations under this Article in a simplified manner.

SECTION 4

Notifying authorities and notified bodies

Article 28

Notifying authorities

1.   Each Member State shall designate or establish at least one notifying authority responsible for setting up and carrying out the necessary procedures for the assessment, designation and notification of conformity assessment bodies and for their monitoring. Those procedures shall be developed in cooperation between the notifying authorities of all Member States.

2.   Member States may decide that the assessment and monitoring referred to in paragraph 1 is to be carried out by a national accreditation body within the meaning of, and in accordance with, Regulation (EC) No 765/2008.

3.   Notifying authorities shall be established, organised and operated in such a way that no conflict of interest arises with conformity assessment bodies, and that the objectivity and impartiality of their activities are safeguarded.

4.   Notifying authorities shall be organised in such a way that decisions relating to the notification of conformity assessment bodies are taken by competent persons different from those who carried out the assessment of those bodies.

5.   Notifying authorities shall offer or provide neither any activities that conformity assessment bodies perform, nor any consultancy services on a commercial or competitive basis.

6.   Notifying authorities shall safeguard the confidentiality of the information that they obtain, in accordance with Article 78.

7. Notifying authorities shall have an adequate number of competent personnel at their disposal for the proper performance of their tasks. Competent personnel shall have the necessary expertise, where applicable, for their function, in fields such as information technologies, AI and law, including the supervision of fundamental rights.

Article 29

Application of a conformity assessment body for notification

1. Conformity assessment bodies shall submit an application for notification to the notifying authority of the Member State in which they are established.

2. The application for notification shall be accompanied by a description of the conformity assessment activities, the conformity assessment module or modules and the types of AI systems for which the conformity assessment body claims to be competent, as well as by an accreditation certificate, where one exists, issued by a national accreditation body attesting that the conformity assessment body fulfils the requirements laid down in Article 31.

Any valid document related to existing designations of the applicant notified body under any other Union harmonisation legislation shall be added.

3. Where the conformity assessment body concerned cannot provide an accreditation certificate, it shall provide the notifying authority with all the documentary evidence necessary for the verification, recognition and regular monitoring of its compliance with the requirements laid down in Article 31.

4. For notified bodies which are designated under any other Union harmonisation legislation, all documents and certificates linked to those designations may be used to support their designation procedure under this Regulation, as appropriate. The notified body shall update the documentation referred to in paragraphs 2 and 3 of this Article whenever relevant changes occur, in order to enable the authority responsible for notified bodies to monitor and verify continuous compliance with all the requirements laid down in Article 31.

Article 30

Notification procedure

1. Notifying authorities may notify only conformity assessment bodies which have satisfied the requirements laid down in Article 31.

2. Notifying authorities shall notify the Commission and the other Member States, using the electronic notification tool developed and managed by the Commission, of each conformity assessment body referred to in paragraph 1.

3. The notification referred to in paragraph 2 of this Article shall include full details of the conformity assessment activities, the conformity assessment module or modules, the types of AI systems concerned, and the relevant attestation of competence. Where a notification is not based on an accreditation certificate as referred to in Article 29(2), the notifying authority shall provide the Commission and the other Member States with documentary evidence which attests to the competence of the conformity assessment body and to the arrangements in place to ensure that that body will be monitored regularly and will continue to satisfy the requirements laid down in Article 31.

4. The conformity assessment body concerned may perform the activities of a notified body only where no objections are raised by the Commission or the other Member States within two weeks of a notification by a notifying authority where it includes an accreditation certificate referred to in Article 29(2), or within two months of a notification by the notifying authority where it includes documentary evidence referred to in Article 29(3).

5. Where objections are raised, the Commission shall, without delay, enter into consultations with the relevant Member States and the conformity assessment body. In view thereof, the Commission shall decide whether the authorisation is justified. The Commission shall address its decision to the Member State concerned and to the relevant conformity assessment body.

Article 31

Requirements relating to notified bodies

1. A notified body shall be established under the national law of a Member State and shall have legal personality.

2. Notified bodies shall satisfy the organisational, quality management, resources and process requirements that are necessary to fulfil their tasks, as well as suitable cybersecurity requirements.

3. The organisational structure, allocation of responsibilities, reporting lines and operation of notified bodies shall ensure confidence in their performance, and in the results of the conformity assessment activities that the notified bodies conduct.

4. Notified bodies shall be independent of the provider of a high-risk AI system in relation to which they perform conformity assessment activities. Notified bodies shall also be independent of any other operator having an economic interest in high-risk AI systems assessed, as well as of any competitors of the provider. This shall not preclude the use of assessed high-risk AI systems that are necessary for the operations of the conformity assessment body, or the use of such high-risk AI systems for personal purposes.

5. Neither a conformity assessment body, its top-level management nor the personnel responsible for carrying out its conformity assessment tasks shall be directly involved in the design, development, marketing or use of high-risk AI systems, nor shall they represent the parties engaged in those activities. They shall not engage in any activity that might conflict with their independence of judgement or integrity in relation to conformity assessment activities for which they are notified. This shall, in particular, apply to consultancy services.

6. Notified bodies shall be organised and operated so as to safeguard the independence, objectivity and impartiality of their activities. Notified bodies shall document and implement a structure and procedures to safeguard impartiality and to promote and apply the principles of impartiality throughout their organisation, personnel and assessment activities.

7. Notified bodies shall have documented procedures in place ensuring that their personnel, committees, subsidiaries, subcontractors and any associated body or personnel of external bodies maintain, in accordance with Article 78, the confidentiality of the information which comes into their possession during the performance of conformity assessment activities, except when its disclosure is required by law. The staff of notified bodies shall be bound to observe professional secrecy with regard to all information obtained in carrying out their tasks under this Regulation, except in relation to the notifying authorities of the Member State in which their activities are carried out.

8. Notified bodies shall have procedures for the performance of activities which take due account of the size of a provider, the sector in which it operates, its structure, and the degree of complexity of the AI system concerned.

9. Notified bodies shall take out appropriate liability insurance for their conformity assessment activities, unless liability is assumed by the Member State in which they are established in accordance with national law or that Member State is itself directly responsible for the conformity assessment.

10. Notified bodies shall be capable of carrying out all their tasks under this Regulation with the highest degree of professional integrity and the requisite competence in the specific field, whether those tasks are carried out by notified bodies themselves or on their behalf and under their responsibility.

11. Notified bodies shall have sufficient internal competences to be able effectively to evaluate the tasks conducted by external parties on their behalf. The notified body shall have permanent availability of sufficient administrative, technical, legal and scientific personnel who possess experience and knowledge relating to the relevant types of AI systems, data and data computing, and relating to the requirements set out in Section 2.

12. Notified bodies shall participate in coordination activities as referred to in Article 38. They shall also take part directly, or be represented in, European standardisation organisations, or ensure that they are aware and up to date in respect of relevant standards.

Article 32

Presumption of conformity with requirements relating to notified bodies

Where a conformity assessment body demonstrates its conformity with the criteria laid down in the relevant harmonised standards or parts thereof, the references of which have been published in the *Official Journal of the European Union*, it shall be presumed to comply with the requirements set out in Article 31 in so far as the applicable harmonised standards cover those requirements.

Article 33

Subsidiaries of notified bodies and subcontracting

1. Where a notified body subcontracts specific tasks connected with the conformity assessment or has recourse to a subsidiary, it shall ensure that the subcontractor or the subsidiary meets the requirements laid down in Article 31, and shall inform the notifying authority accordingly.

2. Notified bodies shall take full responsibility for the tasks performed by any subcontractors or subsidiaries.

3. Activities may be subcontracted or carried out by a subsidiary only with the agreement of the provider. Notified bodies shall make a list of their subsidiaries publicly available.

4. The relevant documents concerning the assessment of the qualifications of the subcontractor or the subsidiary and the work carried out by them under this Regulation shall be kept at the disposal of the notifying authority for a period of five years from the termination date of the subcontracting.

Article 34

Operational obligations of notified bodies

1. Notified bodies shall verify the conformity of high-risk AI systems in accordance with the conformity assessment procedures set out in Article 43.

2. Notified bodies shall avoid unnecessary burdens for providers when performing their activities, and take due account of the size of the provider, the sector in which it operates, its structure and the degree of complexity of the high-risk AI system concerned, in particular in view of minimising administrative burdens and compliance costs for micro- and small enterprises within the meaning of Recommendation 2003/361/EC. The notified body shall, nevertheless, respect the degree of rigour and the level of protection required for the compliance of the high-risk AI system with the requirements of this Regulation.

3. Notified bodies shall make available and submit upon request all relevant documentation, including the providers' documentation, to the notifying authority referred to in Article 28 to allow that authority to conduct its assessment, designation, notification and monitoring activities, and to facilitate the assessment outlined in this Section.

Article 35

Identification numbers and lists of notified bodies

1. The Commission shall assign a single identification number to each notified body, even where a body is notified under more than one Union act.

2. The Commission shall make publicly available the list of the bodies notified under this Regulation, including their identification numbers and the activities for which they have been notified. The Commission shall ensure that the list is kept up to date.

Article 36

Changes to notifications

1. The notifying authority shall notify the Commission and the other Member States of any relevant changes to the notification of a notified body via the electronic notification tool referred to in Article 30(2).

2. The procedures laid down in Articles 29 and 30 shall apply to extensions of the scope of the notification.

For changes to the notification other than extensions of its scope, the procedures laid down in paragraphs (3) to (9) shall apply.

3. Where a notified body decides to cease its conformity assessment activities, it shall inform the notifying authority and the providers concerned as soon as possible and, in the case of a planned cessation, at least one year before ceasing its activities. The certificates of the notified body may remain valid for a period of nine months after cessation of the notified body's activities, on condition that another notified body has confirmed in writing that it will assume responsibilities for the high-risk AI systems covered by those certificates. The latter notified body shall complete a full assessment of the high-risk AI systems affected by the end of that nine-month-period before issuing new certificates for those systems. Where the notified body has ceased its activity, the notifying authority shall withdraw the designation.

4. Where a notifying authority has sufficient reason to consider that a notified body no longer meets the requirements laid down in Article 31, or that it is failing to fulfil its obligations, the notifying authority shall without delay investigate the matter with the utmost diligence. In that context, it shall inform the notified body concerned about the objections raised and give it the possibility to make its views known. If the notifying authority comes to the conclusion that the notified body no longer meets the requirements laid down in Article 31 or that it is failing to fulfil its obligations, it shall restrict, suspend or withdraw the designation as appropriate, depending on the seriousness of the failure to meet those requirements or fulfil those obligations. It shall immediately inform the Commission and the other Member States accordingly.

5. Where its designation has been suspended, restricted, or fully or partially withdrawn, the notified body shall inform the providers concerned within 10 days.

6. In the event of the restriction, suspension or withdrawal of a designation, the notifying authority shall take appropriate steps to ensure that the files of the notified

body concerned are kept, and to make them available to notifying authorities in other Member States and to market surveillance authorities at their request.

7. In the event of the restriction, suspension or withdrawal of a designation, the notifying authority shall:

(a) assess the impact on the certificates issued by the notified body;

(b) submit a report on its findings to the Commission and the other Member States within three months of having notified the changes to the designation;

(c) require the notified body to suspend or withdraw, within a reasonable period of time determined by the authority, any certificates which were unduly issued, in order to ensure the continuing conformity of high-risk AI systems on the market;

(d) inform the Commission and the Member States about certificates the suspension or withdrawal of which it has required;

(e) provide the national competent authorities of the Member State in which the provider has its registered place of business with all relevant information about the certificates of which it has required the suspension or withdrawal; that authority shall take the appropriate measures, where necessary, to avoid a potential risk to health, safety or fundamental rights.

8. With the exception of certificates unduly issued, and where a designation has been suspended or restricted, the certificates shall remain valid in one of the following circumstances:

(a) the notifying authority has confirmed, within one month of the suspension or restriction, that there is no risk to health, safety or fundamental rights in relation to certificates affected by the suspension or restriction, and the notifying authority has outlined a timeline for actions to remedy the suspension or restriction; or

(b) the notifying authority has confirmed that no certificates relevant to the suspension will be issued, amended or re-issued during the course of the suspension or restriction, and states whether the notified body has the capability of continuing to monitor and remain responsible for existing certificates issued for the period of the suspension or restriction; in the event that the notifying authority determines that the notified body does not have the capability to support existing certificates issued, the provider of the system covered by the certificate shall confirm in writing to the national competent authorities of the Member State in which it has its registered place of business, within three months of the suspension or restriction, that another qualified notified body is temporarily assuming the functions of the notified body to monitor and remain responsible for the certificates during the period of suspension or restriction.

9. With the exception of certificates unduly issued, and where a designation has been withdrawn, the certificates shall remain valid for a period of nine months under the following circumstances:

(a) the national competent authority of the Member State in which the provider of the high-risk AI system covered by the certificate has its registered place of business has confirmed that there is no risk to health, safety or fundamental rights associated with the high-risk AI systems concerned; and

(b) another notified body has confirmed in writing that it will assume immediate responsibility for those AI systems and completes its assessment within 12 months of the withdrawal of the designation.

In the circumstances referred to in the first subparagraph, the national competent authority of the Member State in which the provider of the system covered by the certificate has its place of business may extend the provisional validity of the certificates for additional periods of three months, which shall not exceed 12 months in total.

The national competent authority or the notified body assuming the functions of the notified body affected by the change of designation shall immediately inform the Commission, the other Member States and the other notified bodies thereof.

Article 37

Challenge to the competence of notified bodies

1. The Commission shall, where necessary, investigate all cases where there are reasons to doubt the competence of a notified body or the continued fulfilment by a notified body of the requirements laid down in Article 31 and of its applicable responsibilities.

2. The notifying authority shall provide the Commission, on request, with all relevant information relating to the notification or the maintenance of the competence of the notified body concerned.

3. The Commission shall ensure that all sensitive information obtained in the course of its investigations pursuant to this Article is treated confidentially in accordance with Article 78.

4. Where the Commission ascertains that a notified body does not meet or no longer meets the requirements for its notification, it shall inform the notifying Member State accordingly and request it to take the necessary corrective measures, including the suspension or withdrawal of the notification if necessary. Where the Member State fails to take the necessary corrective measures, the Commission may, by means of an implementing act, suspend, restrict or withdraw the designation. That implementing act shall be adopted in accordance with the examination procedure referred to in Article 98(2).

Article 38

Coordination of notified bodies

1. The Commission shall ensure that, with regard to high-risk AI systems, appropriate coordination and cooperation between notified bodies active in the conformity assessment procedures pursuant to this Regulation are put in place and properly operated in the form of a sectoral group of notified bodies.

2. Each notifying authority shall ensure that the bodies notified by it participate in the work of a group referred to in paragraph 1, directly or through designated representatives.

3. The Commission shall provide for the exchange of knowledge and best practices between notifying authorities.

Article 39

Conformity assessment bodies of third countries

Conformity assessment bodies established under the law of a third country with which the Union has concluded an agreement may be authorised to carry out the activities of notified bodies under this Regulation, provided that they meet the requirements laid down in Article 31 or they ensure an equivalent level of compliance.

SECTION 5

Standards, conformity assessment, certificates, registration

Article 40

Harmonised standards and standardisation deliverables

1. High-risk AI systems or general-purpose AI models which are in conformity with harmonised standards or parts thereof the references of which have been published in the *Official Journal of the European Union* in accordance with Regulation (EU) No 1025/2012 shall be presumed to be in conformity with the requirements set out in Section 2 of this Chapter or, as applicable, with the obligations set out in of Chapter V, Sections 2 and 3, of this Regulation, to the extent that those standards cover those requirements or obligations.

2. In accordance with Article 10 of Regulation (EU) No 1025/2012, the Commission shall issue, without undue delay, standardisation requests covering all requirements set out in Section 2 of this Chapter and, as applicable, standardisation requests covering obligations set out in Chapter V, Sections 2 and 3, of this Regulation. The standardisation request shall also ask for deliverables on reporting and documentation processes to improve AI systems' resource performance, such as reducing the high-risk AI system's consumption of energy and of other resources during its lifecycle, and on the energy-efficient development of general-purpose AI models. When preparing a standardisation request, the Commission shall consult the Board and relevant stakeholders, including the advisory forum.

When issuing a standardisation request to European standardisation organisations, the Commission shall specify that standards have to be clear, consistent, including with the standards developed in the various sectors for products covered by the existing Union harmonisation legislation listed in Annex I, and aiming to ensure that high-risk AI systems or general-purpose AI models placed on the market or put into service in the Union meet the relevant requirements or obligations laid down in this Regulation.

The Commission shall request the European standardisation organisations to provide evidence of their best efforts to fulfil the objectives referred to in the first and the second subparagraph of this paragraph in accordance with Article 24 of Regulation (EU) No 1025/2012.

3. The participants in the standardisation process shall seek to promote investment and innovation in AI, including through increasing legal certainty, as well as the competitiveness and growth of the Union market, to contribute to strengthening

global cooperation on standardisation and taking into account existing international standards in the field of AI that are consistent with Union values, fundamental rights and interests, and to enhance multi-stakeholder governance ensuring a balanced representation of interests and the effective participation of all relevant stakeholders in accordance with Articles 5, 6, and 7 of Regulation (EU) No 1025/2012.

Article 41

Common specifications

1. The Commission may adopt, implementing acts establishing common specifications for the requirements set out in Section 2 of this Chapter or, as applicable, for the obligations set out in Sections 2 and 3 of Chapter V where the following conditions have been fulfilled:

(a) the Commission has requested, pursuant to Article 10(1) of Regulation (EU) No 1025/2012, one or more European standardisation organisations to draft a harmonised standard for the requirements set out in Section 2 of this Chapter, or, as applicable, for the obligations set out in Sections 2 and 3 of Chapter V, and:

    (i) the request has not been accepted by any of the European standardisation organisations; or

    (ii) the harmonised standards addressing that request are not delivered within the deadline set in accordance with Article 10(1) of Regulation (EU) No 1025/2012; or

    (iii) the relevant harmonised standards insufficiently address fundamental rights concerns; or

    (iv) the harmonised standards do not comply with the request; and

(b) no reference to harmonised standards covering the requirements referred to in Section 2 of this Chapter or, as applicable, the obligations referred to in Sections 2 and 3 of Chapter V has been published in the *Official Journal of the European Union* in accordance with Regulation (EU) No 1025/2012, and no such reference is expected to be published within a reasonable period.

When drafting the common specifications, the Commission shall consult the advisory forum referred to in Article 67.

The implementing acts referred to in the first subparagraph of this paragraph shall be adopted in accordance with the examination procedure referred to in Article 98(2).

2. Before preparing a draft implementing act, the Commission shall inform the committee referred to in Article 22 of Regulation (EU) No 1025/2012 that it considers the conditions laid down in paragraph 1 of this Article to be fulfilled.

3. High-risk AI systems or general-purpose AI models which are in conformity with the common specifications referred to in paragraph 1, or parts of those specifications, shall be presumed to be in conformity with the requirements set out in Section 2 of this Chapter or, as applicable, to comply with the obligations referred to in Sections 2 and 3 of Chapter V, to the extent those common specifications cover those requirements or those obligations.

4. Where a harmonised standard is adopted by a European standardisation organisation and proposed to the Commission for the publication of its reference in the *Official Journal of the European Union*, the Commission shall assess the harmonised standard in accordance with Regulation (EU) No 1025/2012. When reference to a harmonised standard is published in the *Official Journal of the European Union*, the Commission shall repeal the implementing acts referred to in paragraph 1, or parts thereof which cover the same requirements set out in Section 2 of this Chapter or, as applicable, the same obligations set out in Sections 2 and 3 of Chapter V.

5. Where providers of high-risk AI systems or general-purpose AI models do not comply with the common specifications referred to in paragraph 1, they shall duly justify that they have adopted technical solutions that meet the requirements referred to in Section 2 of this Chapter or, as applicable, comply with the obligations set out in Sections 2 and 3 of Chapter V to a level at least equivalent thereto.

6. Where a Member State considers that a common specification does not entirely meet the requirements set out in Section 2 or, as applicable, comply with obligations set out in Sections 2 and 3 of Chapter V, it shall inform the Commission thereof with a detailed explanation. The Commission shall assess that information and, if appropriate, amend the implementing act establishing the common specification concerned.

Article 42

Presumption of conformity with certain requirements

1. High-risk AI systems that have been trained and tested on data reflecting the specific geographical, behavioural, contextual or functional setting within which they are intended to be used shall be presumed to comply with the relevant requirements laid down in Article 10(4).

2. High-risk AI systems that have been certified or for which a statement of conformity has been issued under a cybersecurity scheme pursuant to Regulation (EU) 2019/881 and the references of which have been published in the *Official Journal of the European Union* shall be presumed to comply with the cybersecurity requirements set out in Article 15 of this Regulation in so far as the cybersecurity certificate or statement of conformity or parts thereof cover those requirements.

Article 43

Conformity assessment

1. For high-risk AI systems listed in point 1 of Annex III, where, in demonstrating the compliance of a high-risk AI system with the requirements set out in Section 2, the provider has applied harmonised standards referred to in Article 40, or, where applicable, common specifications referred to in Article 41, the provider shall opt for one of the following conformity assessment procedures based on:

(a) the internal control referred to in Annex VI; or

(b) the assessment of the quality management system and the assessment of the technical documentation, with the involvement of a notified body, referred to in Annex VII.

In demonstrating the compliance of a high-risk AI system with the requirements set out in Section 2, the provider shall follow the conformity assessment procedure set out in Annex VII where:

(a) harmonised standards referred to in Article 40 do not exist, and common specifications referred to in Article 41 are not available;

(b) the provider has not applied, or has applied only part of, the harmonised standard;

(c) the common specifications referred to in point (a) exist, but the provider has not applied them;

(d) one or more of the harmonised standards referred to in point (a) has been published with a restriction, and only on the part of the standard that was restricted.

For the purposes of the conformity assessment procedure referred to in Annex VII, the provider may choose any of the notified bodies. However, where the high-risk AI system is intended to be put into service by law enforcement, immigration or asylum authorities or by Union institutions, bodies, offices or agencies, the market surveillance authority referred to in Article 74(8) or (9), as applicable, shall act as a notified body.

2. For high-risk AI systems referred to in points 2 to 8 of Annex III, providers shall follow the conformity assessment procedure based on internal control as referred to in Annex VI, which does not provide for the involvement of a notified body.

3. For high-risk AI systems covered by the Union harmonisation legislation listed in Section A of Annex I, the provider shall follow the relevant conformity assessment procedure as required under those legal acts. The requirements set out in Section 2 of this Chapter shall apply to those high-risk AI systems and shall be part of that assessment. Points 4.3., 4.4., 4.5. and the fifth paragraph of point 4.6 of Annex VII shall also apply.

For the purposes of that assessment, notified bodies which have been notified under those legal acts shall be entitled to control the conformity of the high-risk AI systems with the requirements set out in Section 2, provided that the compliance of those notified bodies with requirements laid down in Article 31(4), (5), (10) and (11) has been assessed in the context of the notification procedure under those legal acts.

Where a legal act listed in Section A of Annex I enables the product manufacturer to opt out from a third-party conformity assessment, provided that that manufacturer has applied all harmonised standards covering all the relevant requirements, that manufacturer may use that option only if it has also applied harmonised standards or, where applicable, common specifications referred to in Article 41, covering all requirements set out in Section 2 of this Chapter.

4. High-risk AI systems that have already been subject to a conformity assessment procedure shall undergo a new conformity assessment procedure in the event of a substantial modification, regardless of whether the modified system is intended to be further distributed or continues to be used by the current deployer.

For high-risk AI systems that continue to learn after being placed on the market or put into service, changes to the high-risk AI system and its performance that have been pre-determined by the provider at the moment of the initial conformity assessment and are part of the information contained in the technical documentation referred to in point 2(f) of Annex IV, shall not constitute a substantial modification.

5. The Commission is empowered to adopt delegated acts in accordance with Article 97 in order to amend Annexes VI and VII by updating them in light of technical progress.

6. The Commission is empowered to adopt delegated acts in accordance with Article 97 in order to amend paragraphs 1 and 2 of this Article in order to subject high-risk AI systems referred to in points 2 to 8 of Annex III to the conformity assessment procedure referred to in Annex VII or parts thereof. The Commission shall adopt such delegated acts taking into account the effectiveness of the conformity assessment procedure based on internal control referred to in Annex VI in preventing or minimising the risks to health and safety and protection of fundamental rights posed by such systems, as well as the availability of adequate capacities and resources among notified bodies.

Article 44

Certificates

1. Certificates issued by notified bodies in accordance with Annex VII shall be drawn-up in a language which can be easily understood by the relevant authorities in the Member State in which the notified body is established.

2. Certificates shall be valid for the period they indicate, which shall not exceed five years for AI systems covered by Annex I, and four years for AI systems covered by Annex III. At the request of the provider, the validity of a certificate may be extended for further periods, each not exceeding five years for AI systems covered by Annex I, and four years for AI systems covered by Annex III, based on a re-assessment in accordance with the applicable conformity assessment procedures. Any supplement to a certificate shall remain valid, provided that the certificate which it supplements is valid.

3. Where a notified body finds that an AI system no longer meets the requirements set out in Section 2, it shall, taking account of the principle of proportionality, suspend or withdraw the certificate issued or impose restrictions on it, unless compliance with those requirements is ensured by appropriate corrective action taken by the provider of the system within an appropriate deadline set by the notified body. The notified body shall give reasons for its decision.

An appeal procedure against decisions of the notified bodies, including on conformity certificates issued, shall be available.

Article 45

Information obligations of notified bodies

1. Notified bodies shall inform the notifying authority of the following:

(a) any Union technical documentation assessment certificates, any supplements to those certificates, and any quality management system approvals issued in accordance with the requirements of Annex VII;

(b) any refusal, restriction, suspension or withdrawal of a Union technical documentation assessment certificate or a quality management system approval issued in accordance with the requirements of Annex VII;

(c) any circumstances affecting the scope of or conditions for notification;

(d) any request for information which they have received from market surveillance authorities regarding conformity assessment activities;

(e) on request, conformity assessment activities performed within the scope of their notification and any other activity performed, including cross-border activities and subcontracting.

2. Each notified body shall inform the other notified bodies of:

(a) quality management system approvals which it has refused, suspended or withdrawn, and, upon request, of quality system approvals which it has issued;

(b) Union technical documentation assessment certificates or any supplements thereto which it has refused, withdrawn, suspended or otherwise restricted, and, upon request, of the certificates and/or supplements thereto which it has issued.

3. Each notified body shall provide the other notified bodies carrying out similar conformity assessment activities covering the same types of AI systems with relevant information on issues relating to negative and, on request, positive conformity assessment results.

4. Notified bodies shall safeguard the confidentiality of the information that they obtain, in accordance with Article 78.

Article 46

Derogation from conformity assessment procedure

1. By way of derogation from Article 43 and upon a duly justified request, any market surveillance authority may authorise the placing on the market or the putting into service of specific high-risk AI systems within the territory of the Member State concerned, for exceptional reasons of public security or the protection of life and health of persons, environmental protection or the protection of key industrial and infrastructural assets. That authorisation shall be for a limited period while the necessary conformity assessment procedures are being carried out, taking into account the exceptional reasons justifying the derogation. The completion of those procedures shall be undertaken without undue delay.

2. In a duly justified situation of urgency for exceptional reasons of public security or in the case of specific, substantial and imminent threat to the life or physical safety of natural persons, law-enforcement authorities or civil protection authorities may put a specific high-risk AI system into service without the authorisation referred to in paragraph 1, provided that such authorisation is requested during or after the use without undue delay. If the authorisation referred to in paragraph 1 is refused, the use of the high-risk AI system shall be stopped with immediate effect and all the results and outputs of such use shall be immediately discarded.

3. The authorisation referred to in paragraph 1 shall be issued only if the market surveillance authority concludes that the high-risk AI system complies with the requirements of Section 2. The market surveillance authority shall inform the Commission and the other Member States of any authorisation issued pursuant to paragraphs 1 and 2. This obligation shall not cover sensitive operational data in relation to the activities of law-enforcement authorities.

4. Where, within 15 calendar days of receipt of the information referred to in paragraph 3, no objection has been raised by either a Member State or the Commission in respect of an authorisation issued by a market surveillance authority of a Member State in accordance with paragraph 1, that authorisation shall be deemed justified.

5. Where, within 15 calendar days of receipt of the notification referred to in paragraph 3, objections are raised by a Member State against an authorisation issued by a market surveillance authority of another Member State, or where the Commission considers the authorisation to be contrary to Union law, or the conclusion of the Member States regarding the compliance of the system as referred to in paragraph 3 to be unfounded, the Commission shall, without delay, enter into

consultations with the relevant Member State. The operators concerned shall be consulted and have the possibility to present their views. Having regard thereto, the Commission shall decide whether the authorisation is justified. The Commission shall address its decision to the Member State concerned and to the relevant operators.

6. Where the Commission considers the authorisation unjustified, it shall be withdrawn by the market surveillance authority of the Member State concerned.

7. For high-risk AI systems related to products covered by Union harmonisation legislation listed in Section A of Annex I, only the derogations from the conformity assessment established in that Union harmonisation legislation shall apply.

Article 47

EU declaration of conformity

1. The provider shall draw up a written machine readable, physical or electronically signed EU declaration of conformity for each high-risk AI system, and keep it at the disposal of the national competent authorities for 10 years after the high-risk AI system has been placed on the market or put into service. The EU declaration of conformity shall identify the high-risk AI system for which it has been drawn up. A copy of the EU declaration of conformity shall be submitted to the relevant national competent authorities upon request.

2. The EU declaration of conformity shall state that the high-risk AI system concerned meets the requirements set out in Section 2. The EU declaration of conformity shall contain the information set out in Annex V, and shall be translated into a language that can be easily understood by the national competent authorities of the Member States in which the high-risk AI system is placed on the market or made available.

3. Where high-risk AI systems are subject to other Union harmonisation legislation which also requires an EU declaration of conformity, a single EU declaration of conformity shall be drawn up in respect of all Union law applicable to the high-risk AI system. The declaration shall contain all the information required to identify the Union harmonisation legislation to which the declaration relates.

4. By drawing up the EU declaration of conformity, the provider shall assume responsibility for compliance with the requirements set out in Section 2. The provider shall keep the EU declaration of conformity up-to-date as appropriate.

5.   The Commission is empowered to adopt delegated acts in accordance with Article 97 in order to amend Annex V by updating the content of the EU declaration of conformity set out in that Annex, in order to introduce elements that become necessary in light of technical progress.

Article 48

CE Marking

1.   The CE marking shall be subject to the general principles set out in Article 30 of Regulation (EC) No 765/2008.

2.   For high-risk AI systems provided digitally, a digital CE marking shall be used, only if it can easily be accessed via the interface from which that system is accessed or via an easily accessible machine-readable code or other electronic means.

3.   The CE marking shall be affixed visibly, legibly and indelibly for high-risk AI systems. Where that is not possible or not warranted on account of the nature of the high-risk AI system, it shall be affixed to the packaging or to the accompanying documentation, as appropriate.

4.   Where applicable, the CE marking shall be followed by the identification number of the notified body responsible for the conformity assessment procedures set out in Article 43. The identification number of the notified body shall be affixed by the body itself or, under its instructions, by the provider or by the provider's authorised representative. The identification number shall also be indicated in any promotional material which mentions that the high-risk AI system fulfils the requirements for CE marking.

5.   Where high-risk AI systems are subject to other Union law which also provides for the affixing of the CE marking, the CE marking shall indicate that the high-risk AI system also fulfil the requirements of that other law.

Article 49

Registration

1.   Before placing on the market or putting into service a high-risk AI system listed in Annex III, with the exception of high-risk AI systems referred to in point 2 of Annex III, the provider or, where applicable, the authorised representative shall register themselves and their system in the EU database referred to in Article 71.

2.   Before placing on the market or putting into service an AI system for which the provider has concluded that it is not high-risk according to Article 6(3), that provider or, where applicable, the authorised representative shall register themselves and that system in the EU database referred to in Article 71.

3.   Before putting into service or using a high-risk AI system listed in Annex III, with the exception of high-risk AI systems listed in point 2 of Annex III, deployers that are public authorities, Union institutions, bodies, offices or agencies or persons acting on their behalf shall register themselves, select the system and register its use in the EU database referred to in Article 71.

4.   For high-risk AI systems referred to in points 1, 6 and 7 of Annex III, in the areas of law enforcement, migration, asylum and border control management, the registration referred to in paragraphs 1, 2 and 3 of this Article shall be in a secure non-public section of the EU database referred to in Article 71 and shall include only the following information, as applicable, referred to in:

(a) Section A, points 1 to 10, of Annex VIII, with the exception of points 6, 8 and 9;

(b) Section B, points 1 to 5, and points 8 and 9 of Annex VIII;

(c)   Section C, points 1 to 3, of Annex VIII;

(d)   points 1, 2, 3 and 5, of Annex IX.

Only the Commission and national authorities referred to in Article 74(8) shall have access to the respective restricted sections of the EU database listed in the first subparagraph of this paragraph.

5.   High-risk AI systems referred to in point 2 of Annex III shall be registered at national level.

CHAPTER IV

TRANSPARENCY OBLIGATIONS FOR PROVIDERS AND DEPLOYERS OF CERTAIN AI SYSTEMS

Article 50

Transparency obligations for providers and deployers of certain AI systems

1.   Providers shall ensure that AI systems intended to interact directly with natural persons are designed and developed in such a way that the natural persons concerned

are informed that they are interacting with an AI system, unless this is obvious from the point of view of a natural person who is reasonably well-informed, observant and circumspect, taking into account the circumstances and the context of use. This obligation shall not apply to AI systems authorised by law to detect, prevent, investigate or prosecute criminal offences, subject to appropriate safeguards for the rights and freedoms of third parties, unless those systems are available for the public to report a criminal offence.

2. Providers of AI systems, including general-purpose AI systems, generating synthetic audio, image, video or text content, shall ensure that the outputs of the AI system are marked in a machine-readable format and detectable as artificially generated or manipulated. Providers shall ensure their technical solutions are effective, interoperable, robust and reliable as far as this is technically feasible, taking into account the specificities and limitations of various types of content, the costs of implementation and the generally acknowledged state of the art, as may be reflected in relevant technical standards. This obligation shall not apply to the extent the AI systems perform an assistive function for standard editing or do not substantially alter the input data provided by the deployer or the semantics thereof, or where authorised by law to detect, prevent, investigate or prosecute criminal offences.

3. Deployers of an emotion recognition system or a biometric categorisation system shall inform the natural persons exposed thereto of the operation of the system, and shall process the personal data in accordance with Regulations (EU) 2016/679 and (EU) 2018/1725 and Directive (EU) 2016/680, as applicable. This obligation shall not apply to AI systems used for biometric categorisation and emotion recognition, which are permitted by law to detect, prevent or investigate criminal offences, subject to appropriate safeguards for the rights and freedoms of third parties, and in accordance with Union law.

4. Deployers of an AI system that generates or manipulates image, audio or video content constituting a deep fake, shall disclose that the content has been artificially generated or manipulated. This obligation shall not apply where the use is authorised by law to detect, prevent, investigate or prosecute criminal offence. Where the content forms part of an evidently artistic, creative, satirical, fictional or analogous work or programme, the transparency obligations set out in this paragraph are limited to disclosure of the existence of such generated or manipulated content in an appropriate manner that does not hamper the display or enjoyment of the work.

Deployers of an AI system that generates or manipulates text which is published with the purpose of informing the public on matters of public interest shall disclose that the text has been artificially generated or manipulated. This obligation shall not apply where the use is authorised by law to detect, prevent, investigate or prosecute criminal offences or where the AI-generated content has undergone a process of human review or editorial control and where a natural or legal person holds editorial responsibility for the publication of the content.

5. The information referred to in paragraphs 1 to 4 shall be provided to the natural persons concerned in a clear and distinguishable manner at the latest at the time of the first interaction or exposure. The information shall conform to the applicable accessibility requirements.

6. Paragraphs 1 to 4 shall not affect the requirements and obligations set out in Chapter III, and shall be without prejudice to other transparency obligations laid down in Union or national law for deployers of AI systems.

7. The AI Office shall encourage and facilitate the drawing up of codes of practice at Union level to facilitate the effective implementation of the obligations regarding the detection and labelling of artificially generated or manipulated content. The Commission may adopt implementing acts to approve those codes of practice in accordance with the procedure laid down in Article 56 (6). If it deems the code is not adequate, the Commission may adopt an implementing act specifying common rules for the implementation of those obligations in accordance with the examination procedure laid down in Article 98(2).

CHAPTER V

GENERAL-PURPOSE AI MODELS

SECTION 1

Classification rules

Article 51

Classification of general-purpose AI models as general-purpose AI models with systemic risk

1. A general-purpose AI model shall be classified as a general-purpose AI model with systemic risk if it meets any of the following conditions:

(a) it has high impact capabilities evaluated on the basis of appropriate technical tools and methodologies, including indicators and benchmarks;

(b) based on a decision of the Commission, *ex officio* or following a qualified alert from the scientific panel, it has capabilities or an impact equivalent to those set out in point (a) having regard to the criteria set out in Annex XIII.

2. A general-purpose AI model shall be presumed to have high impact capabilities pursuant to paragraph 1, point (a), when the cumulative amount of computation used for its training measured in floating point operations is greater than 10^{25}.

3. The Commission shall adopt delegated acts in accordance with Article 97 to amend the thresholds listed in paragraphs 1 and 2 of this Article, as well as to supplement benchmarks and indicators in light of evolving technological developments, such as algorithmic improvements or increased hardware efficiency, when necessary, for these thresholds to reflect the state of the art.

Article 52

Procedure

1. Where a general-purpose AI model meets the condition referred to in Article 51(1), point (a), the relevant provider shall notify the Commission without delay and in any event within two weeks after that requirement is met or it becomes known that it will be met. That notification shall include the information necessary to demonstrate that the relevant requirement has been met. If the Commission becomes aware of a general-purpose AI model presenting systemic risks of which it has not been notified, it may decide to designate it as a model with systemic risk.

2. The provider of a general-purpose AI model that meets the condition referred to in Article 51(1), point (a), may present, with its notification, sufficiently substantiated arguments to demonstrate that, exceptionally, although it meets that requirement, the general-purpose AI model does not present, due to its specific characteristics, systemic risks and therefore should not be classified as a general-purpose AI model with systemic risk.

3. Where the Commission concludes that the arguments submitted pursuant to paragraph 2 are not sufficiently substantiated and the relevant provider was not able to demonstrate that the general-purpose AI model does not present, due to its specific characteristics, systemic risks, it shall reject those arguments, and the general-purpose AI model shall be considered to be a general-purpose AI model with systemic risk.

4. The Commission may designate a general-purpose AI model as presenting systemic risks, *ex officio* or following a qualified alert from the scientific panel pursuant to Article 90(1), point (a), on the basis of criteria set out in Annex XIII.

The Commission is empowered to adopt delegated acts in accordance with Article 97 in order to amend Annex XIII by specifying and updating the criteria set out in that Annex.

5. Upon a reasoned request of a provider whose model has been designated as a general-purpose AI model with systemic risk pursuant to paragraph 4, the Commission shall take the request into account and may decide to reassess whether the general-purpose AI model can still be considered to present systemic risks on the basis of the criteria set out in Annex XIII. Such a request shall contain objective, detailed and new reasons that have arisen since the designation decision. Providers may request reassessment at the earliest six months after the designation decision. Where the Commission, following its reassessment, decides to maintain the designation as a general-purpose AI model with systemic risk, providers may request reassessment at the earliest six months after that decision.

6. The Commission shall ensure that a list of general-purpose AI models with systemic risk is published and shall keep that list up to date, without prejudice to the need to observe and protect intellectual property rights and confidential business information or trade secrets in accordance with Union and national law.

SECTION 2

Obligations for providers of general-purpose AI models

Article 53

Obligations for providers of general-purpose AI models

1. Providers of general-purpose AI models shall:

(a) draw up and keep up-to-date the technical documentation of the model, including its training and testing process and the results of its evaluation, which shall contain, at a minimum, the information set out in Annex XI for the purpose of providing it, upon request, to the AI Office and the national competent authorities;

(b) draw up, keep up-to-date and make available information and documentation to providers of AI systems who intend to integrate the general-purpose AI model into their AI systems. Without prejudice to the need to observe and protect intellectual property rights and confidential business information or trade secrets in accordance with Union and national law, the information and documentation shall:

(i) enable providers of AI systems to have a good understanding of the capabilities and limitations of the general-purpose AI model and to comply with their obligations pursuant to this Regulation; and

(ii) contain, at a minimum, the elements set out in Annex XII;

(c) put in place a policy to comply with Union law on copyright and related rights, and in particular to identify and comply with, including through state-of-the-art technologies, a reservation of rights expressed pursuant to Article 4(3) of Directive (EU) 2019/790;

(d) draw up and make publicly available a sufficiently detailed summary about the content used for training of the general-purpose AI model, according to a template provided by the AI Office.

2. The obligations set out in paragraph 1, points (a) and (b), shall not apply to providers of AI models that are released under a free and open-source licence that allows for the access, usage, modification, and distribution of the model, and whose parameters, including the weights, the information on the model architecture, and the information on model usage, are made publicly available. This exception shall not apply to general-purpose AI models with systemic risks.

3. Providers of general-purpose AI models shall cooperate as necessary with the Commission and the national competent authorities in the exercise of their competences and powers pursuant to this Regulation.

4. Providers of general-purpose AI models may rely on codes of practice within the meaning of Article 56 to demonstrate compliance with the obligations set out in paragraph 1 of this Article, until a harmonised standard is published. Compliance with European harmonised standards grants providers the presumption of conformity to the extent that those standards cover those obligations. Providers of general-purpose AI models who do not adhere to an approved code of practice or do not comply with a European harmonised standard shall demonstrate alternative adequate means of compliance for assessment by the Commission.

5. For the purpose of facilitating compliance with Annex XI, in particular points 2 (d) and (e) thereof, the Commission is empowered to adopt delegated acts in accordance with Article 97 to detail measurement and calculation methodologies with a view to allowing for comparable and verifiable documentation.

6. The Commission is empowered to adopt delegated acts in accordance with Article 97(2) to amend Annexes XI and XII in light of evolving technological developments.

7. Any information or documentation obtained pursuant to this Article, including trade secrets, shall be treated in accordance with the confidentiality obligations set out in Article 78.

Article 54

Authorised representatives of providers of general-purpose AI models

1. Prior to placing a general-purpose AI model on the Union market, providers established in third countries shall, by written mandate, appoint an authorised representative which is established in the Union.

2. The provider shall enable its authorised representative to perform the tasks specified in the mandate received from the provider.

3. The authorised representative shall perform the tasks specified in the mandate received from the provider. It shall provide a copy of the mandate to the AI Office upon request, in one of the official languages of the institutions of the Union. For the purposes of this Regulation, the mandate shall empower the authorised representative to carry out the following tasks:

(a) verify that the technical documentation specified in Annex XI has been drawn up and all obligations referred to in Article 53 and, where applicable, Article 55 have been fulfilled by the provider;

(b) keep a copy of the technical documentation specified in Annex XI at the disposal of the AI Office and national competent authorities, for a period of 10 years after the general-purpose AI model has been placed on the market, and the contact details of the provider that appointed the authorised representative;

(c) provide the AI Office, upon a reasoned request, with all the information and documentation, including that referred to in point (b), necessary to demonstrate compliance with the obligations in this Chapter;

(d) cooperate with the AI Office and competent authorities, upon a reasoned request, in any action they take in relation to the general-purpose AI model, including when the model is integrated into AI systems placed on the market or put into service in the Union.

4. The mandate shall empower the authorised representative to be addressed, in addition to or instead of the provider, by the AI Office or the competent authorities, on all issues related to ensuring compliance with this Regulation.

5. The authorised representative shall terminate the mandate if it considers or has reason to consider the provider to be acting contrary to its obligations pursuant to this Regulation. In such a case, it shall also immediately inform the AI Office about the termination of the mandate and the reasons therefor.

6. The obligation set out in this Article shall not apply to providers of general-purpose AI models that are released under a free and open-source licence that allows for the access, usage, modification, and distribution of the model, and whose parameters, including the weights, the information on the model architecture, and the information on model usage, are made publicly available, unless the general-purpose AI models present systemic risks.

SECTION 3

Obligations of providers of general-purpose AI models with systemic risk

Article 55

Obligations of providers of general-purpose AI models with systemic risk

1. In addition to the obligations listed in Articles 53 and 54, providers of general-purpose AI models with systemic risk shall:

(a) perform model evaluation in accordance with standardised protocols and tools reflecting the state of the art, including conducting and documenting adversarial testing of the model with a view to identifying and mitigating systemic risks;

(b) assess and mitigate possible systemic risks at Union level, including their sources, that may stem from the development, the placing on the market, or the use of general-purpose AI models with systemic risk;

(c) keep track of, document, and report, without undue delay, to the AI Office and, as appropriate, to national competent authorities, relevant information about serious incidents and possible corrective measures to address them;

(d) ensure an adequate level of cybersecurity protection for the general-purpose AI model with systemic risk and the physical infrastructure of the model.

2. Providers of general-purpose AI models with systemic risk may rely on codes of practice within the meaning of Article 56 to demonstrate compliance with the obligations set out in paragraph 1 of this Article, until a harmonised standard is published. Compliance with European harmonised standards grants providers the presumption of conformity to the extent that those standards cover those obligations.

Providers of general-purpose AI models with systemic risks who do not adhere to an approved code of practice or do not comply with a European harmonised standard shall demonstrate alternative adequate means of compliance for assessment by the Commission.

3. Any information or documentation obtained pursuant to this Article, including trade secrets, shall be treated in accordance with the confidentiality obligations set out in Article 78.

SECTION 4

Codes of practice

Article 56

Codes of practice

1. The AI Office shall encourage and facilitate the drawing up of codes of practice at Union level in order to contribute to the proper application of this Regulation, taking into account international approaches.

2. The AI Office and the Board shall aim to ensure that the codes of practice cover at least the obligations provided for in Articles 53 and 55, including the following issues:

(a) the means to ensure that the information referred to in Article 53(1), points (a) and (b), is kept up to date in light of market and technological developments;

(b) the adequate level of detail for the summary about the content used for training;

(c) the identification of the type and nature of the systemic risks at Union level, including their sources, where appropriate;

(d) the measures, procedures and modalities for the assessment and management of the systemic risks at Union level, including the documentation thereof, which shall be proportionate to the risks, take into consideration their severity and probability and take into account the specific challenges of tackling those risks in light of the possible ways in which such risks may emerge and materialise along the AI value chain.

3. The AI Office may invite all providers of general-purpose AI models, as well as relevant national competent authorities, to participate in the drawing-up of codes of practice. Civil society organisations, industry, academia and other relevant

stakeholders, such as downstream providers and independent experts, may support the process.

4. The AI Office and the Board shall aim to ensure that the codes of practice clearly set out their specific objectives and contain commitments or measures, including key performance indicators as appropriate, to ensure the achievement of those objectives, and that they take due account of the needs and interests of all interested parties, including affected persons, at Union level.

5. The AI Office shall aim to ensure that participants to the codes of practice report regularly to the AI Office on the implementation of the commitments and the measures taken and their outcomes, including as measured against the key performance indicators as appropriate. Key performance indicators and reporting commitments shall reflect differences in size and capacity between various participants.

6. The AI Office and the Board shall regularly monitor and evaluate the achievement of the objectives of the codes of practice by the participants and their contribution to the proper application of this Regulation. The AI Office and the Board shall assess whether the codes of practice cover the obligations provided for in Articles 53 and 55, and shall regularly monitor and evaluate the achievement of their objectives. They shall publish their assessment of the adequacy of the codes of practice.

The Commission may, by way of an implementing act, approve a code of practice and give it a general validity within the Union. That implementing act shall be adopted in accordance with the examination procedure referred to in Article 98(2).

7. The AI Office may invite all providers of general-purpose AI models to adhere to the codes of practice. For providers of general-purpose AI models not presenting systemic risks this adherence may be limited to the obligations provided for in Article 53, unless they declare explicitly their interest to join the full code.

8. The AI Office shall, as appropriate, also encourage and facilitate the review and adaptation of the codes of practice, in particular in light of emerging standards. The AI Office shall assist in the assessment of available standards.

9. Codes of practice shall be ready at the latest by 2 May 2025. The AI Office shall take the necessary steps, including inviting providers pursuant to paragraph 7.

If, by 2 August 2025, a code of practice cannot be finalised, or if the AI Office deems it is not adequate following its assessment under paragraph 6 of this Article, the Commission may provide, by means of implementing acts, common rules for the implementation of the obligations provided for in Articles 53 and 55, including the issues set out in paragraph 2 of this Article. Those implementing acts shall be adopted in accordance with the examination procedure referred to in Article 98(2).

CHAPTER VI

MEASURES IN SUPPORT OF INNOVATION

Article 57

AI regulatory sandboxes

1. Member States shall ensure that their competent authorities establish at least one AI regulatory sandbox at national level, which shall be operational by 2 August 2026. That sandbox may also be established jointly with the competent authorities of other Member States. The Commission may provide technical support, advice and tools for the establishment and operation of AI regulatory sandboxes.

The obligation under the first subparagraph may also be fulfilled by participating in an existing sandbox in so far as that participation provides an equivalent level of national coverage for the participating Member States.

2. Additional AI regulatory sandboxes at regional or local level, or established jointly with the competent authorities of other Member States may also be established.

3. The European Data Protection Supervisor may also establish an AI regulatory sandbox for Union institutions, bodies, offices and agencies, and may exercise the roles and the tasks of national competent authorities in accordance with this Chapter.

4. Member States shall ensure that the competent authorities referred to in paragraphs 1 and 2 allocate sufficient resources to comply with this Article effectively and in a timely manner. Where appropriate, national competent authorities shall cooperate with other relevant authorities, and may allow for the involvement of other actors within the AI ecosystem. This Article shall not affect other regulatory sandboxes established under Union or national law. Member States shall ensure an appropriate level of cooperation between the authorities supervising those other sandboxes and the national competent authorities.

5. AI regulatory sandboxes established under paragraph 1 shall provide for a controlled environment that fosters innovation and facilitates the development, training, testing and validation of innovative AI systems for a limited time before their being placed on the market or put into service pursuant to a specific sandbox

plan agreed between the providers or prospective providers and the competent authority. Such sandboxes may include testing in real world conditions supervised therein.

6. Competent authorities shall provide, as appropriate, guidance, supervision and support within the AI regulatory sandbox with a view to identifying risks, in particular to fundamental rights, health and safety, testing, mitigation measures, and their effectiveness in relation to the obligations and requirements of this Regulation and, where relevant, other Union and national law supervised within the sandbox.

7. Competent authorities shall provide providers and prospective providers participating in the AI regulatory sandbox with guidance on regulatory expectations and how to fulfil the requirements and obligations set out in this Regulation.

Upon request of the provider or prospective provider of the AI system, the competent authority shall provide a written proof of the activities successfully carried out in the sandbox. The competent authority shall also provide an exit report detailing the activities carried out in the sandbox and the related results and learning outcomes. Providers may use such documentation to demonstrate their compliance with this Regulation through the conformity assessment process or relevant market surveillance activities. In this regard, the exit reports and the written proof provided by the national competent authority shall be taken positively into account by market surveillance authorities and notified bodies, with a view to accelerating conformity assessment procedures to a reasonable extent.

8. Subject to the confidentiality provisions in Article 78, and with the agreement of the provider or prospective provider, the Commission and the Board shall be authorised to access the exit reports and shall take them into account, as appropriate, when exercising their tasks under this Regulation. If both the provider or prospective provider and the national competent authority explicitly agree, the exit report may be made publicly available through the single information platform referred to in this Article.

9. The establishment of AI regulatory sandboxes shall aim to contribute to the following objectives:

(a) improving legal certainty to achieve regulatory compliance with this Regulation or, where relevant, other applicable Union and national law;

(b) supporting the sharing of best practices through cooperation with the authorities involved in the AI regulatory sandbox;

(c) fostering innovation and competitiveness and facilitating the development of an AI ecosystem;

(d) contributing to evidence-based regulatory learning;

(e) facilitating and accelerating access to the Union market for AI systems, in particular when provided by SMEs, including start-ups.

10. National competent authorities shall ensure that, to the extent the innovative AI systems involve the processing of personal data or otherwise fall under the supervisory remit of other national authorities or competent authorities providing or supporting access to data, the national data protection authorities and those other national or competent authorities are associated with the operation of the AI regulatory sandbox and involved in the supervision of those aspects to the extent of their respective tasks and powers.

11. The AI regulatory sandboxes shall not affect the supervisory or corrective powers of the competent authorities supervising the sandboxes, including at regional or local level. Any significant risks to health and safety and fundamental rights identified during the development and testing of such AI systems shall result in an adequate mitigation. National competent authorities shall have the power to temporarily or permanently suspend the testing process, or the participation in the sandbox if no effective mitigation is possible, and shall inform the AI Office of such decision. National competent authorities shall exercise their supervisory powers within the limits of the relevant law, using their discretionary powers when implementing legal provisions in respect of a specific AI regulatory sandbox project, with the objective of supporting innovation in AI in the Union.

12. Providers and prospective providers participating in the AI regulatory sandbox shall remain liable under applicable Union and national liability law for any damage inflicted on third parties as a result of the experimentation taking place in the sandbox. However, provided that the prospective providers observe the specific plan and the terms and conditions for their participation and follow in good faith the guidance given by the national competent authority, no administrative fines shall be imposed by the authorities for infringements of this Regulation. Where other competent authorities responsible for other Union and national law were actively involved in the supervision of the AI system in the sandbox and provided guidance for compliance, no administrative fines shall be imposed regarding that law.

13. The AI regulatory sandboxes shall be designed and implemented in such a way that, where relevant, they facilitate cross-border cooperation between national competent authorities.

14. National competent authorities shall coordinate their activities and cooperate within the framework of the Board.

15. National competent authorities shall inform the AI Office and the Board of the establishment of a sandbox, and may ask them for support and guidance. The AI Office shall make publicly available a list of planned and existing sandboxes and keep it up to date in order to encourage more interaction in the AI regulatory sandboxes and cross-border cooperation.

16. National competent authorities shall submit annual reports to the AI Office and to the Board, from one year after the establishment of the AI regulatory sandbox and every year thereafter until its termination, and a final report. Those reports shall provide information on the progress and results of the implementation of those sandboxes, including best practices, incidents, lessons learnt and recommendations on their setup and, where relevant, on the application and possible revision of this Regulation, including its delegated and implementing acts, and on the application of other Union law supervised by the competent authorities within the sandbox. The national competent authorities shall make those annual reports or abstracts thereof available to the public, online. The Commission shall, where appropriate, take the annual reports into account when exercising its tasks under this Regulation.

17. The Commission shall develop a single and dedicated interface containing all relevant information related to AI regulatory sandboxes to allow stakeholders to interact with AI regulatory sandboxes and to raise enquiries with competent authorities, and to seek non-binding guidance on the conformity of innovative products, services, business models embedding AI technologies, in accordance with Article 62(1), point (c). The Commission shall proactively coordinate with national competent authorities, where relevant.

Article 58

Detailed arrangements for, and functioning of, AI regulatory sandboxes

1. In order to avoid fragmentation across the Union, the Commission shall adopt implementing acts specifying the detailed arrangements for the establishment, development, implementation, operation and supervision of the AI regulatory sandboxes. The implementing acts shall include common principles on the following issues:

(a) eligibility and selection criteria for participation in the AI regulatory sandbox;

(b) procedures for the application, participation, monitoring, exiting from and termination of the AI regulatory sandbox, including the sandbox plan and the exit report;

(c) the terms and conditions applicable to the participants.

Those implementing acts shall be adopted in accordance with the examination procedure referred to in Article 98(2).

2. The implementing acts referred to in paragraph 1 shall ensure:

(a) that AI regulatory sandboxes are open to any applying provider or prospective provider of an AI system who fulfils eligibility and selection criteria, which shall be transparent and fair, and that national competent authorities inform applicants of their decision within three months of the application;

(b) that AI regulatory sandboxes allow broad and equal access and keep up with demand for participation; providers and prospective providers may also submit applications in partnerships with deployers and other relevant third parties;

(c) that the detailed arrangements for, and conditions concerning AI regulatory sandboxes support, to the best extent possible, flexibility for national competent authorities to establish and operate their AI regulatory sandboxes;

(d) that access to the AI regulatory sandboxes is free of charge for SMEs, including start-ups, without prejudice to exceptional costs that national competent authorities may recover in a fair and proportionate manner;

(e) that they facilitate providers and prospective providers, by means of the learning outcomes of the AI regulatory sandboxes, in complying with conformity assessment obligations under this Regulation and the voluntary application of the codes of conduct referred to in Article 95;

(f) that AI regulatory sandboxes facilitate the involvement of other relevant actors within the AI ecosystem, such as notified bodies and standardisation organisations, SMEs, including start-ups, enterprises, innovators, testing and experimentation facilities, research and experimentation labs and European Digital Innovation Hubs, centres of excellence, individual researchers, in order to allow and facilitate cooperation with the public and private sectors;

(g) that procedures, processes and administrative requirements for application, selection, participation and exiting the AI regulatory sandbox are simple, easily intelligible, and clearly communicated in order to facilitate the participation of

SMEs, including start-ups, with limited legal and administrative capacities and are streamlined across the Union, in order to avoid fragmentation and that participation in an AI regulatory sandbox established by a Member State, or by the European Data Protection Supervisor is mutually and uniformly recognised and carries the same legal effects across the Union;

(h) that participation in the AI regulatory sandbox is limited to a period that is appropriate to the complexity and scale of the project and that may be extended by the national competent authority;

(i) that AI regulatory sandboxes facilitate the development of tools and infrastructure for testing, benchmarking, assessing and explaining dimensions of AI systems relevant for regulatory learning, such as accuracy, robustness and cybersecurity, as well as measures to mitigate risks to fundamental rights and society at large.

3. Prospective providers in the AI regulatory sandboxes, in particular SMEs and start-ups, shall be directed, where relevant, to pre-deployment services such as guidance on the implementation of this Regulation, to other value-adding services such as help with standardisation documents and certification, testing and experimentation facilities, European Digital Innovation Hubs and centres of excellence.

4. Where national competent authorities consider authorising testing in real world conditions supervised within the framework of an AI regulatory sandbox to be established under this Article, they shall specifically agree the terms and conditions of such testing and, in particular, the appropriate safeguards with the participants, with a view to protecting fundamental rights, health and safety. Where appropriate, they shall cooperate with other national competent authorities with a view to ensuring consistent practices across the Union.

Article 59

Further processing of personal data for developing certain AI systems in the public interest in the AI regulatory sandbox

1. In the AI regulatory sandbox, personal data lawfully collected for other purposes may be processed solely for the purpose of developing, training and testing certain AI systems in the sandbox when all of the following conditions are met:

(a) AI systems shall be developed for safeguarding substantial public interest by a public authority or another natural or legal person and in one or more of the following areas:

    (i) public safety and public health, including disease detection, diagnosis prevention, control and treatment and improvement of health care systems;

    (ii) a high level of protection and improvement of the quality of the environment, protection of biodiversity, protection against pollution, green transition measures, climate change mitigation and adaptation measures;

    (iii) energy sustainability;

    (iv) safety and resilience of transport systems and mobility, critical infrastructure and networks;

    (v) efficiency and quality of public administration and public services;

(b) the data processed are necessary for complying with one or more of the requirements referred to in Chapter III, Section 2 where those requirements cannot effectively be fulfilled by processing anonymised, synthetic or other non-personal data;

(c) there are effective monitoring mechanisms to identify if any high risks to the rights and freedoms of the data subjects, as referred to in Article 35 of Regulation (EU) 2016/679 and in Article 39 of Regulation (EU) 2018/1725, may arise during the sandbox experimentation, as well as response mechanisms to promptly mitigate those risks and, where necessary, stop the processing;

(d) any personal data to be processed in the context of the sandbox are in a functionally separate, isolated and protected data processing environment under the control of the prospective provider and only authorised persons have access to those data;

(e) providers can further share the originally collected data only in accordance with Union data protection law; any personal data created in the sandbox cannot be shared outside the sandbox;

(f) any processing of personal data in the context of the sandbox neither leads to measures or decisions affecting the data subjects nor does it affect the application of their rights laid down in Union law on the protection of personal data;

(g) any personal data processed in the context of the sandbox are protected by means of appropriate technical and organisational measures and deleted once the participation in the sandbox has terminated or the personal data has reached the end of its retention period;

(h) the logs of the processing of personal data in the context of the sandbox are kept for the duration of the participation in the sandbox, unless provided otherwise by Union or national law;

(i) a complete and detailed description of the process and rationale behind the training, testing and validation of the AI system is kept together with the testing results as part of the technical documentation referred to in Annex IV;

(j) a short summary of the AI project developed in the sandbox, its objectives and expected results is published on the website of the competent authorities; this obligation shall not cover sensitive operational data in relation to the activities of law enforcement, border control, immigration or asylum authorities.

2. For the purposes of the prevention, investigation, detection or prosecution of criminal offences or the execution of criminal penalties, including safeguarding against and preventing threats to public security, under the control and responsibility of law enforcement authorities, the processing of personal data in AI regulatory sandboxes shall be based on a specific Union or national law and subject to the same cumulative conditions as referred to in paragraph 1.

3. Paragraph 1 is without prejudice to Union or national law which excludes processing of personal data for other purposes than those explicitly mentioned in that law, as well as to Union or national law laying down the basis for the processing of personal data which is necessary for the purpose of developing, testing or training of innovative AI systems or any other legal basis, in compliance with Union law on the protection of personal data.

Article 60

Testing of high-risk AI systems in real world conditions outside AI regulatory sandboxes

1. Testing of high-risk AI systems in real world conditions outside AI regulatory sandboxes may be conducted by providers or prospective providers of high-risk AI systems listed in Annex III, in accordance with this Article and the real-world testing plan referred to in this Article, without prejudice to the prohibitions under Article 5.

The Commission shall, by means of implementing acts, specify the detailed elements of the real-world testing plan. Those implementing acts shall be adopted in accordance with the examination procedure referred to in Article 98(2).

This paragraph shall be without prejudice to Union or national law on the testing in real world conditions of high-risk AI systems related to products covered by Union harmonisation legislation listed in Annex I.

2. Providers or prospective providers may conduct testing of high-risk AI systems referred to in Annex III in real world conditions at any time before the placing on the market or the putting into service of the AI system on their own or in partnership with one or more deployers or prospective deployers.

3. The testing of high-risk AI systems in real world conditions under this Article shall be without prejudice to any ethical review that is required by Union or national law.

4. Providers or prospective providers may conduct the testing in real world conditions only where all of the following conditions are met:

(a) The provider or prospective provider has drawn up a real-world testing plan and submitted it to the market surveillance authority in the Member State where the testing in real world conditions is to be conducted;

(b) the market surveillance authority in the Member State where the testing in real world conditions is to be conducted has approved the testing in real world conditions and the real-world testing plan; where the market surveillance authority has not provided an answer within 30 days, the testing in real world conditions and the real-world testing plan shall be understood to have been approved; where national law does not provide for a tacit approval, the testing in real world conditions shall remain subject to an authorisation;

(c) the provider or prospective provider, with the exception of providers or prospective providers of high-risk AI systems referred to in points 1, 6 and 7 of Annex III in the areas of law enforcement, migration, asylum and border control management, and high-risk AI systems referred to in point 2 of Annex III has registered the testing in real world conditions in accordance with Article 71(4) with a Union-wide unique single identification number and with the information specified in Annex IX; the provider or prospective provider of high-risk AI systems referred to in points 1, 6 and 7 of Annex III in the areas of law enforcement, migration, asylum and border control management, has registered the testing in real-world conditions in the secure non-public section of the EU database according to Article 49(4), point (d), with a Union-wide unique single identification number and with the information specified therein; the provider or prospective provider of high-risk AI systems referred to in point 2 of Annex III has registered the testing in real-world conditions in accordance with Article 49(5);

(d) the provider or prospective provider conducting the testing in real world conditions is established in the Union or has appointed a legal representative who is established in the Union;

(e) data collected and processed for the purpose of the testing in real world conditions shall be transferred to third countries only provided that appropriate and applicable safeguards under Union law are implemented;

(f) the testing in real world conditions does not last longer than necessary to achieve its objectives and in any case not longer than six months, which may be extended for an additional period of six months, subject to prior notification by the provider or prospective provider to the market surveillance authority, accompanied by an explanation of the need for such an extension;

(g) the subjects of the testing in real world conditions who are persons belonging to vulnerable groups due to their age or disability, are appropriately protected;

(h) where a provider or prospective provider organises the testing in real world conditions in cooperation with one or more deployers or prospective deployers, the latter have been informed of all aspects of the testing that are relevant to their decision to participate, and given the relevant instructions for use of the AI system referred to in Article 13; the provider or prospective provider and the deployer or prospective deployer shall conclude an agreement specifying their roles and responsibilities with a view to ensuring compliance with the provisions for testing in real world conditions under this Regulation and under other applicable Union and national law;

(i) the subjects of the testing in real world conditions have given informed consent in accordance with Article 61, or in the case of law enforcement, where the seeking of informed consent would prevent the AI system from being tested, the testing itself and the outcome of the testing in the real world conditions shall not have any negative effect on the subjects, and their personal data shall be deleted after the test is performed;

(j) the testing in real world conditions is effectively overseen by the provider or prospective provider, as well as by deployers or prospective deployers through persons who are suitably qualified in the relevant field and have the necessary capacity, training and authority to perform their tasks;

(k) the predictions, recommendations or decisions of the AI system can be effectively reversed and disregarded.

5. Any subjects of the testing in real world conditions, or their legally designated representative, as appropriate, may, without any resulting detriment and without having to provide any justification, withdraw from the testing at any time by revoking their informed consent and may request the immediate and permanent deletion of their personal data. The withdrawal of the informed consent shall not affect the activities already carried out.

6. In accordance with Article 75, Member States shall confer on their market surveillance authorities the powers of requiring providers and prospective providers to provide information, of carrying out unannounced remote or on-site inspections, and of performing checks on the conduct of the testing in real world conditions and the related high-risk AI systems. Market surveillance authorities shall use those powers to ensure the safe development of testing in real world conditions.

7. Any serious incident identified in the course of the testing in real world conditions shall be reported to the national market surveillance authority in accordance with Article 73. The provider or prospective provider shall adopt immediate mitigation measures or, failing that, shall suspend the testing in real world conditions until such mitigation takes place, or otherwise terminate it. The provider or prospective provider shall establish a procedure for the prompt recall of the AI system upon such termination of the testing in real world conditions.

8. Providers or prospective providers shall notify the national market surveillance authority in the Member State where the testing in real world conditions is to be conducted of the suspension or termination of the testing in real world conditions and of the final outcomes.

9. The provider or prospective provider shall be liable under applicable Union and national liability law for any damage caused in the course of their testing in real world conditions.

Article 61

Informed consent to participate in testing in real world conditions outside AI regulatory sandboxes

1.   For the purpose of testing in real world conditions under Article 60, freely-given informed consent shall be obtained from the subjects of testing prior to their participation in such testing and after their having been duly informed with concise, clear, relevant, and understandable information regarding:

(a) the nature and objectives of the testing in real world conditions and the possible inconvenience that may be linked to their participation;

(b) the conditions under which the testing in real world conditions is to be conducted, including the expected duration of the subject or subjects' participation;

(c) their rights, and the guarantees regarding their participation, in particular their right to refuse to participate in, and the right to withdraw from, testing in real world conditions at any time without any resulting detriment and without having to provide any justification;

(d) the arrangements for requesting the reversal or the disregarding of the predictions, recommendations or decisions of the AI system;

(e) the Union-wide unique single identification number of the testing in real world conditions in accordance with Article 60(4) point (c), and the contact details of the provider or its legal representative from whom further information can be obtained.

2.   The informed consent shall be dated and documented and a copy shall be given to the subjects of testing or their legal representative.

Article 62

Measures for providers and deployers, in particular SMEs, including start-ups

1.   Member States shall undertake the following actions:

(a) provide SMEs, including start-ups, having a registered office or a branch in the Union, with priority access to the AI regulatory sandboxes, to the extent that they fulfil the eligibility conditions and selection criteria; the priority access shall not preclude other SMEs, including start-ups, other than those referred to in this paragraph from access to the AI regulatory sandbox, provided that they also fulfil the eligibility conditions and selection criteria;

(b) organise specific awareness raising and training activities on the application of this Regulation tailored to the needs of SMEs including start-ups, deployers and, as appropriate, local public authorities;

(c) utilise existing dedicated channels and where appropriate, establish new ones for communication with SMEs including start-ups, deployers, other innovators and, as appropriate, local public authorities to provide advice and respond to queries about the implementation of this Regulation, including as regards participation in AI regulatory sandboxes;

(d) facilitate the participation of SMEs and other relevant stakeholders in the standardisation development process.

2.   The specific interests and needs of the SME providers, including start-ups, shall be taken into account when setting the fees for conformity assessment under Article 43, reducing those fees proportionately to their size, market size and other relevant indicators.

3.   The AI Office shall undertake the following actions:

(a) provide standardised templates for areas covered by this Regulation, as specified by the Board in its request;

(b) develop and maintain a single information platform providing easy to use information in relation to this Regulation for all operators across the Union;

(c) organise appropriate communication campaigns to raise awareness about the obligations arising from this Regulation;

(d) evaluate and promote the convergence of best practices in public procurement procedures in relation to AI systems.

Article 63

Derogations for specific operators

1. Microenterprises within the meaning of Recommendation 2003/361/EC may comply with certain elements of the quality management system required by Article 17 of this Regulation in a simplified manner, provided that they do not have partner enterprises or linked enterprises within the meaning of that

Recommendation. For that purpose, the Commission shall develop guidelines on the elements of the quality management system which may be complied with in a simplified manner considering the needs of microenterprises, without affecting the level of protection or the need for compliance with the requirements in respect of high-risk AI systems.

2. Paragraph 1 of this Article shall not be interpreted as exempting those operators from fulfilling any other requirements or obligations laid down in this Regulation, including those established in Articles 9, 10, 11, 12, 13, 14, 15, 72 and 73.

CHAPTER VII

GOVERNANCE

SECTION 1

Governance at Union level

Article 64

AI Office

1. The Commission shall develop Union expertise and capabilities in the field of AI through the AI Office.

2. Member States shall facilitate the tasks entrusted to the AI Office, as reflected in this Regulation.

Article 65

Establishment and structure of the European Artificial Intelligence Board

1. A European Artificial Intelligence Board (the 'Board') is hereby established.

2. The Board shall be composed of one representative per Member State. The European Data Protection Supervisor shall participate as observer. The AI Office shall also attend the Board's meetings, without taking part in the votes. Other national and Union authorities, bodies or experts may be invited to the meetings by the Board on a case by case basis, where the issues discussed are of relevance for them.

3. Each representative shall be designated by their Member State for a period of three years, renewable once.

4. Member States shall ensure that their representatives on the Board:

(a) have the relevant competences and powers in their Member State so as to contribute actively to the achievement of the Board's tasks referred to in Article 66;

(b) are designated as a single contact point vis-à-vis the Board and, where appropriate, taking into account Member States' needs, as a single contact point for stakeholders;

(c) are empowered to facilitate consistency and coordination between national competent authorities in their Member State as regards the implementation of this Regulation, including through the collection of relevant data and information for the purpose of fulfilling their tasks on the Board.

5. The designated representatives of the Member States shall adopt the Board's rules of procedure by a two-thirds majority. The rules of procedure shall, in particular, lay down procedures for the selection process, the duration of the mandate of, and specifications of the tasks of, the Chair, detailed arrangements for voting, and the organisation of the Board's activities and those of its sub-groups.

6. The Board shall establish two standing sub-groups to provide a platform for cooperation and exchange among market surveillance authorities and notifying authorities about issues related to market surveillance and notified bodies respectively.

The standing sub-group for market surveillance should act as the administrative cooperation group (ADCO) for this Regulation within the meaning of Article 30 of Regulation (EU) 2019/1020.

The Board may establish other standing or temporary sub-groups as appropriate for the purpose of examining specific issues. Where appropriate, representatives of the advisory forum referred to in Article 67 may be invited to such sub-groups or to specific meetings of those subgroups as observers.

7. The Board shall be organised and operated so as to safeguard the objectivity and impartiality of its activities.

8. The Board shall be chaired by one of the representatives of the Member States. The AI Office shall provide the secretariat for the Board, convene the meetings upon request of the Chair, and prepare the agenda in accordance with the tasks of the Board pursuant to this Regulation and its rules of procedure.

Article 66

Tasks of the Board

The Board shall advise and assist the Commission and the Member States in order to facilitate the consistent and effective application of this Regulation. To that end, the Board may in particular:

(a) contribute to the coordination among national competent authorities responsible for the application of this Regulation and, in cooperation with and subject to the agreement of the market surveillance authorities concerned, support joint activities of market surveillance authorities referred to in Article 74(11);

(b) collect and share technical and regulatory expertise and best practices among Member States;

(c) provide advice on the implementation of this Regulation, in particular as regards the enforcement of rules on general-purpose AI models;

(d) contribute to the harmonisation of administrative practices in the Member States, including in relation to the derogation from the conformity assessment procedures referred to in Article 46, the functioning of AI regulatory sandboxes, and testing in real world conditions referred to in Articles 57, 59 and 60;

(e) at the request of the Commission or on its own initiative, issue recommendations and written opinions on any relevant matters related to the implementation of this Regulation and to its consistent and effective application, including:

(i) on the development and application of codes of conduct and codes of practice pursuant to this Regulation, as well as of the Commission's guidelines;

(ii) the evaluation and review of this Regulation pursuant to Article 112, including as regards the serious incident reports referred to in Article 73, and the functioning of the EU database referred to in Article 71, the preparation of the delegated or implementing acts, and as regards possible alignments of this Regulation with the Union harmonisation legislation listed in Annex I;

(iii) on technical specifications or existing standards regarding the requirements set out in Chapter III, Section 2;

(iv) on the use of harmonised standards or common specifications referred to in Articles 40 and 41;

(v) trends, such as European global competitiveness in AI, the uptake of AI in the Union, and the development of digital skills;

(vi) trends on the evolving typology of AI value chains, in particular on the resulting implications in terms of accountability;

(vii) on the potential need for amendment to Annex III in accordance with Article 7, and on the potential need for possible revision of Article 5 pursuant to Article 112, taking into account relevant available evidence and the latest developments in technology;

(f) support the Commission in promoting AI literacy, public awareness and understanding of the benefits, risks, safeguards and rights and obligations in relation to the use of AI systems;

(g) facilitate the development of common criteria and a shared understanding among market operators and competent authorities of the relevant concepts provided for in this Regulation, including by contributing to the development of benchmarks;

(h) cooperate, as appropriate, with other Union institutions, bodies, offices and agencies, as well as relevant Union expert groups and networks, in particular in the fields of product safety, cybersecurity, competition, digital and media services, financial services, consumer protection, data and fundamental rights protection;

(i) contribute to effective cooperation with the competent authorities of third countries and with international organisations;

(j) assist national competent authorities and the Commission in developing the organisational and technical expertise required for the implementation of this Regulation, including by contributing to the assessment of training needs for staff of Member States involved in implementing this Regulation;

(k) assist the AI Office in supporting national competent authorities in the establishment and development of AI regulatory sandboxes, and facilitate cooperation and information-sharing among AI regulatory sandboxes;

(l) contribute to, and provide relevant advice on, the development of guidance documents;

(m) advise the Commission in relation to international matters on AI;

(n) provide opinions to the Commission on the qualified alerts regarding general-purpose AI models;

(o) receive opinions by the Member States on qualified alerts regarding general-purpose AI models, and on national experiences and practices on the monitoring and enforcement of AI systems, in particular systems integrating the general-purpose AI models.

Article 67

Advisory forum

1. An advisory forum shall be established to provide technical expertise and advise the Board and the Commission, and to contribute to their tasks under this Regulation.

2. The membership of the advisory forum shall represent a balanced selection of stakeholders, including industry, start-ups, SMEs, civil society and academia. The membership of the advisory forum shall be balanced with regard to commercial and non-commercial interests and, within the category of commercial interests, with regard to SMEs and other undertakings.

3. The Commission shall appoint the members of the advisory forum, in accordance with the criteria set out in paragraph 2, from amongst stakeholders with recognised expertise in the field of AI.

4. The term of office of the members of the advisory forum shall be two years, which may be extended by up to no more than four years.

5. The Fundamental Rights Agency, ENISA, the European Committee for Standardization (CEN), the European Committee for Electrotechnical Standardization (CENELEC), and the European Telecommunications Standards Institute (ETSI) shall be permanent members of the advisory forum.

6. The advisory forum shall draw up its rules of procedure. It shall elect two co-chairs from among its members, in accordance with criteria set out in paragraph 2. The term of office of the co-chairs shall be two years, renewable once.

7. The advisory forum shall hold meetings at least twice a year. The advisory forum may invite experts and other stakeholders to its meetings.

8. The advisory forum may prepare opinions, recommendations and written contributions at the request of the Board or the Commission.

9. The advisory forum may establish standing or temporary sub-groups as appropriate for the purpose of examining specific questions related to the objectives of this Regulation.

10. The advisory forum shall prepare an annual report on its activities. That report shall be made publicly available.

Article 68

Scientific panel of independent experts

1. The Commission shall, by means of an implementing act, make provisions on the establishment of a scientific panel of independent experts (the 'scientific panel') intended to support the enforcement activities under this Regulation. That implementing act shall be adopted in accordance with the examination procedure referred to in Article 98(2).

2. The scientific panel shall consist of experts selected by the Commission on the basis of up-to-date scientific or technical expertise in the field of AI necessary for the tasks set out in paragraph 3, and shall be able to demonstrate meeting all of the following conditions:

(a) having particular expertise and competence and scientific or technical expertise in the field of AI;

(b) independence from any provider of AI systems or general-purpose AI models;

(c) an ability to carry out activities diligently, accurately and objectively.

The Commission, in consultation with the Board, shall determine the number of experts on the panel in accordance with the required needs and shall ensure fair gender and geographical representation.

3. The scientific panel shall advise and support the AI Office, in particular with regard to the following tasks:

(a) supporting the implementation and enforcement of this Regulation as regards general-purpose AI models and systems, in particular by:

(i) alerting the AI Office of possible systemic risks at Union level of general-purpose AI models, in accordance with Article 90;

(ii) contributing to the development of tools and methodologies for evaluating capabilities of general-purpose AI models and systems, including through benchmarks;

(iii) providing advice on the classification of general-purpose AI models with systemic risk;

(iv) providing advice on the classification of various general-purpose AI models and systems;

(v) contributing to the development of tools and templates;

(b) supporting the work of market surveillance authorities, at their request;

(c) supporting cross-border market surveillance activities as referred to in Article 74(11), without prejudice to the powers of market surveillance authorities;

(d) supporting the AI Office in carrying out its duties in the context of the Union safeguard procedure pursuant to Article 81.

4. The experts on the scientific panel shall perform their tasks with impartiality and objectivity, and shall ensure the confidentiality of information and data obtained in carrying out their tasks and activities. They shall neither seek nor take instructions from anyone when exercising their tasks under paragraph 3. Each expert shall draw up a declaration of interests, which shall be made publicly available. The AI Office shall establish systems and procedures to actively manage and prevent potential conflicts of interest.

5. The implementing act referred to in paragraph 1 shall include provisions on the conditions, procedures and detailed arrangements for the scientific panel and its members to issue alerts, and to request the assistance of the AI Office for the performance of the tasks of the scientific panel.

Article 69

Access to the pool of experts by the Member States

1. Member States may call upon experts of the scientific panel to support their enforcement activities under this Regulation.

2. The Member States may be required to pay fees for the advice and support provided by the experts. The structure and the level of fees as well as the scale and

structure of recoverable costs shall be set out in the implementing act referred to in Article 68(1), taking into account the objectives of the adequate implementation of this Regulation, cost-effectiveness and the necessity of ensuring effective access to experts for all Member States.

3.   The Commission shall facilitate timely access to the experts by the Member States, as needed, and ensure that the combination of support activities carried out by Union AI testing support pursuant to Article 84 and experts pursuant to this Article is efficiently organised and provides the best possible added value.

SECTION 2

National competent authorities

Article 70

Designation of national competent authorities and single points of contact

1.   Each Member State shall establish or designate as national competent authorities at least one notifying authority and at least one market surveillance authority for the purposes of this Regulation. Those national competent authorities shall exercise their powers independently, impartially and without bias so as to safeguard the objectivity of their activities and tasks, and to ensure the application and implementation of this Regulation. The members of those authorities shall refrain from any action incompatible with their duties. Provided that those principles are observed, such activities and tasks may be performed by one or more designated authorities, in accordance with the organisational needs of the Member State.

2.   Member States shall communicate to the Commission the identity of the notifying authorities and the market surveillance authorities and the tasks of those authorities, as well as any subsequent changes thereto. Member States shall make publicly available information on how competent authorities and single points of contact can be contacted, through electronic communication means by 2 August 2025. Member States shall designate a market surveillance authority to act as the single point of contact for this Regulation, and shall notify the Commission of the identity of the single point of contact. The Commission shall make a list of the single points of contact publicly available.

3.   Member States shall ensure that their national competent authorities are provided with adequate technical, financial and human resources, and with infrastructure to fulfil their tasks effectively under this Regulation. In particular, the national competent authorities shall have a sufficient number of personnel permanently available whose competences and expertise shall include an in-depth

understanding of AI technologies, data and data computing, personal data protection, cybersecurity, fundamental rights, health and safety risks and knowledge of existing standards and legal requirements. Member States shall assess and, if necessary, update competence and resource requirements referred to in this paragraph on an annual basis.

4. National competent authorities shall take appropriate measures to ensure an adequate level of cybersecurity.

5. When performing their tasks, the national competent authorities shall act in accordance with the confidentiality obligations set out in Article 78.

6. By 2 August 2025, and once every two years thereafter, Member States shall report to the Commission on the status of the financial and human resources of the national competent authorities, with an assessment of their adequacy. The Commission shall transmit that information to the Board for discussion and possible recommendations.

7. The Commission shall facilitate the exchange of experience between national competent authorities.

8. National competent authorities may provide guidance and advice on the implementation of this Regulation, in particular to SMEs including start-ups, taking into account the guidance and advice of the Board and the Commission, as appropriate. Whenever national competent authorities intend to provide guidance and advice with regard to an AI system in areas covered by other Union law, the national competent authorities under that Union law shall be consulted, as appropriate.

9. Where Union institutions, bodies, offices or agencies fall within the scope of this Regulation, the European Data Protection Supervisor shall act as the competent authority for their supervision.

CHAPTER VIII

EU DATABASE FOR HIGH-RISK AI SYSTEMS

Article 71

EU database for high-risk AI systems listed in Annex III

1. The Commission shall, in collaboration with the Member States, set up and maintain an EU database containing information referred to in paragraphs 2 and 3

of this Article concerning high-risk AI systems referred to in Article 6(2) which are registered in accordance with Articles 49 and 60 and AI systems that are not considered as high-risk pursuant to Article 6(3) and which are registered in accordance with Article 6(4) and Article 49. When setting the functional specifications of such database, the Commission shall consult the relevant experts, and when updating the functional specifications of such database, the Commission shall consult the Board.

2. The data listed in Sections A and B of Annex VIII shall be entered into the EU database by the provider or, where applicable, by the authorised representative.

3. The data listed in Section C of Annex VIII shall be entered into the EU database by the deployer who is, or who acts on behalf of, a public authority, agency or body, in accordance with Article 49(3) and (4).

4. With the exception of the section referred to in Article 49(4) and Article 60(4), point (c), the information contained in the EU database registered in accordance with Article 49 shall be accessible and publicly available in a user-friendly manner. The information should be easily navigable and machine-readable. The information registered in accordance with Article 60 shall be accessible only to market surveillance authorities and the Commission, unless the prospective provider or provider has given consent for also making the information accessible the public.

5. The EU database shall contain personal data only in so far as necessary for collecting and processing information in accordance with this Regulation. That information shall include the names and contact details of natural persons who are responsible for registering the system and have the legal authority to represent the provider or the deployer, as applicable.

6. The Commission shall be the controller of the EU database. It shall make available to providers, prospective providers and deployers adequate technical and administrative support. The EU database shall comply with the applicable accessibility requirements.

CHAPTER IX

POST-MARKET MONITORING, INFORMATION SHARING AND MARKET SURVEILLANCE

SECTION 1

Post-market monitoring

Article 72

Post-market monitoring by providers and post-market monitoring plan for high-risk AI systems

1. Providers shall establish and document a post-market monitoring system in a manner that is proportionate to the nature of the AI technologies and the risks of the high-risk AI system.

2. The post-market monitoring system shall actively and systematically collect, document and analyse relevant data which may be provided by deployers or which may be collected through other sources on the performance of high-risk AI systems throughout their lifetime, and which allow the provider to evaluate the continuous compliance of AI systems with the requirements set out in Chapter III, Section 2. Where relevant, post-market monitoring shall include an analysis of the interaction with other AI systems. This obligation shall not cover sensitive operational data of deployers which are law-enforcement authorities.

3. The post-market monitoring system shall be based on a post-market monitoring plan. The post-market monitoring plan shall be part of the technical documentation referred to in Annex IV. The Commission shall adopt an implementing act laying down detailed provisions establishing a template for the post-market monitoring plan and the list of elements to be included in the plan by 2 February 2026. That implementing act shall be adopted in accordance with the examination procedure referred to in Article 98(2).

4. For high-risk AI systems covered by the Union harmonisation legislation listed in Section A of Annex I, where a post-market monitoring system and plan are already established under that legislation, in order to ensure consistency, avoid duplications and minimise additional burdens, providers shall have a choice of integrating, as appropriate, the necessary elements described in paragraphs 1, 2 and 3 using the template referred in paragraph 3 into systems and plans already existing under that legislation, provided that it achieves an equivalent level of protection.

The first subparagraph of this paragraph shall also apply to high-risk AI systems referred to in point 5 of Annex III placed on the market or put into service by financial institutions that are subject to requirements under Union financial services law regarding their internal governance, arrangements or processes.

SECTION 2

Sharing of information on serious incidents

Article 73

Reporting of serious incidents

1. Providers of high-risk AI systems placed on the Union market shall report any serious incident to the market surveillance authorities of the Member States where that incident occurred.

2. The report referred to in paragraph 1 shall be made immediately after the provider has established a causal link between the AI system and the serious incident or the reasonable likelihood of such a link, and, in any event, not later than 15 days after the provider or, where applicable, the deployer, becomes aware of the serious incident.

The period for the reporting referred to in the first subparagraph shall take account of the severity of the serious incident.

3. Notwithstanding paragraph 2 of this Article, in the event of a widespread infringement or a serious incident as defined in Article 3, point (49)(b), the report referred to in paragraph 1 of this Article shall be provided immediately, and not later than two days after the provider or, where applicable, the deployer becomes aware of that incident.

4. Notwithstanding paragraph 2, in the event of the death of a person, the report shall be provided immediately after the provider or the deployer has established, or as soon as it suspects, a causal relationship between the high-risk AI system and the serious incident, but not later than 10 days after the date on which the provider or, where applicable, the deployer becomes aware of the serious incident.

5. Where necessary to ensure timely reporting, the provider or, where applicable, the deployer, may submit an initial report that is incomplete, followed by a complete report.

6. Following the reporting of a serious incident pursuant to paragraph 1, the provider shall, without delay, perform the necessary investigations in relation to the serious incident and the AI system concerned. This shall include a risk assessment of the incident, and corrective action.

The provider shall cooperate with the competent authorities, and where relevant with the notified body concerned, during the investigations referred to in the first subparagraph, and shall not perform any investigation which involves altering the AI system concerned in a way which may affect any subsequent evaluation of the causes of the incident, prior to informing the competent authorities of such action.

7. Upon receiving a notification related to a serious incident referred to in Article 3, point (49)(c), the relevant market surveillance authority shall inform the national public authorities or bodies referred to in Article 77(1). The Commission shall develop dedicated guidance to facilitate compliance with the obligations set out in paragraph 1 of this Article. That guidance shall be issued by 2 August 2025, and shall be assessed regularly.

8. The market surveillance authority shall take appropriate measures, as provided for in Article 19 of Regulation (EU) 2019/1020, within seven days from the date it received the notification referred to in paragraph 1 of this Article, and shall follow the notification procedures as provided in that Regulation.

9. For high-risk AI systems referred to in Annex III that are placed on the market or put into service by providers that are subject to Union legislative instruments laying down reporting obligations equivalent to those set out in this Regulation, the notification of serious incidents shall be limited to those referred to in Article 3, point (49)(c).

10. For high-risk AI systems which are safety components of devices, or are themselves devices, covered by Regulations (EU) 2017/745 and (EU) 2017/746, the notification of serious incidents shall be limited to those referred to in Article 3, point (49)(c) of this Regulation, and shall be made to the national competent authority chosen for that purpose by the Member States where the incident occurred.

11. National competent authorities shall immediately notify the Commission of any serious incident, whether or not they have taken action on it, in accordance with Article 20 of Regulation (EU) 2019/1020.

SECTION 3

Enforcement

Article 74

Market surveillance and control of AI systems in the Union market

1. Regulation (EU) 2019/1020 shall apply to AI systems covered by this Regulation. For the purposes of the effective enforcement of this Regulation:

(a) any reference to an economic operator under Regulation (EU) 2019/1020 shall be understood as including all operators identified in Article 2(1) of this Regulation;

(b) any reference to a product under Regulation (EU) 2019/1020 shall be understood as including all AI systems falling within the scope of this Regulation.

2. As part of their reporting obligations under Article 34(4) of Regulation (EU) 2019/1020, the market surveillance authorities shall report annually to the Commission and relevant national competition authorities any information identified in the course of market surveillance activities that may be of potential interest for the application of Union law on competition rules. They shall also annually report to the Commission about the use of prohibited practices that occurred during that year and about the measures taken.

3. For high-risk AI systems related to products covered by the Union harmonisation legislation listed in Section A of Annex I, the market surveillance authority for the purposes of this Regulation shall be the authority responsible for market surveillance activities designated under those legal acts.

By derogation from the first subparagraph, and in appropriate circumstances, Member States may designate another relevant authority to act as a market surveillance authority, provided they ensure coordination with the relevant sectoral market surveillance authorities responsible for the enforcement of the Union harmonisation legislation listed in Annex I.

4. The procedures referred to in Articles 79 to 83 of this Regulation shall not apply to AI systems related to products covered by the Union harmonisation legislation listed in section A of Annex I, where such legal acts already provide for procedures ensuring an equivalent level of protection and having the same objective. In such cases, the relevant sectoral procedures shall apply instead.

5. Without prejudice to the powers of market surveillance authorities under Article 14 of Regulation (EU) 2019/1020, for the purpose of ensuring the effective enforcement of this Regulation, market surveillance authorities may exercise the powers referred to in Article 14(4), points (d) and (j), of that Regulation remotely, as appropriate.

6. For high-risk AI systems placed on the market, put into service, or used by financial institutions regulated by Union financial services law, the market surveillance authority for the purposes of this Regulation shall be the relevant national authority responsible for the financial supervision of those institutions under that legislation in so far as the placing on the market, putting into service, or the use of the AI system is in direct connection with the provision of those financial services.

7. By way of derogation from paragraph 6, in appropriate circumstances, and provided that coordination is ensured, another relevant authority may be identified by the Member State as market surveillance authority for the purposes of this Regulation.

National market surveillance authorities supervising regulated credit institutions regulated under Directive 2013/36/EU, which are participating in the Single Supervisory Mechanism established by Regulation (EU) No 1024/2013, should report, without delay, to the European Central Bank any information identified in the course of their market surveillance activities that may be of potential interest for the prudential supervisory tasks of the European Central Bank specified in that Regulation.

8. For high-risk AI systems listed in point 1 of Annex III to this Regulation, in so far as the systems are used for law enforcement purposes, border management and justice and democracy, and for high-risk AI systems listed in points 6, 7 and 8 of Annex III to this Regulation, Member States shall designate as market surveillance authorities for the purposes of this Regulation either the competent data protection supervisory authorities under Regulation (EU) 2016/679 or Directive (EU) 2016/680, or any other authority designated pursuant to the same conditions laid down in Articles 41 to 44 of Directive (EU) 2016/680. Market surveillance activities shall in no way affect the independence of judicial authorities, or otherwise interfere with their activities when acting in their judicial capacity.

9. Where Union institutions, bodies, offices or agencies fall within the scope of this Regulation, the European Data Protection Supervisor shall act as their market surveillance authority, except in relation to the Court of Justice of the European Union acting in its judicial capacity.

10. Member States shall facilitate coordination between market surveillance authorities designated under this Regulation and other relevant national authorities or bodies which supervise the application of Union harmonisation legislation listed in Annex I, or in other Union law, that might be relevant for the high-risk AI systems referred to in Annex III.

11. Market surveillance authorities and the Commission shall be able to propose joint activities, including joint investigations, to be conducted by either market surveillance authorities or market surveillance authorities jointly with the Commission, that have the aim of promoting compliance, identifying non-compliance, raising awareness or providing guidance in relation to this Regulation with respect to specific categories of high-risk AI systems that are found to present a serious risk across two or more Member States in accordance with Article 9 of Regulation (EU) 2019/1020. The AI Office shall provide coordination support for joint investigations.

12. Without prejudice to the powers provided for under Regulation (EU) 2019/1020, and where relevant and limited to what is necessary to fulfil their tasks, the market surveillance authorities shall be granted full access by providers to the documentation as well as the training, validation and testing data sets used for the development of high-risk AI systems, including, where appropriate and subject to security safeguards, through application programming interfaces (API) or other relevant technical means and tools enabling remote access.

13. Market surveillance authorities shall be granted access to the source code of the high-risk AI system upon a reasoned request and only when both of the following conditions are fulfilled:

(a)access to source code is necessary to assess the conformity of a high-risk AI system with the requirements set out in Chapter III, Section 2; and

(b)testing or auditing procedures and verifications based on the data and documentation provided by the provider have been exhausted or proved insufficient.

14. Any information or documentation obtained by market surveillance authorities shall be treated in accordance with the confidentiality obligations set out in Article 78.

Article 75

Mutual assistance, market surveillance and control of general-purpose AI systems

1. Where an AI system is based on a general-purpose AI model, and the model and the system are developed by the same provider, the AI Office shall have powers to monitor and supervise compliance of that AI system with obligations under this Regulation. To carry out its monitoring and supervision tasks, the AI Office shall have all the powers of a market surveillance authority provided for in this Section and Regulation (EU) 2019/1020.

2. Where the relevant market surveillance authorities have sufficient reason to consider general-purpose AI systems that can be used directly by deployers for at least one purpose that is classified as high-risk pursuant to this Regulation to be non-compliant with the requirements laid down in this Regulation, they shall cooperate with the AI Office to carry out compliance evaluations, and shall inform the Board and other market surveillance authorities accordingly.

3. Where a market surveillance authority is unable to conclude its investigation of the high-risk AI system because of its inability to access certain information related to the general-purpose AI model despite having made all appropriate efforts to obtain that information, it may submit a reasoned request to the AI Office, by which access to that information shall be enforced. In that case, the AI Office shall supply to the applicant authority without delay, and in any event within 30 days, any information that the AI Office considers to be relevant in order to establish whether a high-risk AI system is non-compliant. Market surveillance authorities shall safeguard the confidentiality of the information that they obtain in accordance with Article 78 of this Regulation. The procedure provided for in Chapter VI of Regulation (EU) 2019/1020 shall apply *mutatis mutandis*.

Article 76

Supervision of testing in real world conditions by market surveillance authorities

1. Market surveillance authorities shall have competences and powers to ensure that testing in real world conditions is in accordance with this Regulation.

2. Where testing in real world conditions is conducted for AI systems that are supervised within an AI regulatory sandbox under Article 58, the market surveillance authorities shall verify the compliance with Article 60 as part of their supervisory role for the AI regulatory sandbox. Those authorities may, as appropriate, allow the testing in real world conditions to be conducted by the

provider or prospective provider, in derogation from the conditions set out in Article 60(4), points (f) and (g).

3. Where a market surveillance authority has been informed by the prospective provider, the provider or any third party of a serious incident or has other grounds for considering that the conditions set out in Articles 60 and 61 are not met, it may take either of the following decisions on its territory, as appropriate:

(a) to suspend or terminate the testing in real world conditions;

(b) to require the provider or prospective provider and the deployer or prospective deployer to modify any aspect of the testing in real world conditions.

4. Where a market surveillance authority has taken a decision referred to in paragraph 3 of this Article, or has issued an objection within the meaning of Article 60(4), point (b), the decision or the objection shall indicate the grounds therefor and how the provider or prospective provider can challenge the decision or objection.

5. Where applicable, where a market surveillance authority has taken a decision referred to in paragraph 3, it shall communicate the grounds therefor to the market surveillance authorities of other Member States in which the AI system has been tested in accordance with the testing plan.

Article 77

Powers of authorities protecting fundamental rights

1. National public authorities or bodies which supervise or enforce the respect of obligations under Union law protecting fundamental rights, including the right to non-discrimination, in relation to the use of high-risk AI systems referred to in Annex III shall have the power to request and access any documentation created or maintained under this Regulation in accessible language and format when access to that documentation is necessary for effectively fulfilling their mandates within the limits of their jurisdiction. The relevant public authority or body shall inform the market surveillance authority of the Member State concerned of any such request.

2. By 2 November 2024, each Member State shall identify the public authorities or bodies referred to in paragraph 1 and make a list of them publicly available. Member States shall notify the list to the Commission and to the other Member States, and shall keep the list up to date.

3. Where the documentation referred to in paragraph 1 is insufficient to ascertain whether an infringement of obligations under Union law protecting fundamental rights has occurred, the public authority or body referred to in paragraph 1 may make a reasoned request to the market surveillance authority, to organise testing of the high-risk AI system through technical means. The market surveillance authority shall organise the testing with the close involvement of the requesting public authority or body within a reasonable time following the request.

4. Any information or documentation obtained by the national public authorities or bodies referred to in paragraph 1 of this Article pursuant to this Article shall be treated in accordance with the confidentiality obligations set out in Article 78.

Article 78

Confidentiality

1. The Commission, market surveillance authorities and notified bodies and any other natural or legal person involved in the application of this Regulation shall, in accordance with Union or national law, respect the confidentiality of information and data obtained in carrying out their tasks and activities in such a manner as to protect, in particular:

(a) the intellectual property rights and confidential business information or trade secrets of a natural or legal person, including source code, except in the cases referred to in Article 5 of Directive (EU) 2016/943 of the European Parliament and of the Council;

(b) The effective implementation of this Regulation, in particular for the purposes of inspections, investigations or audits;

(c) public and national security interests;

(d) the conduct of criminal or administrative proceedings;

(e) information classified pursuant to Union or national law.

2. The authorities involved in the application of this Regulation pursuant to paragraph 1 shall request only data that is strictly necessary for the assessment of the risk posed by AI systems and for the exercise of their powers in accordance with this Regulation and with Regulation (EU) 2019/1020. They shall put in place adequate and effective cybersecurity measures to protect the security and confidentiality of the information and data obtained, and shall delete the data

collected as soon as it is no longer needed for the purpose for which it was obtained, in accordance with applicable Union or national law.

3. Without prejudice to paragraphs 1 and 2, information exchanged on a confidential basis between the national competent authorities or between national competent authorities and the Commission shall not be disclosed without prior consultation of the originating national competent authority and the deployer when high-risk AI systems referred to in point 1, 6 or 7 of Annex III are used by law enforcement, border control, immigration or asylum authorities and when such disclosure would jeopardise public and national security interests. This exchange of information shall not cover sensitive operational data in relation to the activities of law enforcement, border control, immigration or asylum authorities.

When the law enforcement, immigration or asylum authorities are providers of high-risk AI systems referred to in point 1, 6 or 7 of Annex III, the technical documentation referred to in Annex IV shall remain within the premises of those authorities. Those authorities shall ensure that the market surveillance authorities referred to in Article 74(8) and (9), as applicable, can, upon request, immediately access the documentation or obtain a copy thereof. Only staff of the market surveillance authority holding the appropriate level of security clearance shall be allowed to access that documentation or any copy thereof.

4. Paragraphs 1, 2 and 3 shall not affect the rights or obligations of the Commission, Member States and their relevant authorities, as well as those of notified bodies, with regard to the exchange of information and the dissemination of warnings, including in the context of cross-border cooperation, nor shall they affect the obligations of the parties concerned to provide information under criminal law of the Member States.

5. The Commission and Member States may exchange, where necessary and in accordance with relevant provisions of international and trade agreements, confidential information with regulatory authorities of third countries with which they have concluded bilateral or multilateral confidentiality arrangements guaranteeing an adequate level of confidentiality.

Article 79

Procedure at national level for dealing with AI systems presenting a risk

1. AI systems presenting a risk shall be understood as a 'product presenting a risk' as defined in Article 3, point 19 of Regulation (EU) 2019/1020, in so far as they present risks to the health or safety, or to fundamental rights, of persons.

---

2.   Where the market surveillance authority of a Member State has sufficient reason to consider an AI system to present a risk as referred to in paragraph 1 of this Article, it shall carry out an evaluation of the AI system concerned in respect of its compliance with all the requirements and obligations laid down in this Regulation. Particular attention shall be given to AI systems presenting a risk to vulnerable groups. Where risks to fundamental rights are identified, the market surveillance authority shall also inform and fully cooperate with the relevant national public authorities or bodies referred to in Article 77(1). The relevant operators shall cooperate as necessary with the market surveillance authority and with the other national public authorities or bodies referred to in Article 77(1).

Where, in the course of that evaluation, the market surveillance authority or, where applicable the market surveillance authority in cooperation with the national public authority referred to in Article 77(1), finds that the AI system does not comply with the requirements and obligations laid down in this Regulation, it shall without undue delay require the relevant operator to take all appropriate corrective actions to bring the AI system into compliance, to withdraw the AI system from the market, or to recall it within a period the market surveillance authority may prescribe, and in any event within the shorter of 15 working days, or as provided for in the relevant Union harmonisation legislation.

The market surveillance authority shall inform the relevant notified body accordingly. Article 18 of Regulation (EU) 2019/1020 shall apply to the measures referred to in the second subparagraph of this paragraph.

3.   Where the market surveillance authority considers that the non-compliance is not restricted to its national territory, it shall inform the Commission and the other Member States without undue delay of the results of the evaluation and of the actions which it has required the operator to take.

4.   The operator shall ensure that all appropriate corrective action is taken in respect of all the AI systems concerned that it has made available on the Union market.

5.   Where the operator of an AI system does not take adequate corrective action within the period referred to in paragraph 2, the market surveillance authority shall take all appropriate provisional measures to prohibit or restrict the AI system's being made available on its national market or put into service, to withdraw the product or the standalone AI system from that market or to recall it. That authority shall without undue delay notify the Commission and the other Member States of those measures.

6.   The notification referred to in paragraph 5 shall include all available details, in particular the information necessary for the identification of the non-compliant AI

system, the origin of the AI system and the supply chain, the nature of the non-compliance alleged and the risk involved, the nature and duration of the national measures taken and the arguments put forward by the relevant operator. In particular, the market surveillance authorities shall indicate whether the non-compliance is due to one or more of the following:

(a) non-compliance with the prohibition of the AI practices referred to in Article 5;

(b) a failure of a high-risk AI system to meet requirements set out in Chapter III, Section 2;

(c) shortcomings in the harmonised standards or common specifications referred to in Articles 40 and 41 conferring a presumption of conformity;

(d)   non-compliance with Article 50.

7.   The market surveillance authorities other than the market surveillance authority of the Member State initiating the procedure shall, without undue delay, inform the Commission and the other Member States of any measures adopted and of any additional information at their disposal relating to the non-compliance of the AI system concerned, and, in the event of disagreement with the notified national measure, of their objections.

8.   Where, within three months of receipt of the notification referred to in paragraph 5 of this Article, no objection has been raised by either a market surveillance authority of a Member State or by the Commission in respect of a provisional measure taken by a market surveillance authority of another Member State, that measure shall be deemed justified. This shall be without prejudice to the procedural rights of the concerned operator in accordance with Article 18 of Regulation (EU) 2019/1020. The three-month period referred to in this paragraph shall be reduced to 30 days in the event of non-compliance with the prohibition of the AI practices referred to in Article 5 of this Regulation.

9.   The market surveillance authorities shall ensure that appropriate restrictive measures are taken in respect of the product or the AI system concerned, such as withdrawal of the product or the AI system from their market, without undue delay.

Article 80

Procedure for dealing with AI systems classified by the provider as non-high-risk in application of Annex III

1. Where a market surveillance authority has sufficient reason to consider that an AI system classified by the provider as non-high-risk pursuant to Article 6(3) is indeed high-risk, the market surveillance authority shall carry out an evaluation of the AI system concerned in respect of its classification as a high-risk AI system based on the conditions set out in Article 6(3) and the Commission guidelines.

2. Where, in the course of that evaluation, the market surveillance authority finds that the AI system concerned is high-risk, it shall without undue delay require the relevant provider to take all necessary actions to bring the AI system into compliance with the requirements and obligations laid down in this Regulation, as well as take appropriate corrective action within a period the market surveillance authority may prescribe.

3. Where the market surveillance authority considers that the use of the AI system concerned is not restricted to its national territory, it shall inform the Commission and the other Member States without undue delay of the results of the evaluation and of the actions which it has required the provider to take.

4. The provider shall ensure that all necessary action is taken to bring the AI system into compliance with the requirements and obligations laid down in this Regulation. Where the provider of an AI system concerned does not bring the AI system into compliance with those requirements and obligations within the period referred to in paragraph 2 of this Article, the provider shall be subject to fines in accordance with Article 99.

5. The provider shall ensure that all appropriate corrective action is taken in respect of all the AI systems concerned that it has made available on the Union market.

6. Where the provider of the AI system concerned does not take adequate corrective action within the period referred to in paragraph 2 of this Article, Article 79(5) to (9) shall apply.

7. Where, in the course of the evaluation pursuant to paragraph 1 of this Article, the market surveillance authority establishes that the AI system was misclassified by the provider as non-high-risk in order to circumvent the application of requirements in Chapter III, Section 2, the provider shall be subject to fines in accordance with Article 99.

———

8. In exercising their power to monitor the application of this Article, and in accordance with Article 11 of Regulation (EU) 2019/1020, market surveillance authorities may perform appropriate checks, taking into account in particular information stored in the EU database referred to in Article 71 of this Regulation.

Article 81

Union safeguard procedure

1. Where, within three months of receipt of the notification referred to in Article 79(5), or within 30 days in the case of non-compliance with the prohibition of the AI practices referred to in Article 5, objections are raised by the market surveillance authority of a Member State to a measure taken by another market surveillance authority, or where the Commission considers the measure to be contrary to Union law, the Commission shall without undue delay enter into consultation with the market surveillance authority of the relevant Member State and the operator or operators, and shall evaluate the national measure. On the basis of the results of that evaluation, the Commission shall, within six months, or within 60 days in the case of non-compliance with the prohibition of the AI practices referred to in Article 5, starting from the notification referred to in Article 79(5), decide whether the national measure is justified and shall notify its decision to the market surveillance authority of the Member State concerned. The Commission shall also inform all other market surveillance authorities of its decision.

2. Where the Commission considers the measure taken by the relevant Member State to be justified, all Member States shall ensure that they take appropriate restrictive measures in respect of the AI system concerned, such as requiring the withdrawal of the AI system from their market without undue delay, and shall inform the Commission accordingly. Where the Commission considers the national measure to be unjustified, the Member State concerned shall withdraw the measure and shall inform the Commission accordingly.

3. Where the national measure is considered justified and the non-compliance of the AI system is attributed to shortcomings in the harmonised standards or common specifications referred to in Articles 40 and 41 of this Regulation, the Commission shall apply the procedure provided for in Article 11 of Regulation (EU) No 1025/2012.

Article 82

Compliant AI systems which present a risk

1. Where, having performed an evaluation under Article 79, after consulting the relevant national public authority referred to in Article 77(1), the market surveillance authority of a Member State finds that although a high-risk AI system complies with this Regulation, it nevertheless presents a risk to the health or safety of persons, to fundamental rights, or to other aspects of public interest protection, it shall require the relevant operator to take all appropriate measures to ensure that the AI system concerned, when placed on the market or put into service, no longer presents that risk without undue delay, within a period it may prescribe.

2. The provider or other relevant operator shall ensure that corrective action is taken in respect of all the AI systems concerned that it has made available on the Union market within the timeline prescribed by the market surveillance authority of the Member State referred to in paragraph 1.

3. The Member States shall immediately inform the Commission and the other Member States of a finding under paragraph 1. That information shall include all available details, in particular the data necessary for the identification of the AI system concerned, the origin and the supply chain of the AI system, the nature of the risk involved and the nature and duration of the national measures taken.

4. The Commission shall without undue delay enter into consultation with the Member States concerned and the relevant operators, and shall evaluate the national measures taken. On the basis of the results of that evaluation, the Commission shall decide whether the measure is justified and, where necessary, propose other appropriate measures.

5. The Commission shall immediately communicate its decision to the Member States concerned and to the relevant operators. It shall also inform the other Member States.

Article 83

Formal non-compliance

1. Where the market surveillance authority of a Member State makes one of the following findings, it shall require the relevant provider to put an end to the non-compliance concerned, within a period it may prescribe:

(a) the CE marking has been affixed in violation of Article 48;

(b)   the CE marking has not been affixed;

(c) the EU declaration of conformity referred to in Article 47 has not been drawn up;

(d) the EU declaration of conformity referred to in Article 47 has not been drawn up correctly;

(e)the registration in the EU database referred to in Article 71 has not been carried out;

(f) where applicable, no authorised representative has been appointed;

(g)   technical documentation is not available.

2.   Where the non-compliance referred to in paragraph 1 persists, the market surveillance authority of the Member State concerned shall take appropriate and proportionate measures to restrict or prohibit the high-risk AI system being made available on the market or to ensure that it is recalled or withdrawn from the market without delay.

Article 84

Union AI testing support structures

1.   The Commission shall designate one or more Union AI testing support structures to perform the tasks listed under Article 21(6) of Regulation (EU) 2019/1020 in the area of AI.

2.   Without prejudice to the tasks referred to in paragraph 1, Union AI testing support structures shall also provide independent technical or scientific advice at the request of the Board, the Commission, or of market surveillance authorities.

SECTION 4

Remedies

Article 85

Right to lodge a complaint with a market surveillance authority

Without prejudice to other administrative or judicial remedies, any natural or legal person having grounds to consider that there has been an infringement of the provisions of this Regulation may submit complaints to the relevant market surveillance authority.

In accordance with Regulation (EU) 2019/1020, such complaints shall be taken into account for the purpose of conducting market surveillance activities, and shall be handled in line with the dedicated procedures established therefor by the market surveillance authorities.

Article 86

Right to explanation of individual decision-making

1. Any affected person subject to a decision which is taken by the deployer on the basis of the output from a high-risk AI system listed in Annex III, with the exception of systems listed under point 2 thereof, and which produces legal effects or similarly significantly affects that person in a way that they consider to have an adverse impact on their health, safety or fundamental rights shall have the right to obtain from the deployer clear and meaningful explanations of the role of the AI system in the decision-making procedure and the main elements of the decision taken.

2. Paragraph 1 shall not apply to the use of AI systems for which exceptions from, or restrictions to, the obligation under that paragraph follow from Union or national law in compliance with Union law.

3. This Article shall apply only to the extent that the right referred to in paragraph 1 is not otherwise provided for under Union law.

Article 87

Reporting of infringements and protection of reporting persons

Directive (EU) 2019/1937 shall apply to the reporting of infringements of this Regulation and the protection of persons reporting such infringements.

———

SECTION 5

Supervision, investigation, enforcement and monitoring in respect of providers of general-purpose AI models

Article 88

Enforcement of the obligations of providers of general-purpose AI models

1. The Commission shall have exclusive powers to supervise and enforce Chapter V, taking into account the procedural guarantees under Article 94. The Commission shall entrust the implementation of these tasks to the AI Office, without prejudice to the powers of organisation of the Commission and the division of competences between Member States and the Union based on the Treaties.

2. Without prejudice to Article 75(3), market surveillance authorities may request the Commission to exercise the powers laid down in this Section, where that is necessary and proportionate to assist with the fulfilment of their tasks under this Regulation.

Article 89

Monitoring actions

1. For the purpose of carrying out the tasks assigned to it under this Section, the AI Office may take the necessary actions to monitor the effective implementation and compliance with this Regulation by providers of general-purpose AI models, including their adherence to approved codes of practice.

2. Downstream providers shall have the right to lodge a complaint alleging an infringement of this Regulation. A complaint shall be duly reasoned and indicate at least:

(a) the point of contact of the provider of the general-purpose AI model concerned;

(b) a description of the relevant facts, the provisions of this Regulation concerned, and the reason why the downstream provider considers that the provider of the general-purpose AI model concerned infringed this Regulation;

(c) any other information that the downstream provider that sent the request considers relevant, including, where appropriate, information gathered on its own initiative.

Article 90

Alerts of systemic risks by the scientific panel

1. The scientific panel may provide a qualified alert to the AI Office where it has reason to suspect that:

(a) a general-purpose AI model poses concrete identifiable risk at Union level; or

(b) a general-purpose AI model meets the conditions referred to in Article 51.

2. Upon such qualified alert, the Commission, through the AI Office and after having informed the Board, may exercise the powers laid down in this Section for the purpose of assessing the matter. The AI Office shall inform the Board of any measure according to Articles 91 to 94.

3. A qualified alert shall be duly reasoned and indicate at least:

(a) the point of contact of the provider of the general-purpose AI model with systemic risk concerned;

(b) a description of the relevant facts and the reasons for the alert by the scientific panel;

(c) any other information that the scientific panel considers to be relevant, including, where appropriate, information gathered on its own initiative.

Article 91

Power to request documentation and information

1. The Commission may request the provider of the general-purpose AI model concerned to provide the documentation drawn up by the provider in accordance with Articles 53 and 55, or any additional information that is necessary for the purpose of assessing compliance of the provider with this Regulation.

2. Before sending the request for information, the AI Office may initiate a structured dialogue with the provider of the general-purpose AI model.

3. Upon a duly substantiated request from the scientific panel, the Commission may issue a request for information to a provider of a general-purpose AI model, where the access to information is necessary and proportionate for the fulfilment of the tasks of the scientific panel under Article 68(2).

---

4. The request for information shall state the legal basis and the purpose of the request, specify what information is required, set a period within which the information is to be provided, and indicate the fines provided for in Article 101 for supplying incorrect, incomplete or misleading information.

5. The provider of the general-purpose AI model concerned, or its representative shall supply the information requested. In the case of legal persons, companies or firms, or where the provider has no legal personality, the persons authorised to represent them by law or by their statutes, shall supply the information requested on behalf of the provider of the general-purpose AI model concerned. Lawyers duly authorised to act may supply information on behalf of their clients. The clients shall nevertheless remain fully responsible if the information supplied is incomplete, incorrect or misleading.

Article 92

Power to conduct evaluations

1. The AI Office, after consulting the Board, may conduct evaluations of the general-purpose AI model concerned:

(a) to assess compliance of the provider with obligations under this Regulation, where the information gathered pursuant to Article 91 is insufficient; or

(b) to investigate systemic risks at Union level of general-purpose AI models with systemic risk, in particular following a qualified alert from the scientific panel in accordance with Article 90(1), point (a).

2. The Commission may decide to appoint independent experts to carry out evaluations on its behalf, including from the scientific panel established pursuant to Article 68. Independent experts appointed for this task shall meet the criteria outlined in Article 68(2).

3. For the purposes of paragraph 1, the Commission may request access to the general-purpose AI model concerned through APIs or further appropriate technical means and tools, including source code.

4. The request for access shall state the legal basis, the purpose and reasons of the request and set the period within which the access is to be provided, and the fines provided for in Article 101 for failure to provide access.

5. The providers of the general-purpose AI model concerned or its representative shall supply the information requested. In the case of legal persons, companies or

firms, or where the provider has no legal personality, the persons authorised to represent them by law or by their statutes, shall provide the access requested on behalf of the provider of the general-purpose AI model concerned.

6. The Commission shall adopt implementing acts setting out the detailed arrangements and the conditions for the evaluations, including the detailed arrangements for involving independent experts, and the procedure for the selection thereof. Those implementing acts shall be adopted in accordance with the examination procedure referred to in Article 98(2).

7. Prior to requesting access to the general-purpose AI model concerned, the AI Office may initiate a structured dialogue with the provider of the general-purpose AI model to gather more information on the internal testing of the model, internal safeguards for preventing systemic risks, and other internal procedures and measures the provider has taken to mitigate such risks.

Article 93

Power to request measures

1. Where necessary and appropriate, the Commission may request providers to:

(a) take appropriate measures to comply with the obligations set out in Articles 53 and 54;

(b) implement mitigation measures, where the evaluation carried out in accordance with Article 92 has given rise to serious and substantiated concern of a systemic risk at Union level;

(c) restrict the making available on the market, withdraw or recall the model.

2. Before a measure is requested, the AI Office may initiate a structured dialogue with the provider of the general-purpose AI model.

3. If, during the structured dialogue referred to in paragraph 2, the provider of the general-purpose AI model with systemic risk offers commitments to implement mitigation measures to address a systemic risk at Union level, the Commission may, by decision, make those commitments binding and declare that there are no further grounds for action.

Article 94

Procedural rights of economic operators of the general-purpose AI model

Article 18 of Regulation (EU) 2019/1020 shall apply *mutatis mutandis* to the providers of the general-purpose AI model, without prejudice to more specific procedural rights provided for in this Regulation.

CHAPTER X

CODES OF CONDUCT AND GUIDELINES

Article 95

Codes of conduct for voluntary application of specific requirements

1. The AI Office and the Member States shall encourage and facilitate the drawing up of codes of conduct, including related governance mechanisms, intended to foster the voluntary application to AI systems, other than high-risk AI systems, of some or all of the requirements set out in Chapter III, Section 2 taking into account the available technical solutions and industry best practices allowing for the application of such requirements.

2. The AI Office and the Member States shall facilitate the drawing up of codes of conduct concerning the voluntary application, including by deployers, of specific requirements to all AI systems, on the basis of clear objectives and key performance indicators to measure the achievement of those objectives, including elements such as, but not limited to:

(a) applicable elements provided for in Union ethical guidelines for trustworthy AI;

(b) assessing and minimising the impact of AI systems on environmental sustainability, including as regards energy-efficient programming and techniques for the efficient design, training and use of AI;

(c) promoting AI literacy, in particular that of persons dealing with the development, operation and use of AI;

(d) facilitating an inclusive and diverse design of AI systems, including through the establishment of inclusive and diverse development teams and the promotion of stakeholders' participation in that process;

(e) assessing and preventing the negative impact of AI systems on vulnerable persons or groups of vulnerable persons, including as regards accessibility for persons with a disability, as well as on gender equality.

3.   Codes of conduct may be drawn up by individual providers or deployers of AI systems or by organisations representing them or by both, including with the involvement of any interested stakeholders and their representative organisations, including civil society organisations and academia. Codes of conduct may cover one or more AI systems taking into account the similarity of the intended purpose of the relevant systems.

4.   The AI Office and the Member States shall take into account the specific interests and needs of SMEs, including start-ups, when encouraging and facilitating the drawing up of codes of conduct.

Article 96

Guidelines from the Commission on the implementation of this Regulation

1.   The Commission shall develop guidelines on the practical implementation of this Regulation, and in particular on:

(a) the application of the requirements and obligations referred to in Articles 8 to 15 and in Article 25;

(b)   the prohibited practices referred to in Article 5;

(c) the practical implementation of the provisions related to substantial modification;

(d) the practical implementation of transparency obligations laid down in Article 50;

(e) detailed information on the relationship of this Regulation with the Union harmonisation legislation listed in Annex I, as well as with other relevant Union law, including as regards consistency in their enforcement;

(f) the application of the definition of an AI system as set out in Article 3, point (1).

When issuing such guidelines, the Commission shall pay particular attention to the needs of SMEs including start-ups, of local public authorities and of the sectors most likely to be affected by this Regulation.

The guidelines referred to in the first subparagraph of this paragraph shall take due account of the generally acknowledged state of the art on AI, as well as of relevant harmonised standards and common specifications that are referred to in Articles 40 and 41, or of those harmonised standards or technical specifications that are set out pursuant to Union harmonisation law.

2. At the request of the Member States or the AI Office, or on its own initiative, the Commission shall update guidelines previously adopted when deemed necessary.

CHAPTER XI

DELEGATION OF POWER AND COMMITTEE PROCEDURE

Article 97

Exercise of the delegation

1. The power to adopt delegated acts is conferred on the Commission subject to the conditions laid down in this Article.

2. The power to adopt delegated acts referred to in Article 6(6) and (7), Article 7(1) and (3), Article 11(3), Article 43(5) and (6), Article 47(5), Article 51(3), Article 52(4) and Article 53(5) and (6) shall be conferred on the Commission for a period of five years from 1 August 2024. The Commission shall draw up a report in respect of the delegation of power not later than nine months before the end of the five-year period. The delegation of power shall be tacitly extended for periods of an identical duration, unless the European Parliament or the Council opposes such extension not later than three months before the end of each period.

3. The delegation of power referred to in Article 6(6) and (7), Article 7(1) and (3), Article 11(3), Article 43(5) and (6), Article 47(5), Article 51(3), Article 52(4) and Article 53(5) and (6) may be revoked at any time by the European Parliament or by the Council. A decision of revocation shall put an end to the delegation of power specified in that decision. It shall take effect the day following that of its publication in the *Official Journal of the European Union* or at a later date specified therein. It shall not affect the validity of any delegated acts already in force.

4. Before adopting a delegated act, the Commission shall consult experts designated by each Member State in accordance with the principles laid down in the Interinstitutional Agreement of 13 April 2016 on Better Law-Making.

5. As soon as it adopts a delegated act, the Commission shall notify it simultaneously to the European Parliament and to the Council.

6. Any delegated act adopted pursuant to Article 6(6) or (7), Article 7(1) or (3), Article 11(3), Article 43(5) or (6), Article 47(5), Article 51(3), Article 52(4) or Article 53(5) or (6) shall enter into force only if no objection has been expressed by either the European Parliament or the Council within a period of three months of notification of that act to the European Parliament and the Council or if, before the expiry of that period, the European Parliament and the Council have both informed the Commission that they will not object. That period shall be extended by three months at the initiative of the European Parliament or of the Council.

Article 98

Committee procedure

1. The Commission shall be assisted by a committee. That committee shall be a committee within the meaning of Regulation (EU) No 182/2011.

2. Where reference is made to this paragraph, Article 5 of Regulation (EU) No 182/2011 shall apply.

CHAPTER XII

PENALTIES

Article 99

Penalties

1. In accordance with the terms and conditions laid down in this Regulation, Member States shall lay down the rules on penalties and other enforcement measures, which may also include warnings and non-monetary measures, applicable to infringements of this Regulation by operators, and shall take all measures necessary to ensure that they are properly and effectively implemented, thereby taking into account the guidelines issued by the Commission pursuant to Article 96. The penalties provided for shall be effective, proportionate and dissuasive. They shall take into account the interests of SMEs, including start-ups, and their economic viability.

2. The Member States shall, without delay and at the latest by the date of entry into application, notify the Commission of the rules on penalties and of other

enforcement measures referred to in paragraph 1, and shall notify it, without delay, of any subsequent amendment to them.

3.   Non-compliance with the prohibition of the AI practices referred to in Article 5 shall be subject to administrative fines of up to EUR 35 000 000 or, if the offender is an undertaking, up to 7 % of its total worldwide annual turnover for the preceding financial year, whichever is higher.

4.   Non-compliance with any of the following provisions related to operators or notified bodies, other than those laid down in Articles 5, shall be subject to administrative fines of up to EUR 15 000 000 or, if the offender is an undertaking, up to 3 % of its total worldwide annual turnover for the preceding financial year, whichever is higher:

(a)   obligations of providers pursuant to Article 16;

(b) obligations of authorised representatives pursuant to Article 22;

(c)   obligations of importers pursuant to Article 23;

(d)   obligations of distributors pursuant to Article 24;

(e)   obligations of deployers pursuant to Article 26;

(f) requirements and obligations of notified bodies pursuant to Article 31, Article 33(1), (3) and (4) or Article 34;

(g)   transparency obligations for providers and deployers pursuant to Article 50.

5.   The supply of incorrect, incomplete or misleading information to notified bodies or national competent authorities in reply to a request shall be subject to administrative fines of up to EUR 7 500 000 or, if the offender is an undertaking, up to 1 % of its total worldwide annual turnover for the preceding financial year, whichever is higher.

6.   In the case of SMEs, including start-ups, each fine referred to in this Article shall be up to the percentages or amount referred to in paragraphs 3, 4 and 5, whichever thereof is lower.

7.   When deciding whether to impose an administrative fine and when deciding on the amount of the administrative fine in each individual case, all relevant circumstances of the specific situation shall be taken into account and, as appropriate, regard shall be given to the following:

(a) the nature, gravity and duration of the infringement and of its consequences, taking into account the purpose of the AI system, as well as, where appropriate, the number of affected persons and the level of damage suffered by them;

(b) whether administrative fines have already been applied by other market surveillance authorities to the same operator for the same infringement;

(c) whether administrative fines have already been applied by other authorities to the same operator for infringements of other Union or national law, when such infringements result from the same activity or omission constituting a relevant infringement of this Regulation;

(d) the size, the annual turnover and market share of the operator committing the infringement;

(e) any other aggravating or mitigating factor applicable to the circumstances of the case, such as financial benefits gained, or losses avoided, directly or indirectly, from the infringement;

(f) the degree of cooperation with the national competent authorities, in order to remedy the infringement and mitigate the possible adverse effects of the infringement;

(g) the degree of responsibility of the operator taking into account the technical and organisational measures implemented by it;

(h) the manner in which the infringement became known to the national competent authorities, in particular whether, and if so to what extent, the operator notified the infringement;

(i) the intentional or negligent character of the infringement;

(j) any action taken by the operator to mitigate the harm suffered by the affected persons.

8. Each Member State shall lay down rules on to what extent administrative fines may be imposed on public authorities and bodies established in that Member State.

9. Depending on the legal system of the Member States, the rules on administrative fines may be applied in such a manner that the fines are imposed by competent national courts or by other bodies, as applicable in those Member States. The application of such rules in those Member States shall have an equivalent effect.

10. The exercise of powers under this Article shall be subject to appropriate procedural safeguards in accordance with Union and national law, including effective judicial remedies and due process.

11. Member States shall, on an annual basis, report to the Commission about the administrative fines they have issued during that year, in accordance with this Article, and about any related litigation or judicial proceedings.

Article 100

Administrative fines on Union institutions, bodies, offices and agencies

1. The European Data Protection Supervisor may impose administrative fines on Union institutions, bodies, offices and agencies falling within the scope of this Regulation. When deciding whether to impose an administrative fine and when deciding on the amount of the administrative fine in each individual case, all relevant circumstances of the specific situation shall be taken into account and due regard shall be given to the following:

(a) the nature, gravity and duration of the infringement and of its consequences, taking into account the purpose of the AI system concerned, as well as, where appropriate, the number of affected persons and the level of damage suffered by them;

(b) the degree of responsibility of the Union institution, body, office or agency, taking into account technical and organisational measures implemented by them;

(c) any action taken by the Union institution, body, office or agency to mitigate the damage suffered by affected persons;

(d) the degree of cooperation with the European Data Protection Supervisor in order to remedy the infringement and mitigate the possible adverse effects of the infringement, including compliance with any of the measures previously ordered by the European Data Protection Supervisor against the Union institution, body, office or agency concerned with regard to the same subject matter;

(e) any similar previous infringements by the Union institution, body, office or agency;

(f) the manner in which the infringement became known to the European Data Protection Supervisor, in particular whether, and if so to what extent, the Union institution, body, office or agency notified the infringement;

(g) the annual budget of the Union institution, body, office or agency.

2. Non-compliance with the prohibition of the AI practices referred to in Article 5 shall be subject to administrative fines of up to EUR 1 500 000.

3. The non-compliance of the AI system with any requirements or obligations under this Regulation, other than those laid down in Article 5, shall be subject to administrative fines of up to EUR 750 000.

4. Before taking decisions pursuant to this Article, the European Data Protection Supervisor shall give the Union institution, body, office or agency which is the subject of the proceedings conducted by the European Data Protection Supervisor the opportunity of being heard on the matter regarding the possible infringement. The European Data Protection Supervisor shall base his or her decisions only on elements and circumstances on which the parties concerned have been able to comment. Complainants, if any, shall be associated closely with the proceedings.

5. The rights of defence of the parties concerned shall be fully respected in the proceedings. They shall be entitled to have access to the European Data Protection Supervisor's file, subject to the legitimate interest of individuals or undertakings in the protection of their personal data or business secrets.

6. Funds collected by imposition of fines in this Article shall contribute to the general budget of the Union. The fines shall not affect the effective operation of the Union institution, body, office or agency fined.

7. The European Data Protection Supervisor shall, on an annual basis, notify the Commission of the administrative fines it has imposed pursuant to this Article and of any litigation or judicial proceedings it has initiated.

Article 101

Fines for providers of general-purpose AI models

1. The Commission may impose on providers of general-purpose AI models fines not exceeding 3 % of their annual total worldwide turnover in the preceding financial year or EUR 15 000 000, whichever is higher., when the Commission finds that the provider intentionally or negligently:

(a) infringed the relevant provisions of this Regulation;

(b) failed to comply with a request for a document or for information pursuant to Article 91, or supplied incorrect, incomplete or misleading information;

(c) failed to comply with a measure requested under Article 93;

(d) failed to make available to the Commission access to the general-purpose AI model or general-purpose AI model with systemic risk with a view to conducting an evaluation pursuant to Article 92.

In fixing the amount of the fine or periodic penalty payment, regard shall be had to the nature, gravity and duration of the infringement, taking due account of the principles of proportionality and appropriateness. The Commission shall also into account commitments made in accordance with Article 93(3) or made in relevant codes of practice in accordance with Article 56.

2. Before adopting the decision pursuant to paragraph 1, the Commission shall communicate its preliminary findings to the provider of the general-purpose AI model and give it an opportunity to be heard.

3. Fines imposed in accordance with this Article shall be effective, proportionate and dissuasive.

4. Information on fines imposed under this Article shall also be communicated to the Board as appropriate.

5. The Court of Justice of the European Union shall have unlimited jurisdiction to review decisions of the Commission fixing a fine under this Article. It may cancel, reduce or increase the fine imposed.

6. The Commission shall adopt implementing acts containing detailed arrangements and procedural safeguards for proceedings in view of the possible adoption of decisions pursuant to paragraph 1 of this Article. Those implementing acts shall be adopted in accordance with the examination procedure referred to in Article 98(2).

CHAPTER XIII

FINAL PROVISIONS

Article 102

Amendment to Regulation (EC) No 300/2008

In Article 4(3) of Regulation (EC) No 300/2008, the following subparagraph is added:

'When adopting detailed measures related to technical specifications and procedures for approval and use of security equipment concerning Artificial Intelligence systems within the meaning of Regulation (EU) 2024/1689 of the European Parliament and of the Council (*), the requirements set out in Chapter III, Section 2, of that Regulation shall be taken into account.

Article 103

Amendment to Regulation (EU) No 167/2013

In Article 17(5) of Regulation (EU) No 167/2013, the following subparagraph is added:

'When adopting delegated acts pursuant to the first subparagraph concerning artificial intelligence systems which are safety components within the meaning of Regulation (EU) 2024/1689 of the European Parliament and of the Council (*), the requirements set out in Chapter III, Section 2, of that Regulation shall be taken into account.

Article 104

Amendment to Regulation (EU) No 168/2013

In Article 22(5) of Regulation (EU) No 168/2013, the following subparagraph is added:

'When adopting delegated acts pursuant to the first subparagraph concerning Artificial Intelligence systems which are safety components within the meaning of Regulation (EU) 2024/1689 of the European Parliament and of the Council (*), the requirements set out in Chapter III, Section 2, of that Regulation shall be taken into account.

Article 105

Amendment to Directive 2014/90/EU

In Article 8 of Directive 2014/90/EU, the following paragraph is added:

'5. For Artificial Intelligence systems which are safety components within the meaning of Regulation (EU) 2024/1689 of the European Parliament and of the Council (*), when carrying out its activities pursuant to paragraph 1 and when adopting technical specifications and testing standards in accordance with paragraphs 2 and 3, the Commission shall take into account the requirements set out in Chapter III, Section 2, of that Regulation.

Article 106

Amendment to Directive (EU) 2016/797

In Article 5 of Directive (EU) 2016/797, the following paragraph is added:

'12. When adopting delegated acts pursuant to paragraph 1 and implementing acts pursuant to paragraph 11 concerning Artificial Intelligence systems which are safety components within the meaning of Regulation (EU) 2024/1689 of the European Parliament and of the Council (*), the requirements set out in Chapter III, Section 2, of that Regulation shall be taken into account.

Article 107

Amendment to Regulation (EU) 2018/858

In Article 5 of Regulation (EU) 2018/858 the following paragraph is added:

'4. When adopting delegated acts pursuant to paragraph 3 concerning Artificial Intelligence systems which are safety components within the meaning of Regulation (EU) 2024/1689 of the European Parliament and of the Council (*), the requirements set out in Chapter III, Section 2, of that Regulation shall be taken into account.

Article 108

Amendments to Regulation (EU) 2018/1139

Regulation (EU) 2018/1139 is amended as follows:

(1) in Article 17, the following paragraph is added:

'3. Without prejudice to paragraph 2, when adopting implementing acts pursuant to paragraph 1 concerning Artificial Intelligence systems which are safety components within the meaning of Regulation (EU) 2024/1689 of the European Parliament and of the Council (*), the requirements set out in Chapter III, Section 2, of that Regulation shall be taken into account.

(*) Regulation (EU) 2024/1689 of the European Parliament and of the Council of 13 June 2024 laying down harmonised rules on artificial intelligence and amending Regulations (EC) No 300/2008, (EU) No 167/2013, (EU) No 168/2013, (EU) 2018/858, (EU) 2018/1139 and (EU) 2019/2144 and Directives 2014/90/EU, (EU) 2016/797 and (EU) 2020/1828 (Artificial Intelligence Act) (OJ L, 2024/1689, 12.7.2024, ELI: http://data.europa.eu/eli/reg/2024/1689/oj).';"

(2) in Article 19, the following paragraph is added:

'4. When adopting delegated acts pursuant to paragraphs 1 and 2 concerning Artificial Intelligence systems which are safety components within the meaning of Regulation (EU) 2024/1689, the requirements set out in Chapter III, Section 2, of that Regulation shall be taken into account.'

;

(3) in Article 43, the following paragraph is added:

'4. When adopting implementing acts pursuant to paragraph 1 concerning Artificial Intelligence systems which are safety components within the meaning of Regulation (EU) 2024/1689, the requirements set out in Chapter III, Section 2, of that Regulation shall be taken into account.'

;

(4) in Article 47, the following paragraph is added:

'3. When adopting delegated acts pursuant to paragraphs 1 and 2 concerning Artificial Intelligence systems which are safety components within the meaning of Regulation (EU) 2024/1689, the requirements set out in Chapter III, Section 2, of that Regulation shall be taken into account.'

;

(5) in Article 57, the following subparagraph is added:

'When adopting those implementing acts concerning Artificial Intelligence systems which are safety components within the meaning of Regulation (EU) 2024/1689, the requirements set out in Chapter III, Section 2, of that Regulation shall be taken into account.'

;

(6) in Article 58, the following paragraph is added:

'3. When adopting delegated acts pursuant to paragraphs 1 and 2 concerning Artificial Intelligence systems which are safety components within the meaning of Regulation (EU) 2024/1689, the requirements set out in Chapter III, Section 2, of that Regulation shall be taken into account.'.

Article 109

Amendment to Regulation (EU) 2019/2144

In Article 11 of Regulation (EU) 2019/2144, the following paragraph is added:

'3. When adopting the implementing acts pursuant to paragraph 2, concerning artificial intelligence systems which are safety components within the meaning of Regulation (EU) 2024/1689 of the European Parliament and of the Council , the requirements set out in Chapter III, Section 2, of that Regulation shall be taken into account.

Article 110

Amendment to Directive (EU) 2020/1828

In Annex I to Directive (EU) 2020/1828 of the European Parliament and of the Council ([58]), the following point is added:

'(68) Regulation (EU) 2024/1689 of the European Parliament and of the Council of 13 June 2024 laying down harmonised rules on artificial intelligence and amending Regulations (EC) No 300/2008, (EU) No 167/2013, (EU) No 168/2013, (EU) 2018/858, (EU) 2018/1139 and (EU) 2019/2144 and Directives 2014/90/EU, (EU) 2016/797 and (EU) 2020/1828 (Artificial Intelligence Act) (OJ L, 2024/1689, 12.7.2024, ELI: http://data.europa.eu/eli/reg/2024/1689/oj).'.

Article 111

AI systems already placed on the market or put into service and general-purpose AI models already placed on the marked

1. Without prejudice to the application of Article 5 as referred to in Article 113(3), point (a), AI systems which are components of the large-scale IT systems established by the legal acts listed in Annex X that have been placed on the market or put into service before 2 August 2027 shall be brought into compliance with this Regulation by 31 December 2030.

The requirements laid down in this Regulation shall be taken into account in the evaluation of each large-scale IT system established by the legal acts listed in Annex X to be undertaken as provided for in those legal acts and where those legal acts are replaced or amended.

2. Without prejudice to the application of Article 5 as referred to in Article 113(3), point (a), this Regulation shall apply to operators of high-risk AI systems, other than the systems referred to in paragraph 1 of this Article, that have been placed on the market or put into service before 2 August 2026, only if, as from that date, those systems are subject to significant changes in their designs. In any case, the providers and deployers of high-risk AI systems intended to be used by public authorities shall take the necessary steps to comply with the requirements and obligations of this Regulation by 2 August 2030.

3. Providers of general-purpose AI models that have been placed on the market before 2 August 2025 shall take the necessary steps in order to comply with the obligations laid down in this Regulation by 2 August 2027.

Article 112

Evaluation and review

1. The Commission shall assess the need for amendment of the list set out in Annex III and of the list of prohibited AI practices laid down in Article 5, once a year following the entry into force of this Regulation, and until the end of the period of the delegation of power laid down in Article 97. The Commission shall submit the findings of that assessment to the European Parliament and the Council.

2. By 2 August 2028 and every four years thereafter, the Commission shall evaluate and report to the European Parliament and to the Council on the following:

(a) the need for amendments extending existing area headings or adding new area headings in Annex III;

(b) amendments to the list of AI systems requiring additional transparency measures in Article 50;

(c) amendments enhancing the effectiveness of the supervision and governance system.

3. By 2 August 2029 and every four years thereafter, the Commission shall submit a report on the evaluation and review of this Regulation to the European Parliament and to the Council. The report shall include an assessment with regard to the structure of enforcement and the possible need for a Union agency to resolve any identified shortcomings. On the basis of the findings, that report shall, where appropriate, be accompanied by a proposal for amendment of this Regulation. The reports shall be made public.

4. The reports referred to in paragraph 2 shall pay specific attention to the following:

(a) the status of the financial, technical and human resources of the national competent authorities in order to effectively perform the tasks assigned to them under this Regulation;

(b) the state of penalties, in particular administrative fines as referred to in Article 99(1), applied by Member States for infringements of this Regulation;

(c) adopted harmonised standards and common specifications developed to support this Regulation;

(d) the number of undertakings that enter the market after the entry into application of this Regulation, and how many of them are SMEs.

5. By 2 August 2028, the Commission shall evaluate the functioning of the AI Office, whether the AI Office has been given sufficient powers and competences to fulfil its tasks, and whether it would be relevant and needed for the proper implementation and enforcement of this Regulation to upgrade the AI Office and its enforcement competences and to increase its resources. The Commission shall submit a report on its evaluation to the European Parliament and to the Council.

6. By 2 August 2028 and every four years thereafter, the Commission shall submit a report on the review of the progress on the development of standardisation deliverables on the energy-efficient development of general-purpose AI models, and

asses the need for further measures or actions, including binding measures or actions. The report shall be submitted to the European Parliament and to the Council, and it shall be made public.

7. By 2 August 2028 and every three years thereafter, the Commission shall evaluate the impact and effectiveness of voluntary codes of conduct to foster the application of the requirements set out in Chapter III, Section 2 for AI systems other than high-risk AI systems and possibly other additional requirements for AI systems other than high-risk AI systems, including as regards environmental sustainability.

8. For the purposes of paragraphs 1 to 7, the Board, the Member States and national competent authorities shall provide the Commission with information upon its request and without undue delay.

9. In carrying out the evaluations and reviews referred to in paragraphs 1 to 7, the Commission shall take into account the positions and findings of the Board, of the European Parliament, of the Council, and of other relevant bodies or sources.

10. The Commission shall, if necessary, submit appropriate proposals to amend this Regulation, in particular taking into account developments in technology, the effect of AI systems on health and safety, and on fundamental rights, and in light of the state of progress in the information society.

11. To guide the evaluations and reviews referred to in paragraphs 1 to 7 of this Article, the AI Office shall undertake to develop an objective and participative methodology for the evaluation of risk levels based on the criteria outlined in the relevant Articles and the inclusion of new systems in:

(a) the list set out in Annex III, including the extension of existing area headings or the addition of new area headings in that Annex;

(b) the list of prohibited practices set out in Article 5; and

(c) the list of AI systems requiring additional transparency measures pursuant to Article 50.

12. Any amendment to this Regulation pursuant to paragraph 10, or relevant delegated or implementing acts, which concerns sectoral Union harmonisation legislation listed in Section B of Annex I shall take into account the regulatory specificities of each sector, and the existing governance, conformity assessment and enforcement mechanisms and authorities established therein.

13. By 2 August 2031, the Commission shall carry out an assessment of the enforcement of this Regulation and shall report on it to the European Parliament, the Council and the European Economic and Social Committee, taking into account the first years of application of this Regulation. On the basis of the findings, that report shall, where appropriate, be accompanied by a proposal for amendment of this Regulation with regard to the structure of enforcement and the need for a Union agency to resolve any identified shortcomings.

Article 113

Entry into force and application

This Regulation shall enter into force on the twentieth day following that of its publication in the *Official Journal of the European Union.*

It shall apply from 2 August 2026.

However:

(a) Chapters I and II shall apply from 2 February 2025;

(b) Chapter III Section 4, Chapter V, Chapter VII and Chapter XII and Article 78 shall apply from 2 August 2025, with the exception of Article 101;

(c) Article 6(1) and the corresponding obligations in this Regulation shall apply from 2 August 2027.

This Regulation shall be binding in its entirety and directly applicable in all Member States.

Done at Brussels, 13 June 2024.

For the European Parliament

The President

R. METSOLA

For the Council

The President

M. MICHEL

Council of Europe, Framework Convention on Artificial Intelligence and Human Rights, Democracy and the Rule of Law (2024)

Preamble

The member States of the Council of Europe and the other signatories hereto,

Considering that the aim of the Council of Europe is to achieve greater unity between its members, based in particular on the respect for human rights, democracy and the rule of law;

Recognising the value of fostering co-operation between the Parties to this Convention and of extending such co-operation to other States that share the same values;

Conscious of the accelerating developments in science and technology and the profound changes brought about through activities within the lifecycle of artificial intelligence systems, which have the potential to promote human prosperity as well as individual and societal wellbeing, sustainable development, gender equality and the empowerment of all women and girls, as well as other important goals and interests, by enhancing progress and innovation;

Recognising that activities within the lifecycle of artificial intelligence systems may offer unprecedented opportunities to protect and promote human rights, democracy and the rule of law;

Concerned that certain activities within the lifecycle of artificial intelligence systems may undermine human dignity and individual autonomy, human rights, democracy and the rule of law;

Concerned about the risks of discrimination in digital contexts, particularly those involving artificial intelligence systems, and their potential effect of creating or aggravating inequalities, including those experienced by women and individuals in vulnerable situations, regarding the enjoyment of their human rights and their full, equal and effective participation in economic, social, cultural and political affairs;

Concerned by the misuse of artificial intelligence systems and opposing the use of such systems for repressive purposes in violation of international human rights law, including through arbitrary or unlawful surveillance and censorship practices that erode privacy and individual autonomy;

Conscious of the fact that human rights, democracy and the rule of law are inherently interwoven;

Convinced of the need to establish, as a matter of priority, a globally applicable legal framework setting out common general principles and rules governing the activities within the lifecycle of artificial intelligence systems that effectively preserves shared values and harnesses the benefits of artificial intelligence for the promotion of these values in a manner conducive to responsible innovation;

Recognising the need to promote digital literacy, knowledge about, and trust in the design, development, use and decommissioning of artificial intelligence systems; Recognising the framework character of this Convention, which may be supplemented by further instruments to address specific issues relating to the activities within the lifecycle of artificial intelligence systems;

Underlining that this Convention is intended to address specific challenges which arise throughout the lifecycle of artificial intelligence systems and encourage the consideration of the wider risks and impacts related to these technologies including, but not limited to, human health and the environment, and socio-economic aspects, such as employment and labour;

Noting relevant efforts to advance international understanding and co-operation on artificial intelligence by other international and supranational organisations and fora; Mindful of applicable international human rights instruments, such as the 1948 Universal Declaration of Human Rights, the 1950 Convention for the Protection of Human Rights and Fundamental Freedoms (ETS No. 5), the 1966 International Covenant on Civil and Political Rights, the 1966 International Covenant on Economic, Social and Cultural Rights, the 1961 European Social Charter (ETS No. 35), as well as their respective protocols, and the 1996 European Social Charter (Revised) (ETS No. 163);

Mindful also of the 1989 United Nations Convention on the Rights of the Child and the 2006 United Nations Convention on the Rights of Persons with Disabilities; Mindful also of the privacy rights of individuals and the protection of personal data, as applicable and conferred, for example, by the 1981 Convention for the Protection of Individuals with regard to Automatic Processing of Personal Data (ETS No. 108) and its protocols;

Affirming the commitment of the Parties to protecting human rights, democracy and the rule of law, and fostering trustworthiness of artificial intelligence systems through this Convention,

Have agreed as follows:

Chapter I – General provisions

Article 1 – Object and purpose

1. The provisions of this Convention aim to ensure that activities within the lifecycle of artificial intelligence systems are fully consistent with human rights, democracy and the rule of law.

2. Each Party shall adopt or maintain appropriate legislative, administrative or other measures to give effect to the provisions set out in this Convention. These measures shall be graduated and differentiated as may be necessary in view of the severity and probability of the occurrence of adverse impacts on human rights, democracy and the rule of law throughout the lifecycle of artificial intelligence systems. This may include specific or horizontal measures that apply irrespective of the type of technology used.

3. In order to ensure effective implementation of its provisions by the Parties, this Convention establishes a follow-up mechanism and provides for international co-operation.

Article 2 – Definition of artificial intelligence systems

For the purposes of this Convention, "artificial intelligence system" means a machine-based system that, for explicit or implicit objectives, infers, from the input it receives, how to generate outputs such as predictions, content, recommendations or decisions that may influence physical or virtual environments. Different artificial intelligence systems vary in their levels of autonomy and adaptiveness after deployment.

Article 3 – Scope

1. The scope of this Convention covers the activities within the lifecycle of artificial intelligence systems that have the potential to interfere with human rights, democracy and the rule of law as follows:

> a. Each Party shall apply this Convention to the activities within the lifecycle of artificial intelligence systems undertaken by public authorities, or private actors acting on their behalf.

> b. Each Party shall address risks and impacts arising from activities within the lifecycle of artificial intelligence systems by private actors to the extent not covered in subparagraph a in a manner conforming with the object and purpose of this Convention.

Each Party shall specify in a declaration submitted to the Secretary General of the Council of Europe at the time of signature, or when depositing its instrument of ratification, acceptance, approval or accession, how it intends to implement this obligation, either by applying the principles and obligations set forth in Chapters II to VI of this Convention to activities of private actors or by taking other appropriate measures to fulfil the obligation set out in this subparagraph. Parties may, at any time and in the same manner, amend their declarations.

When implementing the obligation under this subparagraph, a Party may not derogate from or limit the application of its international obligations undertaken to protect human rights, democracy and the rule of law.

2. A Party shall not be required to apply this Convention to activities within the lifecycle of artificial intelligence systems related to the protection of its national security interests, with the understanding that such activities are conducted in a manner consistent with applicable international law, including international human rights law obligations, and with respect for its democratic institutions and processes.

3. Without prejudice to Article 13 and Article 25, paragraph 2, this Convention shall not apply to research and development activities regarding artificial intelligence systems not yet made available for use, unless testing or similar activities are undertaken in such a way that they have the potential to interfere with human rights, democracy and the rule of law.

4. Matters relating to national defence do not fall within the scope of this Convention.

Chapter II – General obligations

Article 4 – Protection of human rights

Each Party shall adopt or maintain measures to ensure that the activities within the lifecycle of artificial intelligence systems are consistent with obligations to protect human rights, as enshrined in applicable international law and in its domestic law.

Article 5 – Integrity of democratic processes and respect for the rule of law

1. Each Party shall adopt or maintain measures that seek to ensure that artificial intelligence systems are not used to undermine the integrity, independence and effectiveness of democratic institutions and processes, including the principle of the separation of powers, respect for judicial independence and access to justice.

2. Each Party shall adopt or maintain measures that seek to protect its democratic processes in the context of activities within the lifecycle of artificial intelligence systems, including individuals' fair access to and participation in public debate, as well as their ability to freely form opinions.

Chapter III – Principles related to activities within the lifecycle of artificial intelligence systems

Article 6 – General approach

This chapter sets forth general common principles that each Party shall implement in regard to artificial intelligence systems in a manner appropriate to its domestic legal system and the other obligations of this Convention.

Article 7 – Human dignity and individual autonomy

Each Party shall adopt or maintain measures to respect human dignity and individual autonomy in relation to activities within the lifecycle of artificial intelligence systems.

Article 8 – Transparency and oversight

Each Party shall adopt or maintain measures to ensure that adequate transparency and oversight requirements tailored to the specific contexts and risks are in place in respect of activities within the lifecycle of artificial intelligence systems, including with regard to the identification of content generated by artificial intelligence systems.

Article 9 – Accountability and responsibility

Each Party shall adopt or maintain measures to ensure accountability and responsibility for adverse impacts on human rights, democracy and the rule of law resulting from activities within the lifecycle of artificial intelligence systems.

Article 10 – Equality and non-discrimination

1. Each Party shall adopt or maintain measures with a view to ensuring that activities within the lifecycle of artificial intelligence systems respect equality, including gender equality, and the prohibition of discrimination, as provided under applicable international and domestic law.

2. Each Party undertakes to adopt or maintain measures aimed at overcoming inequalities to achieve fair, just and equitable outcomes, in line with its applicable

domestic and international human rights obligations, in relation to activities within the lifecycle of artificial intelligence systems.

Article 11 – Privacy and personal data protection

Each Party shall adopt or maintain measures to ensure that, with regard to activities within the lifecycle of artificial intelligence systems:

> a. privacy rights of individuals and their personal data are protected, including through applicable domestic and international laws, standards and frameworks; and

> b. effective guarantees and safeguards have been put in place for individuals, in accordance with applicable domestic and international legal obligations.

Article 12 – Reliability

Each Party shall take, as appropriate, measures to promote the reliability of artificial intelligence systems and trust in their outputs, which could include requirements related to adequate quality and security throughout the lifecycle of artificial intelligence systems.

Article 13 – Safe innovation

With a view to fostering innovation while avoiding adverse impacts on human rights, democracy and the rule of law, each Party is called upon to enable, as appropriate, the establishment of controlled environments for developing, experimenting and testing artificial intelligence systems under the supervision of its competent authorities.

Chapter IV – Remedies

Article 14 – Remedies

1. Each Party shall, to the extent remedies are required by its international obligations and consistent with its domestic legal system, adopt or maintain measures to ensure the availability of accessible and effective remedies for violations of human rights resulting from the activities within the lifecycle of artificial intelligence systems.

2. With the aim of supporting paragraph 1 above, each Party shall adopt or maintain measures including:

a. measures to ensure that relevant information regarding artificial intelligence systems which have the potential to significantly affect human rights and their relevant usage is documented, provided to bodies authorised to access that information and, where appropriate and applicable, made available or communicated to affected persons;

b. measures to ensure that the information referred to in subparagraph a is sufficient for the affected persons to contest the decision(s) made or substantially informed by the use of the system, and, where relevant and appropriate, the use of the system itself; and

c. an effective possibility for persons concerned to lodge a complaint to competent authorities.

Article 15 – Procedural safeguards

1. Each Party shall ensure that, where an artificial intelligence system significantly impacts upon the enjoyment of human rights, effective procedural guarantees, safeguards and rights, in accordance with the applicable international and domestic law, are available to persons affected thereby.

2. Each Party shall seek to ensure that, as appropriate for the context, persons interacting with artificial intelligence systems are notified that they are interacting with such systems rather than with a human.

Chapter V – Assessment and mitigation of risks and adverse impacts

Article 16 – Risk and impact management framework

1. Each Party shall, taking into account the principles set forth in Chapter III, adopt or maintain measures for the identification, assessment, prevention and mitigation of risks posed by artificial intelligence systems by considering actual and potential impacts to human rights, democracy and the rule of law.

2. Such measures shall be graduated and differentiated, as appropriate, and:

a. take due account of the context and intended use of artificial intelligence systems, in particular as concerns risks to human rights, democracy, and the rule of law;

b. take due account of the severity and probability of potential impacts;

c. consider, where appropriate, the perspectives of relevant stakeholders, in particular persons whose rights may be impacted;

d. apply iteratively throughout the activities within the lifecycle of the artificial intelligence system;

e. include monitoring for risks and adverse impacts to human rights, democracy, and the rule of law;

f. include documentation of risks, actual and potential impacts, and the risk management approach; and

g. require, where appropriate, testing of artificial intelligence systems before making them available for first use and when they are significantly modified. Each Party shall adopt or maintain measures that seek to ensure that adverse impacts of artificial intelligence systems to human rights, democracy, and the rule of law are adequately addressed. Such adverse impacts and measures to address them should be documented and inform the relevant risk management measures described in paragraph

3. Each Party shall assess the need for a moratorium or ban or other appropriate measures in respect of certain uses of artificial intelligence systems where it considers such uses incompatible with the respect for human rights, the functioning of democracy or the rule of law.

Chapter VI – Implementation of the Convention

Article 17 – Non-discrimination

The implementation of the provisions of this Convention by the Parties shall be secured without discrimination on any ground, in accordance with their international human rights obligations.

Article 18 – Rights of persons with disabilities and of children

Each Party shall, in accordance with its domestic law and applicable international obligations, take due account of any specific needs and vulnerabilities in relation to respect for the rights of persons with disabilities and of children.

Article 19 – Public consultation

Each Party shall seek to ensure that important questions raised in relation to artificial intelligence systems are, as appropriate, duly considered through public discussion and multistakeholder consultation in the light of social, economic, legal, ethical, environmental and other relevant implications.

Article 20 – Digital literacy and skills

Each Party shall encourage and promote adequate digital literacy and digital skills for all segments of the population, including specific expert skills for those responsible for the identification, assessment, prevention and mitigation of risks posed by artificial intelligence systems.

Article 21 – Safeguard for existing human rights

Nothing in this Convention shall be construed as limiting, derogating from or otherwise affecting the human rights or other related legal rights and obligations which may be guaranteed under the relevant laws of a Party or any other relevant international agreement to which it is party.

Article 22 – Wider protection

None of the provisions of this Convention shall be interpreted as limiting or otherwise affecting the possibility for a Party to grant a wider measure of protection than is stipulated in this Convention.

Chapter VII – Follow-up mechanism and co-operation

Article 23 – Conference of the

1. The Conference of the Parties shall be composed of representatives of the Parties to this Convention.

2. The Parties shall consult periodically with a view to:

    a. facilitating the effective application and implementation of this Convention, including the identification of any problems and the effects of any reservation made in pursuance of Article 34, paragraph 1, or any declaration made under this Convention;

    b. considering the possible supplementation to or amendment of this Convention;

    c. considering matters and making specific recommendations concerning the interpretation and application of this Convention;

d. facilitating the exchange of information on significant legal, policy or technological developments of relevance, including in pursuit of the objectives defined in Article 25, for the implementation of this Convention;

e. facilitating, where necessary, the friendly settlement of disputes related to the application of this Convention; and

f. facilitating co-operation with relevant stakeholders concerning pertinent aspects of the implementation of this Convention, including through public hearings where appropriate.

3, The Conference of the Parties shall be convened by the Secretary General of the Council of Europe whenever necessary and, in any case, when a majority of the Parties or the Committee of Ministers requests its convocation.

4. The Conference of the Parties shall adopt its own rules of procedure by consensus within twelve months of the entry into force of this Convention.

5. The Parties shall be assisted by the Secretariat of the Council of Europe in carrying out their functions pursuant to this article.

6. The Conference of the Parties may propose to the Committee of Ministers appropriate ways to engage relevant expertise in support of the effective implementation of this Convention.

7. Any Party which is not a member of the Council of Europe shall contribute to the funding of the activities of the Conference of the Parties. The contribution of a non-member of the Council of Europe shall be established jointly by the Committee of Ministers and that non-member.

8. The Conference of the Parties may decide to restrict the participation in its work of a Party that has ceased to be a member of the Council of Europe under Article 8 of the Statute of the Council of Europe (ETS No. 1) for a serious violation of Article 3 of the Statute. Similarly, measures can be taken in respect of any Party that is not a member State of the Council of Europe by a decision of the Committee of Ministers to cease its relations with that State on grounds similar to those mentioned in Article 3 of the Statute.

Article 24 – Reporting obligation

1. Each Party shall provide a report to the Conference of the Parties within the first two years after becoming a Party, and then periodically thereafter with details of the

activities undertaken to give effect to Article 3, paragraph 1, sub-paragraphs a and b.

2, The Conference of the Parties shall determine the format and the process for the report in accordance with its rules of procedure.

Article 25 – International co-operation

1. The Parties shall co-operate in the realisation of the purpose of this Convention. Parties are further encouraged, as appropriate, to assist States that are not Parties to this Convention in acting consistently with the terms of this Convention and becoming a Party to it.

2. The Parties shall, as appropriate, exchange relevant and useful information between themselves concerning aspects related to artificial intelligence which may have significant positive or negative effects on the enjoyment of human rights, the functioning of democracy and the observance of the rule of law, including risks and effects that have arisen in research contexts and in relation to the private sector. Parties are encouraged to involve, as appropriate, relevant stakeholders and States that are not Parties to this Convention in such exchanges of information.

3. The Parties are encouraged to strengthen co-operation, including with relevant stakeholders where appropriate, to prevent and mitigate risks and adverse impacts on human rights, democracy and the rule of law in the context of activities within the lifecycle of artificial intelligence systems.

Article 26 – Effective oversight mechanisms

1. Each Party shall establish or designate one or more effective mechanisms to oversee compliance with the obligations in this Convention.

2. Each Party shall ensure that such mechanisms exercise their duties independently and impartially and that they have the necessary powers, expertise and resources to effectively fulfil their tasks of overseeing compliance with the obligations in this Convention, as given effect by the Parties. I

3. If a Party has provided for more than one such mechanism, it shall take measures, where practicable, to facilitate effective cooperation among them.

4. If a Party has provided for mechanisms different from existing human rights structures, it shall take measures, where practicable, to promote effective cooperation between the mechanisms referred to in paragraph 1 and those existing domestic human rights structures.

Chapter VIII – Final clauses

Article 27 – Effects of the Convention

1. If two or more Parties have already concluded an agreement or treaty on the matters dealt with in this Convention, or have otherwise established relations on such matters, they shall also be entitled to apply that agreement or treaty or to regulate those relations accordingly, so long as they do so in a manner which is not inconsistent with the object and purpose of this Convention.

2. Parties which are members of the European Union shall, in their mutual relations, apply European Union rules governing the matters within the scope of this Convention without prejudice to the object and purpose of this Convention and without prejudice to its full application with other Parties. The same applies to other Parties to the extent that they are bound by such rules.

Article 28 – Amendments

1. Amendments to this Convention may be proposed by any Party, the Committee of Ministers of the Council of Europe or the Conference of the Parties.

2. Any proposal for amendment shall be communicated by the Secretary General of the Council of Europe to the Parties.

3. Any amendment proposed by a Party, or the Committee of Ministers, shall be communicated to the Conference of the Parties, which shall submit to the Committee of Ministers its opinion on the proposed amendment.

4. The Committee of Ministers shall consider the proposed amendment and the opinion submitted by the Conference of the Parties and may approve the amendment.

5. The text of any amendment approved by the Committee of Ministers in accordance with paragraph 4 shall be forwarded to the Parties for acceptance.

6. Any amendment approved in accordance with paragraph 4 shall come into force on the thirtieth day after all Parties have informed the Secretary General of their acceptance thereof.

Article 29 – Dispute settlement

In the event of a dispute between Parties as to the interpretation or application of this Convention, these Parties shall seek a settlement of the dispute through negotiation or any other peaceful means of their choice, including through the Conference of the Parties, as provided for in Article 23, paragraph 2, sub-paragraph e.

Article 30 – Signature and entry into force

1. This Convention shall be open for signature by the member States of the Council of Europe, the non-member States which have participated in its elaboration and the European Union.

2. This Convention is subject to ratification, acceptance or approval. Instruments of ratification, acceptance or approval shall be deposited with the Secretary General of the Council of Europe.

3. This Convention shall enter into force on the first day of the month following the expiration of a period of three months after the date on which five signatories, including at least three member States of the Council of Europe, have expressed their consent to be bound by this Convention in accordance with paragraph 2.

4. In respect of any signatory which subsequently expresses its consent to be bound by it, this Convention shall enter into force on the first day of the month following the expiration of a period of three months after the date of the deposit of its instrument of ratification, acceptance or approval.

Article 31 – Accession

1. After the entry into force of this Convention, the Committee of Ministers of the Council of Europe may, after consulting the Parties to this Convention and obtaining their unanimous consent, invite any non-member State of the Council of Europe which has not participated in the elaboration of this Convention to accede to this Convention by a decision taken by the majority provided for in Article 20.d of the Statute of the Council of Europe, and by unanimous vote of the representatives of the Parties entitled to sit on the Committee of Ministers.

2. In respect of any acceding State, this Convention shall enter into force on the first day of the month following the expiration of a period of three months after the date of deposit of the instrument of accession with the Secretary General of the Council of Europe.

Article 32 – Territorial application

1. Any State or the European Union may, at the time of signature or when depositing its instrument of ratification, acceptance, approval or accession, specify the territory

or territories to which this Convention shall apply. Any Party may, at a later date, by a declaration addressed to the Secretary General of the Council of Europe, extend the application of this Convention to any other territory specified in the declaration.

2. In respect of such territory, this Convention shall enter into force on the first day of the month following the expiration of a period of three months after the date of receipt of the declaration by the Secretary General.

3. Any declaration made under the two preceding paragraphs may, in respect of any territory specified in said declaration, be withdrawn by a notification addressed to the Secretary General of the Council of Europe. The withdrawal shall become effective on the first day of the month following the expiration of a period of three months after the date of receipt of such notification by the Secretary General.

Article 33 – Federal clause

1. A federal State may reserve the right to assume obligations under this Convention consistent with its fundamental principles governing the relationship between its central government and constituent states or other similar territorial entities, provided that this Convention shall apply to the central government of the federal State.

2. With regard to the provisions of this Convention, the application of which comes under the jurisdiction of constituent states or other similar territorial entities that are not obliged by the constitutional system of the federation to take legislative measures, the federal government shall inform the competent authorities of such states of the said provisions with its favourable opinion, and encourage them to take appropriate action to give them effect.

Article 34 – Reservations

1. By a written notification addressed to the Secretary General of the Council of Europe, any State may, at the time of signature or when depositing its instrument of ratification, acceptance, approval or accession, declare that it avails itself of the reservation provided for in Article 33, paragraph 1.

2. No other reservation may be made in respect of this Convention.

Article 35 – Denunciation

1. Any Party may, at any time, denounce this Convention by means of a notification addressed to the Secretary General of the Council of Europe.

2. Such denunciation shall become effective on the first day of the month following the expiration of a period of three months after the date of receipt of the notification by the Secretary General.

Article 36 – Notification

The Secretary General of the Council of Europe shall notify the member States of the Council of Europe, the non-member States which have participated in the elaboration of this Convention, the European Union, any signatory, any contracting State, any Party and any other State which has been invited to accede to this Convention, of:

a. any signature;

b. the deposit of any instrument of ratification, acceptance, approval or accession;

c. any date of entry into force of this Convention, in accordance with Article 30, paragraphs 3 and 4, and Article 31, paragraph 2;

d. any amendment adopted in accordance with Article 28 and the date on which such an amendment enters into force;

e. any declaration made in pursuance of Article 3, paragraph 1, sub-paragraph b;

f. any reservation and withdrawal of a reservation made in pursuance of;

g. any denunciation made in pursuance of Article 35;

h. any other act, declaration, notification or communication relating to this Convention.

In witness whereof the undersigned, being duly authorised thereto, have signed this Convention.

Done in Vilnius, this 5th day of September 2024, in English and in French, both texts being equally authentic, in a single copy which shall be deposited in the archives of the Council of Europe. The Secretary General of the Council of Europe shall transmit certified copies to each member State of the Council of Europe, to the non-member States which have participated in the elaboration of this Convention, to the European Union and to any State invited to accede to this Convention.

Declaration of Montevideo
(2024)

"For the construction of a regional approach on the governance of Artificial Intelligence and its impacts on our society"

In Montevideo, Oriental Republic of Uruguay, the Ministers and High-Level Authorities representing the countries gathered at the Second Ministerial and High-Level Authorities Summit on the Ethics of Artificial Intelligence in Latin America and the Caribbean, on October 3rd and 4th, 2024.

REAFFIRMING the region's commitment to the development and deployment of Artificial Intelligence (AI) based on the promotion, respect for and protection of human rights, fundamental freedoms, rule of law and democracy, focused on the human being, wellbeing and dignity of people, fostering innovation and ensuring inclusive and sustainable technological progress.

RECOGNIZING the milestone constituted by the first Ministerial and High-Level Authorities Summit on the Ethics of Artificial Intelligence in Latin America and the Caribbean in Santiago, Chile, on October 23rd and 24th, 2023 with the support of the United Nations Educational, Scientific and Cultural Organization (UNESCO) and CAF development bank of Latin America and the Caribbean- and its declaration, which laid the foundations towards the process of building an ethical AI with a special regional focus. APPRECIATING the contributions and efforts of initiatives led by the different countries of the region, international organizations and by other international and domestic actors that have contributed to the regional debate on AI governance, and

REAFFIRMING the importance of consolidating collective instances that include multiple stakeholders, encouraging cooperation and ongoing dialogue among them to ensure the interoperability of the various instruments, with a view to avoiding duplication of efforts.

EMPHASIZING the power of each country to establish its policies according to its interests, national priorities and circumstances, and internal regulations and international commitments, without ignoring the importance of regional and global governance, based on the impacts of AI in all areas, including ethics, safety and equity; promoting international cooperation to maximize the benefits and mitigate the risks associated with the development and use of AI, particularly in the enjoyment of human rights.

REAFFIRMING that AI governance must take into account the special features that characterize the region -including their socio-economic and cultural contexts-, in relation to the opportunities, as well as the current and potential risks and adverse impacts arising from AI technologies, at all stages of their life cycle, as well as the need to promote digital inclusion and equity in access to such technologies.

RECOGNIZING that the construction of this regional approach requires an effective coordination and consolidation of spaces for political and technical discussion that will make it possible to reach consensus and HIGHLIGHTING in this process the importance of the Working Group created as a result of the Ministerial and High-Level Authorities Summit on the Ethics of Artificial Intelligence in Latin America and the Caribbean in Santiago, Chile.

EMPHASIZING the contribution of UNESCO and CAF to this regional process in line with the UNESCO Recommendation on the ethics of AI, the accompaniment of the Working Group, and the technical support provided to countries for the development and implementation of national AI policies.

HIGHLIGHTING the usefulness of the UNESCO Readiness Assessment Methodology (RAM) to understand and diagnose the situation and degree of progress of countries in the region in the implementation of the UNESCO Recommendation on the ethics of AI and the interest to proceed in all countries of Latin America and the Caribbean to reach a regional diagnosis.

EMPHASIZING the usefulness of CAF's Practical Guide "Design of public policies on artificial intelligence. Development of enablers for their implementation in Latin America and the Caribbean", by providing methodological frameworks for the participatory design of AI public policies and for strengthening the enablers for their implementation, with a focus on our region.

UNDERLINING the importance of having a common and actionable Roadmap, built by our countries that takes into account regional objectives and priorities, as an instrument to strengthen the digital sovereignty of our region and its development.

UNDERTAKING THE COMMITMENT to continue working to address the priority challenges contemplated in the Roadmap, which are guidelines to the development and effective implementation of public policies on AI in Latin America and the Caribbean.

Therefore, RESOLVE:

To approve the Roadmap agreed at this Summit, its prioritized areas and associated initiatives as an instrument to strengthen regional technical and political dialogues relating to governance and to the development of capacities for the use of AI in Latin America and the Caribbean, with the technical support and contribution of UNESCO and CAF in our region, among others.

To include in the approved Roadmap the following prioritized areas: Governance and Regulation; Talent and future of work; Protection of vulnerable groups; Environment, Sustainability and Climate Change; and Infrastructure. Each of them has associated products and specific activities that seek to implement AI ethically.

To consolidate the creation of the Working Group on the Ethics of Artificial Intelligence in Latin America and the Caribbean as a space for permanent dialogue and periodic meetings, with a regional approach, entrusting it with the preparation of the terms of reference for its operation and the coordination of the necessary actions for the implementation of the approved Roadmap, as well as formulating proposals for subsequent reviews.

To establish the Ministerial and High-Level Authorities Summit on the Ethics of Artificial Intelligence in Latin America and the Caribbean as an annual meeting to analyze and discuss on the development of regional policies on AI, and to follow up on the implementation of the approved Roadmap and its reviews.

To acknowledge the work of the national focal points in contributing to the creation of the first Roadmap and continue with the commitment to advance in the regional process and in the implementation of the prioritized areas, ensuring coordination and convergence, where appropriate, with other regional initiatives such as the Digital Agenda for Latin America and the Caribbean (eLAC) and the e-Government Network of Latin America and the Caribbean (Red Gealc), thus contributing to the strengthening of regional dialogue in this area.

To thank UNESCO and CAF for their support and collaboration in the construction and permanence of this regional process, and their offer to continue supporting the work of the Working Group, as technical secretary, within the framework of its commitment to provide technical and financial assistance for the implementation of the approved Roadmap and the actions that are defined in subsequent reviews.

To thank the Government of the Republic of Chile for its leadership role in the Working Group established in Santiago, Chile, thus far, and to entrust the Oriental Republic of Uruguay with the continuation of its work and the determination of the procedural aspects necessary to ensure adequate progress of the initiatives provided for in the Roadmap.

To thank the Government of the Oriental Republic of Uruguay and especially the Agency for Electronic Government and the Information and Knowledge Society

(AGESIC) of the Presidency of the Republic, UNESCO and CAF for the organization of this Summit.

Vatican Guidelines on AI
N. DCCII - Decree of the Pontifical Commission for the State of City bearing "Guidelines on Artificial Intelligence" (2025)

Chapter I

(General Provisions)

Article 1

(Purpose and scope)

§ 1 These guidelines recite general principles aimed at enhancing and promoting the ethical and transparent use of artificial intelligence, in an anthropocentric and trustworthy dimension, with respect for human dignity and the common good.

§ 2 The research, testing and development of artificial intelligence systems and models must comply with these guidelines.

§ 3 The following provisions shall take effect for the Governorate of the Vatican City State, limited to the territory of the Vatican City State, and for the activities carried out by the Governorate in the areas referred to in Articles 15 and 16 of the Lateran Treaty, and shall apply:

a) To Operational Bodies;

b) To Scientific Bodies;

c) To the Judicial Organs of the Vatican City State;

d) to the staff of the Governorate of the State of Vatican City, as identified in Article 3 of the *General Regulations for the Staff of the Governorate of the State of Vatican City*, November 21, 2010, as amended;

e) to economic operators, suppliers or bidders, and temporary professional appointees, as defined in Article 2 (c), (e), (f) of Decree No. CCCLXXXVII, on norms on transparency, control and competition in public contracts of the Holy See and the Vatican City State, December 1, 2020, as amended.

Article 2

(Definitions)

For the purposes of these guidelines:

a) "*Artificial intelligence*" means the set of computational systems and models that, through automated processes, are able to analyze data, learn from it, make decisions and perform tasks that would normally require human intelligence;

b) "*Artificial intelligence system*" means an automated system designed to operate with varying levels of autonomy and which may exhibit adaptability after deployment and which, for explicit or implicit purposes, deduces from the analyzed data it receives how to generate responses such as predictions, content, recommendations, or decisions that may influence physical or virtual environments;

c) "*Artificial intelligence models*" means software and hardware systems that, through the use of machine learning techniques and the ability to identify recurring structures in data collections, are able to perform tasks and activities typically associated with human intelligence;

d) "*Data*" means any information, act or fact represented in digital form;

e) "Biometric data" means personal data obtained by specific technical processing relating to physical, physiological or behavioral characteristics of a natural person, such as facial images or dactyloscopic data;

f) "*Risk*" means the use of data within the artificial intelligence system and artificial intelligence models that exposes to a high probability of the occurrence of harm.

Article 3

(Basic Principles)

§ 1 The entities referred to in Article 1, § 3 above must ensure that the conduct of their activities in the field of experimentation, development, adoption and use of artificial intelligence systems and models is in accordance with respect for human dignity, the common good and is inspired by the principles of ethical responsibility, transparency and proportionality of administrative action.

§ 2 Artificial intelligence systems and models must be developed and applied while ensuring the security of the Vatican City State, protection and confidentiality of personal data, non-discrimination of human beings, economic sustainability, and care for Creation.

§ 3 The subjects mentioned in Article 1, § 3 above, to the extent of their competence, must verify and supervise the data processing and management processes in the development and application of artificial intelligence systems and models, so that the results are correct, reliable, appropriate and obtained in accordance with the principles of transparency and proportionality.

§ 4 Operational, scientific, and judicial bodies, while respecting the anthropocentric dimension in the use of artificial intelligence systems and models, must ensure the vocation of artificial intelligence to serve humans, while preserving respect for the autonomy and decision-making power of humans.

§ 5 The Operational, Scientific and Judicial Bodies, must be vigilant to ensure that the development and application of artificial intelligence systems and models do not harm the pastoral mission of the Supreme Pontiff, the integrity of the Catholic Church and the proper conduct of the institutional activities of the Governorate of Vatican City State.

Article 4

(Prohibitions on the use of artificial intelligence)

The following practices are prohibited:

a) The use of an artificial intelligence system to draw general anthropological inferences with discriminatory effects on the individual;

b) The use of an artificial intelligence system that uses subliminal manipulation techniques suitable for causing the person or a group of people physical or psychological harm;

c) the use of an artificial intelligence system that precludes people with disabilities from accessing artificial intelligence and related features and applications;

d) the use of an artificial intelligence system that, through the processing of data, creates social inequality, degrading human dignity and violating fundamental human principles;

e) the use of an artificial intelligence system likely to compromise the security of the Vatican City State and the areas referred to in Articles 15 and 16 of the Lateran Treaty and the maintenance of public order, and to encourage the proliferation of criminal conduct;

f) the use of an artificial intelligence system whose purposes conflict with the mission of the Supreme Pontiff, the integrity of the Catholic Church and the proper conduct of the institutional activities of the Governorate of Vatican City State;

g) The use of an artificial intelligence system that conflicts with the provisions of these guidelines.

Chapter II

(General principles by subject matter)

Article 5

(Principles on information and data processing)

§ 1 The use of artificial intelligence systems and models in the circulation of information, data processing, and the processing of personal data must be in compliance with the principles set forth in Decree No. DCLVII, containing the *General Regulations on the Protection of Personal Data*, April 30, 2024.

§ 2 Circulation of information, data processing, and processing of personal data through the use of artificial intelligence, shall not be detrimental to truthfulness, freedom of expression, impartiality, and completeness.

§ 3 The circulation of information, data processing, and the processing of personal data by means of artificial intelligence systems and models must not be aimed at producing discriminatory effects, harming human dignity, or damaging the image of the Vatican City State and the Catholic Church.

§ 4 Information and communications related to the use of artificial intelligence systems must be through the use of clear and simple language, such as to ensure full knowability and the subject's ability to object to incorrect processing of his or her personal data.

§ 5 The provisions of this article shall apply mutatis mutandis to the use of biometric data through artificial intelligence systems and models.

Article 6

(Principles in scientific research and health care)

§ 1 The Governorate of Vatican City State, through the Directorate of Health and Hygiene, may encourage the introduction of artificial intelligence systems and models that contribute to the improvement of personal health care and the protection of public health and hygiene.

§ 2 The use of artificial intelligence systems and models referred to in the preceding paragraph must ensure respect for human rights, fundamental freedoms and protection in the processing of personal data.

§ 3 Any person who uses health care services provided by the Directorate of Health and Hygiene shall be informed about the use of artificial intelligence systems applied.

§ 4 Artificial intelligence systems developed and applied to the field of scientific research and health care must not impair or limit the decision-making assessment of medical professionals.

§ 5 Artificial intelligence systems and models used in health care, and the related data used, should be periodically verified and updated to ensure their reliability and minimize risk.

Article 7

(Principles of copyright protection)

§ 1 The use of artificial intelligence systems and models in the reproduction, extraction, and creation of textual, musical, photographic, audiovisual, and radio and visual arts content must comply with the provisions of the *Law on the Protection of Copyright on Intellectual Works and Related Rights,* No. CXCVII, Sept. 1, 2017.

§ 2 The contents referred to in the preceding paragraph shall be identified by the acronym "IA."

§ 3 The Governorate of the State of Vatican City is the exclusive owner of the right of paternity and rights of economic use over textual, musical, photographic, audiovisual and radio content and figurative arts created, through the use of artificial intelligence, within the territory of the State of Vatican City and in the areas referred to in Articles 15 and 16 of the Lateran Treaty.

§ 4 The use of artificial intelligence systems and models in the reproduction, extraction and creation of textual, musical, photographic, audiovisual and radio content and the figurative arts must not cause harm to the honor, reputation, decorum and prestige of the Supreme Pontiff, the Catholic Church and the Vatican City State.

Article 8

(Principles on cultural property)

§ 1 The Governorate of Vatican City State, through the Directorate of Museums and Cultural Heritage, may encourage the introduction of systems and models of artificial intelligence, which contribute to the improvement of the conservation, management, enhancement and enjoyment of the artistic - museum heritage of the Vatican City State.

§ 2 The development and application of artificial intelligence systems and models to the subject matter of the cultural property of the Vatican City State, must be carried out in accordance with the institutional purposes of the Governorate and without harming the integrity and preservation of the property itself.

§ 3 The use of artificial intelligence within the scope of cultural heritage restoration activities must be carried out in accordance with the internationally recognized principles of method and the functions of coordination and technical guidance of the *Directorate of Museums and Cultural Heritage,* as stipulated in Decree No. CCCLVI, July 25, 2001, on *Regulations for the Execution of Law* No. CCCLV, July 25, 2001, on the Protection of Cultural Heritage.

§ 4 For the reproduction and economic exploitation of cultural property by means of artificial intelligence systems and models, the provisions of Article 7 above and the requirements contained in the *Law on the Protection of Cultural Property*, No. CCCLV, July 25, 2001, shall be observed.

Article 9

(Principles on infrastructure and services)

§ 1 The Governorate of the State of Vatican City, through the Directorate of Infrastructure and Services, may make use of artificial intelligence systems and models in order to incentivize economic and environmental sustainability in the execution of infrastructure interventions and service delivery.

§ 2 Artificial intelligence systems developed and applied to the scope of the competencies of the Directorate of Infrastructure and Services, referred to in Article 9, *Law on the Government of the State of Vatican City*, No. CCLXXIV, November 25, 2018, must not prejudice or restrict the decision-making evaluations of the individuals identified by the Administration as responsible for carrying out the activities.

§ 3 The processing and treatment of data and technical information through the use of artificial intelligence, as part of the design, maintenance and execution of works, must be carried out in compliance with the provisions of Article 5 above and in a manner that does not compromise the security of the Vatican City State and the areas referred to in Articles 15 and 16 of the Lateran Treaty.

Article 10

(Principles on administrative procedures)

§ 1 The Governorate of the State of Vatican City, through the Bodies in charge of the implementation of the administrative guidelines, may make use of artificial intelligence in order to facilitate the process of simplification of procedures, to reduce the time required for the definition of the procedures themselves, to raise the performance levels of administrative action, ensuring that the interested parties are aware of the regulatory interventions.

§ 2 The use of artificial intelligence in the matter of administrative procedures, must conform to compliance with the basic principles set forth in these guidelines and also with the following principles:

a) ethicality in guiding administrative choices;

b) transparency, cost-effectiveness, effectiveness, efficiency and result;

c) Segregation of functions and good performance of administrative action.

§ 3 The use of artificial intelligence in the matter of administrative procedures, must be done while respecting the autonomy and decision-making power of the person who remains solely responsible for measures and procedures in which artificial intelligence has been used.

§ 4 The use of artificial intelligence in the subject matter of this article, shall play an instrumental and supporting function in administrative activities in order to enhance the skills and aptitudes of human resources.

§ 5 The development and application of artificial intelligence systems and models in administrative procedures, should be preceded by a report containing the analysis of the impact of the regulation and followed, six months after the entry into force of the regulatory intervention, by a report containing the evaluation of the effects produced on the Vatican system.

Article 11

(Principles in labor matters)

§ 1 The Governorate of the Vatican City State may make use of artificial intelligence systems and models to augment personnel training processes, To improve safety conditions in workplaces and protect the health of workers.

§ 2 As part of personnel selection procedures, the testing and application of artificial intelligence systems and models must be carried out in accordance with the principle of transparency, preventing any violation of human dignity and potential discriminatory effects among participants in the selection process.

§ 3 In the matter referred to in this Article, the use of artificial intelligence shall not limit the decision-making power of the persons assigned by the Administration to the organization, operation, and coordination of the personnel of the Governorate of the State of Vatican City.

§ 4 The processing of labor-related data by means of artificial intelligence systems and models shall be carried out in accordance with the principles set forth in Decree No. DCLVII, containing the *General Regulations on the Protection of Personal Data*, April 30, 2024, and the provisions of these guidelines.

Article 12

(Principles on judicial activity)

§ 1 Artificial intelligence systems may be used exclusively for the organization and simplification of judicial work, as well as for jurisprudential and doctrinal research.

§ 2 It is reserved exclusively for the magistrate to decide on the interpretation of the law, the evaluation of facts and evidence, and the adoption of any measure.

Article 13

(Principles on security)

The general principles regarding the use of artificial intelligence in the context of the security and information security of the Vatican City State and the areas referred to in Articles 15 and 16 of the Lateran Treaty will be regulated, together with the relevant regulations, by special implementing regulations.

Chapter III (Final Provisions)

Article 14

(Commission on Artificial Intelligence)

§ 1 The President of the Governorate of Vatican City State appoints by his own measure the Commission on Artificial Intelligence, composed of five members and chaired by the Secretary General.

§ 2 Members of the Commission will be identified as two officials from the Legal Department, two officials from the Directorate of Telecommunications and Information Systems, and one official from the Directorate of Security and Civil Defense Services.

§ 3 The Commission on Artificial Intelligence carries out the following tasks:

a) Prepares implementing laws and regulations referred to in Article 15 below;

b) Delivers opinions on proposals for experimentation and application of artificial intelligence systems and models within the territory of the Vatican City State and the areas referred to in Articles 15 and 16 of the Lateran Treaty;

c) carries out monitoring activities on the application of artificial intelligence systems and models, reporting potential risks to the Governing Bodies;

d) Prepares a semi-annual report on the impact of the use of artificial intelligence in the Vatican City State and the areas referred to in Articles 15 and 16 of the Lateran Treaty.

§ 4 The members of the Commission serve for a three-year term; at the end of the three years, the President of the Governorate may renew the term of office of the members or replace them.

Article 15

(Implementing Laws and Regulations)

Within 12 months of the entry into force of this Decree, implementing laws and regulations implementing these guidelines will be adopted.

Domestic Laws

Interim Measures for the Management of Generative Artificial Intelligence Services
(China 2023)

Chapter I: General Provisions

Article 1: These Provisions are drafted on the basis of the Cybersecurity Law of the PRC, the PRC Data Security Law, the Personal Information Protection Law of the PRC, the PRC Law on the Scientific and Technological Progress, and other relevant laws and administrative regulations, so as promote the healthy development and regulated use of generative AI, preserve national security and the societal public interest, and protect the lawful rights and interests of citizens, legal persons, and other organizations.

Article 2: These measures apply to the use of generative AI technologies to provide services to the public in the [mainland] PRC for the generation of text, images, audio, video, or other content (hereinafter generative AI services).

Where the state has other provisions on the use of generative AI services to engage in activities such as news and publication, film and television production, and artistic creation, those provisions are to be followed.

These Measures do not apply where industry associations, enterprises, education and research institutions, public cultural bodies, and related professional bodies, etc., research, develop, and use generative AI technology, but have not provided generative AI services to the (mainland) public.

Article 3: The state is to adhere to the principle of placing equal emphasis on development and security, merging the promotion of innovation with governance in accordance with law; employing effective measures to encourage innovation and development in generative AI, and carrying out tolerant and cautious graded management by category of generative AI services.

Article 4: The provision and use of generative AI services shall comply with the requirements of laws and administrative regulations, respect social mores, ethics, and morality, and obey the following provisions:

(1) Uphold the Core Socialist Values; content that is prohibited by laws and administrative regulations such as that inciting subversion of national sovereignty or the overturn of the socialist system, endangering national security and interests or harming the nation's image, inciting separatism or undermining national unity and social stability, advocating terrorism or extremism, promoting ethnic hatred and ethnic discrimination, violence and obscenity, as well as fake and harmful information;

(2) During processes such as algorithm design, the selection of training data, model generation and optimization, and the provision of services, effective measures are to be employed to prevent the creation of discrimination such as by race, ethnicity, faith, nationality, region, sex, age, profession, or health;

(3) Respect intellectual property rights and commercial ethics, and protect commercial secrets, advantages in algorithms, data, platforms, and so forth must not be used for monopolies or to carry out unfair competition;

(4) Respect the lawful rights and interests of others, the physical and psychological well-being of others must not be endangered, and the rights and interests of others, such as in their image, reputation, honor, privacy, and personal information, must not be infringed;

(5) Based on the characteristics of the service type, employ effective measures to increase transparency in generative AI services and to increase the accuracy and reliability of generated content.

Chapter II: Development and Governance of Technology

Article 5: Encourage the innovative application of generative AI technology in each industry and field, generate exceptional content that is positive, healthy, and uplifting, and explore the optimization of usage scenarios in building an application ecosystem.

Support industry associations, enterprises, education and research institutions, public cultural bodies, and relevant professional bodies, etc. to coordinate in areas such as innovation in generative AI technology, the establishment of data resources, applications, and risk prevention.

Article 6: Encourage independent innovation in basic technologies for generative AI such as algorithms, frameworks, chips, and supporting software platforms, carry out international exchanges and cooperation in an equal and mutually beneficial way, and participate in the formulation of international rules related to generative AI.

Promote the establishment of generative AI infrastructure and public training data resource platforms. Promote collaboration and sharing of algorithm resources, increasing efficiency in the use of computing resources. Promote the orderly opening of public data by type and grade, expanding high-quality public training data resources. Encourage the adoption of safe and reliable chips, software, tools, computational power, and data resources.

Article 7: The providers of generative AI services (hereinafter "providers") shall carry out pre-training, optimization training, and other activities handling training data in accordance with law, and comply with the following provisions:

(1) Use data and foundational models that have lawful sources;

(2) Where intellectual property rights are involved, the intellectual property rights that are lawfully enjoyed by others must not be infringed;

(3) Where personal information is involved, the consent of the personal information subject shall be obtained or it shall comply with other situations provided by laws and administrative regulations;

(4) Employ effective measures to increase the quality of training data, and increase the truth, accuracy, objectivity, and diversity of training data;

(5) Other provisions in laws and administrative regulations such as the PRC Cybersecurity Law, The PRC Data Security Law, and the PRC Personal Information Protection Law, and the regulatory requirements of relevant departments in charge.

Article 8: When manual tagging is conducted in the course of researching and developing generative AI technology, the providers shall formulate clear, specific, and feasible tagging rules that meet the requirements of these Measures; carry out assessments of the quality of data tagging, with spot checks to verify the accuracy of tagging content; and conduct necessary training for tagging personnel to increase their awareness of legal compliance and oversee and guide them to carry out tagging efforts in a standardized way.

Chapter III: Service Specifications

Article 9: Providers shall bear responsibility as the producers of online information content in accordance with law and are to fulfill the online information security obligations. Where personal information is involved, they are to bear responsibility as personal information handlers and fulfill obligations to protect personal information.

Providers shall sign service agreements with users who register for their generative AI services (hereinafter "users"), clarifying the rights and obligations of both parties.

Article 10: Providers shall clarify and disclose the user groups, occasions, and uses of their services, guide users' scientific understanding and lawful use of generative AI technology, and employ effective measures to prevent minor users from overreliance or addiction to generative AI services.

Article 11: Providers shall fulfill confidentiality obligations towards information input by users and users' usage records in accordance with law; they must not collect unnecessary personal information, must not illegally retain user input information and usage records from which users' identities can be determined, and must not illegally provide user input information and usage records to others.

Providers shall lawfully and promptly accept and address requests from individuals such as to access, reproduce, modify, supplement, or delete their personal information.

Article 12: Providers shall label generated content such as images and video in accordance with the Provisions on the Administration of Deep Synthesis Internet Information Services.

Article 13: Providers shall provide safe, stable, and sustained services throughout the course of services to ensure users' normal usage.

Article 14: Where providers discover illegal content they shall promptly employ measures to address it such as stopping generation, stopping transmission, and removal, employ measures such as model optimization training to make corrections and report to the relevant departments in charge.

Where providers discover that users are using generative AI services to engage in illegal activities, they shall employ measures in accordance with laws and agreements to address it, including warnings, limiting functions, and suspending or concluding the provision of services, and store the relevant records and report to the relevant departments in charge.

Article 15: Providers shall establish and complete mechanisms for making complaints and reports, setting up easy complaint and reporting portals, disclosing the process for handling them and the time limits for giving responses, and promptly accepting and handling complaints and reports from the public and giving feedback on the outcome.

Chapter IV: Oversight Inspections and Legal Responsibility

Article 16: Based on their respective duties, departments such as for internet information, reform and development, education, science and technology, industry and informatization, public security, radio and television, and press and publication are to strengthen the management of generative AI services in accordance with law.

In light of the characteristics of generative AI technology and its service applications in relevant industries and fields, the state's competent departments are to improve scientific regulatory methods that are compatible with innovation and development, and formulate rules or guidelines for corresponding regulation by type and grade.

Article 17: Those providing generative AI services with public opinion properties or the capacity for social mobilization shall carry out security assessments in accordance with relevant state provisions and perform formalities for the filing, modification, or canceling of filings on algorithms in accordance with the "Provisions on the Management of Algorithmic Recommendations in Internet Information Services".

Article 18: Where users discover that generative AI services do not comply with laws, administrative regulations, or these Measures, they have the right to make a complaint or report to the relevant departments in charge.

Article 19: Providers shall cooperate with the relevant departments in charge that are carrying out oversight inspections of generative AI services on the basis of their duties, explaining the sources, models, types, tagging rules, algorithm mechanisms, etc. for training data as required, and providing necessary technical, data, and other supports and assistance.

Relevant bodies and personnel participating in security assessments and oversight inspections of generative AI services shall strictly keep the confidentiality of state secrets, commercial secrets, personal privacy, and personal information that they learn of in performing their duties, and must not leak or unlawfully provide it to others.

Article 20: Where generative AI services provided from outside the [mainland] PRC do not meet the requirements of laws, administrative regulations, or these Measures, the state internet information department shall notify the relevant organs to employ technical measures and other necessary measures to address it.

Article 21: Where providers violate these Measures, penalties are to be given by the relevant regulatory departments in accordance with the provisions of the PRC Cybersecurity Law, The PRC Data Security Law, the PRC Law on the Protection of

Personal Information, The PRC Law on Scientific and Technological Progress, and other such laws and administrative regulations; and where laws and administrative regulations are silent, the relevant departments in charge are to give warnings, circulate criticism, or order corrections in a set period of time on the basis of their duties, and if corrections are refused or the circumstances are serious, an order is to be given to suspend the provision of the related services.

Where violations of public security are constituted, a public security administrative sanction is lawfully given; where a crime is constituted, criminal responsibility is to be pursued in accordance with law.

Chapter V: Supplementary Provisions

Article 22: The meanings of the following terms used in these Measures are:

(1) "Generative AI technology" refers to models and relevant technologies that have the ability to generate content such as texts, images, audio, or video.

(2) "Generative AI service providers" refers to organizations and individuals that use generative AI technology to provide generative AI services (including providing generative AI services through programmable interfaces and other means).

(3) "Generative AI service users" refers to organizations and individuals that use generative AI services to generate content.

Article 23: Where laws and administrative regulations provide that administrative permits shall be acquired for the provision of generative AI services, the provides shall obtain permits in accordance with law.

Foreign investment in generative AI services shall comply with laws and administrative regulations related to foreign investment.

Article 24: These measures take effect on August 15, 2023.

Cybersecurity Administration

People's Republic of China National Development and Reform Commission

Ministry of Education of the People's Republic of China

Ministry of Science and Technology of the People's Republic of China

Ministry of Industry and Information Technology of the People's Republic of China

Ministry of Public Security of the People's Republic of China

State Administration of Radio and Television

Directive on Automated Decision-Making
(Canada 2023)

1. Effective date

1.1 This directive takes effect on April 1, 2019, with compliance required by no later than April 1, 2020.

1.2 This directive applies to all automated decision systems developed or procured after April 1, 2020. However,

> 1.2.1 existing systems developed or procured prior to April 25, 2023 will have until April 25, 2024 to fully transition to the requirements in subsections 6.2.3, 6.3.1, 6.3.4, 6.3.5 and 6.3.6 in this directive;

> 1.2.2 new systems developed or procured after April 25, 2023 will have until October 25, 2023 to meet the requirements in this directive.

1.3 This directive will be reviewed every two years, and as determined by the Chief Information Officer of Canada.

2. Authorities

2.1 This directive is issued pursuant to the same authority indicated in section 2 of the *Policy on Service and Digital*.

3. Definitions

3.1 Definitions to be used in the interpretation of this directive are listed in Appendix A.

4. Objectives and expected results

4.1 The objective of this directive is to ensure that automated decision systems are deployed in a manner that reduces risks to clients, federal institutions and Canadian society, and leads to more efficient, accurate, consistent and interpretable decisions made pursuant to Canadian law.

4.2 The expected results of this directive are as follows:

> 4.2.1 Decisions made by federal institutions are data-driven, responsible and comply with procedural fairness and due process requirements.

4.2.2 Impacts of algorithms on administrative decisions are assessed and negative outcomes are reduced, when encountered.

4.2.3 Data and information on the use of automated decision systems in federal institutions are made available to the public, where appropriate.

5. Scope

5.1 This directive applies to any system, tool, or statistical model used to make an administrative decision or a related assessment about a client.

5.2 This directive applies only to automated decision systems in production and excludes systems operating in test environments.

6. Requirements

The Assistant Deputy Minister responsible for the program using the automated decision system, or any other person named by the Deputy Head, is responsible for:

6.1 Algorithmic Impact Assessment

6.1.1 Completing and releasing the final results of an Algorithmic Impact Assessment prior to the production of any automated decision system.

6.1.2 Applying the relevant requirements prescribed in Appendix C as determined by the Algorithmic Impact Assessment.

6.1.3 Reviewing and updating the Algorithmic Impact Assessment on a scheduled basis, including when the functionality or scope of the automated decision system changes.

6.1.4 Releasing the final results of the Algorithmic Impact Assessment in an accessible format via Government of Canada websites and any other services designated by the Treasury Board of Canada Secretariat pursuant to the *Directive on Open Government*.

6.2 Transparency

Providing notice before decisions

6.2.1 Providing notice through all service delivery channels in use that the decision rendered will be undertaken in whole or in part by an automated decision system, as prescribed in Appendix C.

6.2.2 Providing notices prominently and in plain language, pursuant to the Canada.ca Content Style Guide.

Providing explanations after decisions

6.2.3 Providing a meaningful explanation to affected individuals of how and why the decision was made, as prescribed in Appendix C.

Access to components

6.2.4 Determining the appropriate licence for software components, including consideration of open source software in accordance with the measures specified in the *Government of Canada Enterprise Architecture Framework*.

6.2.5 If using a proprietary licence, ensuring that:

> 6.2.5.1 All released versions of proprietary software components used for automated decision systems are delivered to, and safeguarded by, the department.
>
> 6.2.5.2 The Government of Canada retains the right to access and test automated decision systems, including all released versions of proprietary software components, in case it is necessary for a specific audit, investigation, inspection, examination, enforcement action, or judicial proceeding, subject to safeguards against unauthorized disclosure.
>
> 6.2.5.3 As part of this access, the Government of Canada retains the right to authorize external parties to review and audit these components as necessary.

Release of source code

6.2.6 Releasing custom source code owned by the Government of Canada in accordance with the measures specified in the *Government of Canada Enterprise Architecture Framework*, unless:

> 6.2.6.1 the source code is processing data classified as Secret, Top Secret or Protected C; or

6.2.6.2 disclosure would otherwise be exempted or excluded under the *Access to Information Act*, if the *Access to Information Act* were to apply.

6.2.7 Determining the appropriate access restrictions to the released source code.

Documenting decisions

6.2.8 Documenting the decisions of automated decision systems in accordance with the *Directive on Service and Digital*, and in support of the monitoring (6.3.2), data governance (6.3.4) and reporting requirements (6.5.1).

6.3 Quality assurance

Testing and monitoring outcomes

6.3.1 Before launching into production, developing processes so that the data and information used by the automated decision system, as well as the system's underlying model, are tested for unintended biases and other factors that may unfairly impact the outcomes.

6.3.2 Developing processes to monitor the outcomes of the automated decision system to safeguard against unintentional outcomes and to verify compliance with institutional and program legislation, as well as this directive, on a scheduled basis.

Data quality

6.3.3 Validating that the data collected for, and used by, the automated decision system is relevant, accurate, up-to-date, and in accordance with the *Policy on Service and Digital* and the *Privacy Act*.

Data governance

6.3.4 Establishing measures to ensure that data used and generated by the automated decision system are traceable, protected and accessed appropriately, and lawfully collected, used, retained and disposed of in accordance with the *Directive on Service and Digital*, *Directive on Privacy Practices*, and *Directive on Security Management*.

Peer review

6.3.5 Consulting the appropriate qualified experts to review the automated decision system and publishing the complete review or a plain language summary of the findings prior to the system's production, as prescribed in Appendix C.

Gender-based Analysis Plus

6.3.6 Completing a Gender-based Analysis Plus during the development or modification of the automated decision system, as prescribed in Appendix C.

Employee training

6.3.7 Providing adequate employee training in the design, function, and implementation of the automated decision system to be able to review, explain and oversee its operations, as prescribed in Appendix C.

IT and business continuity management

6.3.8 Establishing strategies, plans and/or measures to support IT and business continuity management, as prescribed in Appendix C and in accordance with the *Directive on Security Management.*

Security

6.3.9 Conducting risk assessments during the development of the automated decision system and establishing appropriate safeguards, in accordance with the *Policy on Government Security.*

Legal

6.3.10 Consulting with the institution's legal services from the concept stage of an automation project to ensure that the use of the automated decision system is compliant with applicable legal requirements.

Ensuring human intervention

6.3.11 Ensuring that the automated decision system allows for human intervention, when appropriate, as prescribed in Appendix C.

6.3.12 Obtaining the appropriate level of approvals prior to the production of an automated decision system, as prescribed in Appendix C.

6.4 Recourse

6.4.1 Providing clients with any applicable recourse options that are available to them to challenge the administrative decision.

6.5 Reporting

6.5.1 Publishing information on the effectiveness and efficiency of the automated decision system in meeting program objectives on a website or service designated by the Treasury Board of Canada Secretariat.

7. Consequences

7.1 Consequences of non-compliance with this directive can include any

7.2 For an outline of the consequences of non-compliance, refer to the *Framework for the Management of Compliance*, Appendix C: Consequences for Institutions and Appendix D: Consequences for Individuals.

8. Roles and responsibilities of Treasury Board of Canada Secretariat

Subject to the necessary delegations, the Chief Information Officer of Canada is responsible for:

8.1 Providing government-wide guidance on the use of automated decision systems.

8.2 Developing and maintaining the Algorithmic Impact Assessment and any supporting documentation.

8.3 Communicating and engaging government-wide and with partners in other jurisdictions and sectors to develop common strategies, approaches, and processes to support the responsible use of automated decision systems.

9. Application

9.1 This directive applies to all institutions subject to the *Policy on Service and Digital*, unless excluded by specific acts, regulations or Orders-in-Council;

9.1.1 Agents of Parliament are excluded from this directive, including the:

- Office of the Auditor General of Canada,

- Office of the Chief Electoral Officer,

- Office of the Commissioner of Lobbying of Canada,

- Office of the Commissioner of Official Languages,

- Office of the Information Commissioner of Canada,

- Office of the Privacy Commissioner of Canada, and

- Office of the Public Sector Integrity Commissioner of Canada.

9.2 Agencies, Crown Corporations, or Agents of Parliament may enter into Specific Agreements with the Treasury Board of Canada Secretariat to adopt the requirements of this directive and apply them to their organization, as required.

10. References

10.1 Legislation

- ○ *Financial Administration Act*

- ○ *Access to Information Act*

- ○ *Canadian Human Rights Act*

- ○ *Privacy Act*

- ○ *Security of Information Act*

- ○ *Accessible Canada Act*

10.2 Related policy instruments

- ○ *Policy on Access to Information*

- ○ *Policy on Service and Digital*

- ○ *Policy on Government Security*

- ○ *Policy on Privacy Protection*

- ○ *Policy on People Management*

- ○ *Directive on Open Government*

    o   *Standard on Security Screening*

11. Enquiries

11.1 For interpretation of any aspect of this directive, contact Treasury Board of Canada Secretariat Public Enquiries.

11.2 Individuals from federal institutions may contact ai-ia@tbs-sct.gc.ca for any questions regarding this directive, including the Algorithmic Impact Assessment.

An Act Concerning Consumer Protections in Interactions with Artificial Intelligence Systems
(Colorado, U.S. 2024)

Part 17 Artificial Intelligence

6-1-1701. Definitions. As used in this part 17, unless the context otherwise requires:

(1) (a) "algorithmic discrimination" means any condition in which the use of an artificial intelligence system results in an unlawful differential treatment or impact that disfavors an individual or group of individuals on the basis of their actual or perceived age, color, disability, ethnicity, genetic information, limited proficiency in the English language, national origin, race, religion, reproductive health, sex, veteran status, or other classification protected under the laws of this state or federal law.

(b) "algorithmic discrimination" does not include:

(i) the offer, license, or use of a high-risk artificial intelligence system by a developer or deployer for the sole purpose of:

(a) the developer's or deployer's self-testing to identify, mitigate, or prevent discrimination or otherwise ensure compliance with state and federal law; or

(b) expanding an applicant, customer, or participant pool to increase diversity or redress historical discrimination; or

(ii) an act or omission by or on behalf of a private club or other establishment that is not in fact open to the public, as set forth in title ii of the federal "civil rights act of 1964", 42 u.s.c. Sec. 2000a (e), as amended.

(2) "artificial intelligence system" means any machine-based system that, for any explicit or implicit objective, infers from the inputs the system receives how to generate outputs, including content, decisions, predictions, or recommendations, that can influence physical or virtual environments.

(3) "consequential decision" means a decision that has a material legal or similarly significant effect on the provision or denial to any consumer of, or the cost or terms of:

(a) education enrollment or an education opportunity;

(b) employment or an employment opportunity;

(c) a financial or lending service;

(d) an essential government service; (e) i iealth-care services;

(f) housing;

(g) insurance; or

 (h) a legal service.

(4) "consumer" means an individual who is a Colorado resident.

(5) "deploy" means to use a high-risk artificial intelligence system.

(6) "deployer" means a person doing business in this state that deploys a high-risk artificial intelligence system.

(7) "developer" means a person doing business in this state that develops or intentionally and substantially modifies an artificial intelligence system.

(8) "i iealth-care services" has the same meaning as provided in 42 u.s.c. Sec. 234 (d)(2).

(9) (a) "high-risk artificial intelligence system" means any artificial intelligence system that, when deployed, makes, or is a substantial factor in making, a consequential decision.

(b) include:

 (i) "high-risk artificial intelligence system" does not an artificial intelligence system if the artificial intelligence system is intended to:

(a) perform a narrow procedural task; or

(b) detect decision-making patterns or deviations from prior decision-making patterns and is not intended to replace or influence a previously completed human assessment without sufficient human review; or

(ii) the following technologies, unless the technologies, when deployed, make, or are a substantial factor in making, a consequential decision:

(a) anti-fraud technology that does not use facial recognition technology;

(b) anti-malware;

(c) anti-virus;

(d) artificial intelligence-enabled video games;

(e) calculators;

(f) cybersecurity;

(g) databases;

(h) data storage;

(i) firewall;

(j) internet domain registration;

(k) internet website loading;

(l) networking;

(m) spam- and robocall-filtering;

(n) spell-checking;

(o) spreadsheets;

(p) web caching;

(q) web hosting or any similar technology; or (r) technology that communicates with consumers in natural language for the purpose of providing users with information, making referrals or recommendations, and answering questions and is subject to an accepted use policy that prohibits generating content that is discriminatory or harmful.

(10) (a) "intentional and substantial modification" or "intentionally and substantially modifies" means a deliberate change made to an artificial intelligence system that results in any new reasonably foreseeable risk of algorithmic discrimination.

(b) "intentional and substantial modification" or "intentionally and substantially modifies" does not include a change made to a high-risk artificial intelligence system, or the performance of a high-risk artificial intelligence system, if:

(i) the high-risk artificial intelligence system continues to learn after the high-risk artificial intelligence system is:

(a) offered, sold, leased, licensed, given, or otherwise made available to a deployer; or

(b) deployed; (ii) the change is made to the high-risk artificial intelligence system as a result of any learning described in subsection (10)(b)(i) of this section;

(iii) the change was predetermined by the deployer, or a third party contracted by the deployer, when the deployer or third party completed an initial impact assessment of such high-risk artificial intelligence system pursuant to section 6-1-1703 (3); and

(iv) the change is included in technical documentation for the high-risk artificial intelligence system.

(11) (a) "substantial factor" means a factor that:

(i) assists in making a consequential decision;

(ii) is capable of altering the outcome of a consequential decision; and

(iii) is generated by an artificial intelligence system.

(b) "substantial factor" includes any use of an artificial intelligence system to generate any content, decision, prediction, or recommendation concerning a consumer that is used as a basis to make a consequential decision concerning the consumer.

(12) "trade secret" has the meaning set forth in section 7-74-102 (4).

6-1-1702. Developer duty to avoid algorithmic discrimination required documentation.

(1) on and after February 1, 2026, a developer of a high-risk artificial intelligence system shall use reasonable care to protect consumers from any known or reasonably foreseeable risks of algorithmic discrimination arising from the intended and contracted uses of the high-risk artificial intelligence system. In any

enforcement action brought on or after February 1, 2026, by the attorney general pursuant to section 6-1-1706, there is a rebuttable presumption that a developer used reasonable care as required under this section if the developer complied with this section and any additional requirements or obligations as set forth in rules promulgated by the attorney general pursuant to section 6-1-1707.

(2) on and after February 1, 2026, and except as provided in subsection (6) of this section, a developer of a high-risk artificial intelligence system shall make available to the deployer or other developer of the high-risk artificial intelligence system:

(a) a general statement describing the reasonably foreseeable uses and known harmful or inappropriate uses of the high-risk artificial intelligence system;

(b) documentation disclosing:

(i) high-level summaries of the type of data used to train the high-risk artificial intelligence system;

(ii) known or reasonably foreseeable limitations of the high-risk artificial intelligence system, including known or reasonably foreseeable risks of algorithmic discrimination arising from the intended uses of the high-risk artificial intelligence system;

(iii) the purpose of the high-risk artificial intelligence system;

(iv) the intended benefits and uses of the high-risk artificial intelligence system; and (v) all other information necessary to allow the deployer to comply with the requirements of section 6-1-1703;

(c) documentation describing:

(i) how the high-risk artificial intelligence system was evaluated for performance and mitigation of algorithmic discrimination before the high-risk artificial intelligence system was offered, sold, leased, licensed, given, or otherwise made available to the deployer;

(ii) the data governance measures used to cover the training datasets and the measures used to examine the suitability of data sources, possible biases, and appropriate mitigation;

(iii) the intended outputs of the high-risk artificial intelligence system;

(iv) the measures the developer has taken to mitigate known or reasonably foreseeable risks of algorithmic discrimination that may arise from the reasonably foreseeable deployment of the high-risk artificial intelligence system; and

(v) how the high-risk artificial intelligence system should be used, not be used, and be monitored by an individual when the high-risk artificial intelligence system is used to make, or is a substantial factor in making, a consequential decision; and

(d) any additional documentation that is reasonably necessary to assist the deployer in understanding the outputs and monitor the performance of the high-risk artificial intelligence system for risks of algorithmic discrimination.

(3) (a) except as provided in subsection (6) of this section, a developer that offers, sells, leases, licenses, gives, or otherwise makes available to a deployer or other developer a high-risk artificial intelligence system on or after February 1, 2026, shall make available to the deployer or other developer, to the extent feasible, the documentation and information, through artifacts such as model cards, dataset cards, or other impact assessments, necessary for a deployer, or for a third party contracted by a deployer, to complete an impact assessment pursuant to section 6-1-1703 (3).

(b) a developer that also serves as a deployer for a high-risk artificial intelligence system is not required to generate the documentation required by this section unless the high-risk artificial intelligence system is provided to an unaffiliated entity acting as a deployer.

(4) (a) on and after February 1, 2026, a developer shall make available, in a manner that is clear and readily available on the developer's website or in a public use case inventory, a statement summarizing:

(i) the types of high-risk artificial intelligence systems that the developer has developed or intentionally and substantially modified and currently makes available to a deployer or other developer; and

(ii) how the developer manages known or reasonably foreseeable risks of algorithmic discrimination that may arise from the development or intentional and substantial modification of the types of high-risk artificial intelligence systems described in accordance with subsection (4)(a)(i) of this section.

(b) a developer shall update the statement described in subsection (4)(a) of this section:

(i) as necessary to ensure that the statement remains accurate; and

(ii) no later than ninety days after the developer intentionally and substantially modifies any high-risk artificial intelligence system described in subsection (4)(a)(i) of this section.

(5) on and after February 1,2026, a developer of a high-risk artificial intelligence system shall disclose to the attorney general, in a form and manner prescribed by the attorney general, and to all known deployers or other developers of the high-risk artificial intelligence system, any known or reasonably foreseeable risks of algorithmic discrimination arising from the intended uses of the high-risk artificial intelligence system without unreasonable delay but no later than ninety days after the date on which:

(a) the developer discovers through the developer's ongoing testing and analysis that the developer's high-risk artificial intelligence system has been deployed and has caused or is reasonably likely to have caused algorithmic discrimination; or

(b) the developer receives from a deployer a credible report that the high-risk artificial intelligence system has been deployed and has caused algorithmic discrimination.

(6) nothing in subsections (2) to (5) of this section requires a developer to disclose a trade secret, information protected from disclosure by state or federal law, or information that would create a security risk to the developer.

(7) on and after February 1,2026, the attorney general may require that a developer disclose to the attorney general, no later than ninety days after the request and in a form and manner prescribed by the attorney general, the statement or documentation described in subsection (2) of this section. The attorney general may evaluate such statement or documentation to ensure compliance with this part 17, and the statement or documentation is not subject to disclosure under the "Colorado open records act", part 2 of article 72 of title 24. In a disclosure pursuant to this subsection (7), a developer may designate the statement or documentation as including proprietary information or a trade secret. To the extent that any information contained in the statement or documentation includes information subject to attorney-client privilege or work-product protection, the disclosure does not constitute a waiver of the privilege or protection.

6-1-1703. Deployer duty to avoid algorithmic discrimination risk management policy and program. (1) on and after February 1, 2026, a deployer of a high-risk artificial intelligence system shall use reasonable care to protect consumers from any known or reasonably foreseeable risks of algorithmic discrimination. In any enforcement action brought on or after February 1, 2026, by the attorney general pursuant to section 6-1-1706, there is a rebuttable presumption that a deployer of a high-risk artificial intelligence system used reasonable care as required under this section if the deployer complied with this section and any additional requirements or obligations as set forth in rules promulgated by the attorney general pursuant to section 6-1-1707.

(2) (a) on and after February 1, 2026, and except as provided in subsection (6) of this section, a deployer of a high-risk artificial intelligence system shall implement a risk management policy and program to govern the deployer's deployment of the high-risk artificial intelligence system. The risk management policy and program must specify and incorporate the principles, processes, and personnel that the deployer uses to identify, document, and mitigate known or reasonably foreseeable risks of algorithmic discrimination. The risk management policy and program must be an iterative process planned, implemented, and regularly and systematically reviewed and updated over the life cycle of a high-risk artificial intelligence system, requiring regular, systematic review and updates. A risk management policy and program implemented and maintained pursuant to this subsection (2) must be reasonable considering:

(i) (a) the guidance and standards set forth in the latest version of the "artificial intelligence risk management framework" published by the national institute of standards and technology in the united states department of commerce, standard iso/iec 42001 of the international organization for standardization, or another nationally or internationally recognized risk management framework for artificial intelligence systems, if the standards are substantially equivalent to or more stringent than the requirements of this part 17; or

(b) any risk management framework for artificial intelligence systems that the attorney general, in the attorney general's discretion, may designate;

(ii) the size and complexity of the deployer;

(iii) the nature and scope of the high-risk artificial intelligence systems deployed by the deployer, including the intended uses of the high-risk artificial intelligence systems; and (iv) the sensitivity and volume of data processed in connection with the high-risk artificial intelligence systems deployed by the deployer.

(b) a risk management policy and program implemented pursuant to subsection (2)(a) of this section may cover multiple high-risk artificial intelligence systems deployed by the deployer.

___

(3) (a) except as provided in subsections (3)(d), (3)(e), and (6) of this section:

(i) a deployer, or a third party contracted by the deployer, that deploys a high-risk artificial intelligence system on or after February 1, 2026, shall complete an impact assessment for the high-risk artificial intelligence system; and

(ii) on and after February 1, 2026, a deployer, or a third party contracted by the deployer, shall complete an impact assessment for a deployed high-risk artificial intelligence system at least annually and within ninety days after any intentional and substantial modification to the high-risk artificial intelligence system is made available.

(b) an impact assessment completed pursuant to this subsection (3) must include, at a minimum, and to the extent reasonably known by or available to the deployer:

(i) a statement by the deployer disclosing the purpose, intended use cases, and deployment context of, and benefits afforded by, the high-risk artificial intelligence system;

(ii) an analysis of whether the deployment of the high-risk artificial intelligence system poses any known or reasonably foreseeable risks of algorithmic discrimination and, if so, the nature of the algorithmic discrimination and the steps that have been taken to mitigate the risks;

(iii) a description of the categories of data the high-risk artificial intelligence system processes as inputs and the outputs the high-risk artificial intelligence system produces;

(iv) if the deployer used data to customize the high-risk artificial intelligence system, an overview of the categories of data the deploy er used to customize the high-risk artificial intelligence system;

(v) any metrics used to evaluate the performance and known limitations of the high-risk artificial intelligence system;

(vi) a description of any transparency measures taken concerning the high-risk artificial intelligence system, including any measures taken to disclose to a consumer that the high-risk artificial intelligence system is in use when the high-risk artificial intelligence system is in use; and

(vii) a description of the post-deployment monitoring and user safeguards provided concerning the high-risk artificial intelligence system, including the oversight, use, and learning process established by the deployer to address issues arising from the deployment of the high-risk artificial intelligence system.

(c) in addition to the information required under subsection (3)(b) of this section, an impact assessment completed pursuant to this subsection (3) following an intentional and substantial modification to a high-risk artificial intelligence system on or after February 1, 2026, must include a statement disclosing the extent to which the high-risk artificial intelligence system was used in a manner that was consistent with, or varied from, the developer's intended uses of the high-risk artificial intelligence system.

(d) a single impact assessment may address a comparable set of high-risk artificial intelligence systems deployed by a deployer.

(e) if a deployer, or a third party contracted by the deployer, completes an impact assessment for the purpose of complying with another applicable law or regulation, the impact assessment satisfies the requirements established in this subsection (3) if the impact assessment is reasonably similar in scope and effect to the impact assessment that would otherwise be completed pursuant to this subsection (3).

(f) a deployer shall maintain the most recently completed impact assessment for a high-risk artificial intelligence system as required under this subsection (3), all records concerning each impact assessment, and all prior impact assessments, if any, for at least three years following the final deployment of the high-risk artificial intelligence system.

(g) on or before February 1, 2026, and at least annually thereafter, a deployer, or a third party contracted by the deployer, must review the deployment of each high-risk artificial intelligence system deployed by the deployer to ensure that the high-risk artificial intelligence system is not causing algorithmic discrimination.

(4) (a) on and after February 1, 2026, and no later than the time that a deployer deploys a high-risk artificial intelligence system to make, or be a substantial factor in making, a consequential decision concerning a consumer, the deployer shall:

(i) notify the consumer that the deployer has deployed a high-risk artificial intelligence system to make, or be a substantial factor in making, a consequential decision before the decision is made;

(ii) provide to the consumer a statement disclosing the purpose of the high-risk artificial intelligence system and the nature of the consequential decision; the contact information for the deployer; a description, in plain language, of the high-risk artificial intelligence system; and instructions on how to access the statement required by subsection (5)(a) of this section; and

(iii) provide to the consumer information, if applicable, regarding the consumer's right to opt out of the processing of personal data concerning the consumer for purposes of profiling in furtherance of decisions that produce legal or similarly significant effects concerning the consumer under section 6-1-1306 (1)(a)(i)(c).

(b) on and after February 1, 2026, a deployer that has deployed a high-risk artificial intelligence system to make, or be a substantial factor in making, a consequential decision concerning a consumer shall, if the consequential decision is adverse to the consumer, provide to the consumer:

(i) a statement disclosing the principal reason or reasons for the consequential decision, including:

(a) the degree to which, and manner in which, the high-risk artificial intelligence system contributed to the consequential decision;

(b) the type of data that was processed by the high-risk artificial intelligence system in making the consequential decision; and

(c) the source or sources of the data described in subsection (4)(b)(i)(b) of this section;

(ii) an opportunity to correct any incorrect personal data that the high-risk artificial intelligence system processed in making, or as a substantial factor in making, the consequential decision; and

(iii) an opportunity to appeal an adverse consequential decision concerning the consumer arising from the deployment of a high-risk artificial intelligence system, which appeal must, if technically feasible, allow for human review unless providing the opportunity for appeal is not in the best interest of the consumer, including in instances in which any delay might pose a risk to the life or safety of such consumer.

(c) (i) except as provided in subsection (4)(c)(ii) of this section, a deployer shall provide the notice, statement, contact information, and description required by subsections (4)(a) and (4)(b) of this section:

(a) directly to the consumer;

(b) in plain language;

(c) in all languages in which the deployer, in the ordinary course of the deployer's business, provides contracts, disclaimers, sale announcements, and other information to consumers; and

(d) in a format that is accessible to consumers with disabilities.

(ii) if the deployer is unable to provide the notice, statement, contact information, and description required by subsections (4)(a) and (4)(b) of this section directly to the consumer, the deployer shall make the notice, statement, contact information, and description available in a manner that is reasonably calculated to ensure that the consumer receives the notice, statement, contact information, and description.

(5) (a) on and after February 1, 2026, and except as provided in subsection (6) of this section, a deployer shall make available, in a manner that is clear and readily available on the deployer's website, a statement summarizing:

(i) the types of high-risk artificial intelligence systems that are currently deployed by the deployer;

(ii) how the deployer manages known or reasonably foreseeable risks of algorithmic discrimination that may arise from the deployment of each high-risk artificial intelligence system described pursuant to subsection (5)(a)(i) of this section; and

(iii) in detail, the nature, source, and extent of the information collected and used by the deployer.

(b) a deployer shall periodically update the statement described in subsection (5)(a) of this section.

(6) subsections (2), (3), and (5) of this section do not apply to a deployer if, at the time the deployer deploys a high-risk artificial intelligence system and at all times while the high-risk artificial intelligence system is deployed:

(a) the deployer:

(i) employs fewer than fifty full-time equivalent employees; and

(ii) does not use the deployer's own data to train the high-risk artificial intelligence system;

(b) the high-risk artificial intelligence system:

(i) is used for the intended uses that are disclosed to the deployer as required by section 6-1-1702 (2)(a); and

(ii) continues learning based on data derived from sources other than the deployer's own data; and

(c) the deployer makes available to consumers any impact assessment that:

(i) the developer of the high-risk artificial intelligence system has completed and provided to the deployer; and

(ii) includes information that is substantially similar to the information in the impact assessment required under subsection (3)(b) of this section.

(7) if a deployer deploys a high-risk artificial intelligence system on or after February 1, 2026, and subsequently discovers that the high-risk artificial intelligence system has caused algorithmic discrimination, the deployer, without unreasonable delay, but no later than ninety days after the date of the discovery, shall send to the attorney general, in a form and manner prescribed by the attorney general, a notice disclosing the discovery.

(8) nothing in subsections (2) to (5) and (7) of this section requires a deployer to disclose a trade secret or information protected from disclosure by state or federal law. To the extent that a deployer withholds information pursuant to this subsection (8) or section 6-1-1705 (5), the deployer shall notify the consumer and provide a basis for the withholding.

(9) on and after February 1, 2026, the attorney general may require that a deployer, or a third party contracted by the deployer, disclose to the attorney general, no later than ninety days after the request and in a form and manner prescribed by the attorney general, the risk management policy implemented pursuant to subsection (2) of this section, the impact assessment completed pursuant to subsection (3) of this section, or the records maintained pursuant to subsection (3)(f) of this section. The attorney general may evaluate the risk management policy, impact assessment, or records to ensure compliance with this part 17, and the risk management policy, impact assessment, and records are not subject to disclosure under the "Colorado open records act", part 2 of article 72 of title 24. In a disclosure pursuant to this subsection (9), a deployer may designate the statement or

documentation as including proprietary information or a trade secret. To the extent that any information contained in the risk management policy, impact assessment, or records include information subject to attorney-client privilege or work-product protection, the disclosure does not constitute a waiver of the privilege or protection.

6-1-1704. Disclosure of an artificial intelligence system to consumer. (1) on and after February 1, 2026, and except as provided in subsection (2) of this section, a deployer or other developer that deploys, offers, sells, leases, licenses, gives, or otherwise makes available an artificial intelligence system that is intended to interact with consumers shall ensure the disclosure to each consumer who interacts with the artificial intelligence system that the consumer is interacting with an artificial intelligence system.

(2) disclosure is not required under subsection (1) of this section under circumstances in which it would be obvious to a reasonable person that the person is interacting with an artificial intelligence system.

6-1-1705. Compliance with other legal obligations - definitions. (1) nothing in this part 17 restricts a developer's, a deployer's, or other person's ability to:

(a) comply with federal, state, or municipal laws, ordinances, or regulations;

(b) comply with a civil, criminal, or regulatory inquiry, investigation, subpoena, or summons by a federal, a state, a municipal, or other governmental authority;

(c) cooperate with a law enforcement agency concerning conduct or activity that the developer, deployer, or other person reasonably and in good faith believes may violate federal, state, or municipal laws, ordinances, or regulations;

(d) investigate, establish, exercise, prepare for, or defend legal claims;

(e) take immediate steps to protect an interest that is essential for the life or physical safety of a consumer or another individual;

(f) by any means other than the use of facial recognition technology, prevent, detect, protect against, or respond to security incidents, identity theft, fraud, harassment, malicious or deceptive activities, or illegal activity; investigate, report, or prosecute the persons responsible for any such action; or preserve the integrity or security of systems;

(g) engage in public or peer-reviewed scientific or statistical research in the public interest that adheres to all other applicable ethics and privacy laws and is conducted in accordance with 45 cfr46, as amended, or relevant requirements established by the federal food and drug administration;

(h) conduct research, testing, and development activities regarding an artificial intelligence system or model, other than testing conducted under real-world conditions, before the artificial intelligence system or model is placed on the market, deployed, or put into service, as applicable; or

(i) assist another developer, deployer, or other person with any of the obligations imposed under this part 17.

(2) the obligations imposed on developers, deployers, or other persons under this part 17 do not restrict a developer's, a deployer's, or other person's ability to:

(a) effectuate a product recall; or

(b) identify and repair technical errors that impair existing or intended functionality.

(3) the obligations imposed on developers, deployers, or other persons under this part 17 do not apply where compliance with this part 17 by the developer, deployer, or other person would violate an evidentiary privilege under the laws of this state.

(4) nothing in this part 17 imposes any obligation on a developer, a deployer, or other person that adversely affects the rights or freedoms of a person, including the rights of a person to freedom of speech or freedom of the press that are guaranteed in:

(a) the first amendment to the united states constitution; or

(b) section 10 of article ii of the state constitution.

(5) nothing in this part 17 applies to a developer, a deployer, or other person:

(a) insofar as the developer, deployer, or other person develops, deploys, puts into service, or intentionally and substantially modifies, as applicable, a high-risk artificial intelligence system:

(i) that has been approved, authorized, certified, cleared, developed, or granted by a federal agency, such as the federal food and drug administration or the federal aviation administration, acting within the scope of the federal agency's authority, or by a regulated entity subject to the supervision and regulation of the federal housing finance agency; or

(ii) in compliance with standards established by a federal agency, including standards established by the federal office of the national coordinator for health information technology, or by a regulated entity subject to the supervision and regulation of the federal housing finance agency, if the standards are substantially equivalent or more stringent than the requirements of this part 17;

(b) conducting research to support an application for approval or certification from a federal agency, including the federal aviation administration, the federal communications commission, or the federal food and drug administration or research to support an application otherwise subject to review by the federal agency;

(c) performing work under, or in connection with, a contract with the united states department of commerce, the united states department of defense, or the national aeronautics and space administration, unless the developer, deployer, or other person is performing the work on a high-risk artificial intelligence system that is used to make, or is a substantial factor in making, a decision concerning employment or housing; or

(d) that is a covered entity within the meaning of the federal "health insurance portability and accountability act of 1996", 42 u.s.c. Secs. 1320d to 1320d-9, and the regulations promulgated under the federal act, as both may be amended from time to time, and is providing health-care recommendations that:

(i) are generated by an artificial intelligence system;

(ii) require a health-care provider to take action to implement the recommendations; and

(iii) are not considered to be high risk.

(6) nothing in this part 17 applies to any artificial intelligence system that is acquired by or for the federal government or any federal agency or department, including the united states department of commerce, the united states department of defense, or the national aeronautics and space administration, unless the artificial intelligence system is a high-risk artificial intelligence system that is used to make, or is a substantial factor in making, a decision concerning employment or housing.

(7) an insurer, as defined in section 10-1-102 (13), a fraternal benefit society, as described in section 10-14-102, or a developer of an artificial intelligence system used by an insurer is in full compliance with this part 17 if the insurer, the fraternal

benefit society, or the developer is subject to the requirements of section 10-3-1104.9 and any rules adopted by the commissioner of insurance pursuant to section 10-3-1104.9.

(8) (a) a bank, out-of-state bank, credit union chartered by the state of Colorado, federal credit union, out-of-state credit union, or any affiliate or subsidiary thereof, is in full compliance with this part 17 if the bank, out-of-state bank, credit union chartered by the state of Colorado, federal credit union, out-of-state credit union, or affiliate or subsidiary is subject to examination by a state or federal prudential regulator under any published guidance or regulations that apply to the use of high-risk artificial intelligence systems and the guidance or regulations:

(i) impose requirements that are substantially equivalent to or more stringent than the requirements imposed in this part 17; and

(ii) at a minimum, require the bank, out-of-state bank, credit union chartered by the state of Colorado, federal credit union, out-of-state credit union, or affiliate or subsidiary to:

(a) regularly audit the bank's, out-of-state bank's, credit union chartered by the state of Colorado's, federal credit union's, out-of-state credit union's, or affiliate's or subsidiary's use of high-risk artificial intelligence systems for compliance with state and federal anti-discrimination laws and regulations applicable to the bank, out-of-state bank, credit union chartered by the state of Colorado, federal credit union, out-of-state credit union, or affiliate or subsidiary; and

(b) mitigate any algorithmic discrimination caused by the use of a high-risk artificial intelligence system or any risk of algorithmic discrimination that is reasonably foreseeable as a result of the use of a high-risk artificial intelligence system.

(b) as used in this subsection (8): (i) "affiliate" has the meaning set forth in section 11-101-401 (3.5).

(ii) "bank" has the meaning set forth in section 11-101-401 (5).

(iii) "credit union" has the meaning set forth in section 11-30-101 (1)(a).

(iv) "out-of-state bank" has the meaning set forth in section 11-101-401 (50).

(9) if a developer, a deployer, or other person engages in an action pursuant to an exemption set forth in this section, the developer, deployer, or other person bears the burden of demonstrating that the action qualifies for the exemption.

———

6-1-1706. Enforcement by attorney general. (1) notwithstanding section 6-1-103, the attorney general has exclusive authority to enforce this part 17.

(2) except as provided in subsection (3) of this section, a violation of the requirements established in this part 17 constitutes an unfair trade practice pursuant to section 6-1-105 (1) (hhhh).

(3) in any action commenced by the attorney general to enforce this part 17, it is an affirmative defense that the developer, deployer, or other person:

(a) discovers and cures a violation of this part 17 as a result of:

(i) feedback that the developer, deployer, or other person encourages deployers or users to provide to the developer, deployer, or other person;

(ii) adversarial testing or red teaming, as those terms are defined or used by the national institute of standards and technology; or

(iii) an internal review process; and

(b) is otherwise in compliance with:

(i) the latest version of the "artificial intelligence risk management framework" published by the national institute of standards and technology in the united states department of commerce and standard iso/iec 42001 of the international organization for standardization;

(ii) another nationally or internationally recognized risk management framework for artificial intelligence systems, if the standards are substantially equivalent to or more stringent than the requirements of this part 17; or

(iii) any risk management framework for artificial intelligence systems that the attorney general, in the attorney general's discretion, may designate and, if designated, shall publicly disseminate.

(4) a developer, a deployer, or other person bears the burden of demonstrating to the attorney general that the requirements established in subsection (3) of this section have been satisfied.

(5) nothing in this part 17, including the enforcement authority granted to the attorney general under this section, preempts or otherwise affects any right, claim, remedy, presumption, or defense available at law or in equity. A rebuttable presumption or affirmative defense established under this part 17 applies only to an

enforcement action brought by the attorney general pursuant to this section and does not apply to any right, claim, remedy, presumption, or defense available at law or in equity.

(6) this part 17 does not provide the basis for, and is not subject to, a private right of action for violations of this part 17 or any other law.

6-1-1707. Rules. (1) the attorney general may promulgate rules as necessary for the purpose of implementing and enforcing this part 17, including:

(a) the documentation and requirements for developers pursuant to section 6-1-1702 (2);

(b) the contents of and requirements for the notices and disclosures required by sections 6-1-1702 (5) and (7); 6-1-1703 (4), (5), (7), and (9); and 6-1-1704;

(c) the content and requirements of the risk management policy and program required by section 6-1-1703 (2);

(d) the content and requirements of the impact assessments required by section 6-1-1703 (3);

(e) the requirements for the rebuttable presumptions set forth in sections 6-1-1702 and 6-1-1703; and

(f) the requirements for the affirmative defense set forth in section 6-1-1706 (3), including the process by which the attorney general will recognize any other nationally or internationally recognized risk management framework for artificial intelligence systems.

California AI Transparency Act
(California, U.S. 2024)

CHAPTER 25. AI Transparency Act

22757. This chapter shall be known as the California AI Transparency Act.

22757.1. As used in this chapter:

(a) "Artificial intelligence" or "AI" means an engineered or machine-based system that varies in its level of autonomy and that can, for explicit or implicit objectives, infer from the input it receives how to generate outputs that can influence physical or virtual environments.

(b) "Covered provider" means a person that creates, codes, or otherwise produces a generative artificial intelligence system that has over 1,000,000 monthly visitors or users and is publicly accessible within the geographic boundaries of the state.

(c) "Generative artificial intelligence system" or "GenAI system" means an artificial intelligence that can generate derived synthetic content, including text, images, video, and audio, that emulates the structure and characteristics of the system's training data.

(d) "Latent" means present but not manifest.

(e) "Manifest" means easily perceived, understood, or recognized by a natural person.

(f) "Metadata" means structural or descriptive information about data.

(g) "Personal information" has the same meaning as defined in Section 1798.140 of the Civil Code.

(h) "Personal provenance data" means provenance data that contains either of the following:

> (1) Personal information.

> (2) Unique device, system, or service information that is reasonably capable of being associated with a particular user.

(i) "Provenance data" means data that is embedded into digital content, or that is included in the digital content's metadata, for the purpose of verifying the digital content's authenticity, origin, or history of modification.

(j) "System provenance data" means provenance data that is not reasonably capable of being associated with a particular user and that contains either of the following:

> (1) Information regarding the type of device, system, or service that was used to generate a piece of digital content.

> (2) Information related to content authenticity.

22757.2. (a) A covered provider shall make available an AI detection tool at no cost to the user that meets all of the following criteria:

> (1) The tool allows a user to assess whether image, video, or audio content, or content that is any combination thereof, was created or altered by the covered provider's GenAI system.

> (2) The tool outputs any system provenance data that is detected in the content.

> (3) The tool does not output any personal provenance data that is detected in the content.

> (4)

>> (A) Subject to subparagraph (B), the tool is publicly accessible.

>> (B) A covered provider may impose reasonable limitations on access to the tool to prevent, or respond to, demonstrable risks to the security or integrity of its GenAI system.

> (5) The tool allows a user to upload content or provide a uniform resource locator (URL) linking to online content.

> (6) The tool supports an application programming interface that allows a user to invoke the tool without visiting the covered provider's internet website.

(b) A covered provider shall collect user feedback related to the efficacy of the covered provider's AI detection tool and incorporate relevant feedback into any attempt to improve the efficacy of the tool.

(c) A covered provider shall not do any of the following:

    (1)

        (A) Except as provided in subparagraph (B), collect or retain personal information from users of the covered provider's AI detection tool.

        (B)

            (i) A covered provider may collect and retain the contact information of a user who submits feedback pursuant to subdivision (b) if the user opts in to being contacted by the covered provider.

            (ii) User information collected pursuant to clause (i) shall be used only to evaluate and improve the efficacy of the covered provider's AI detection tool.

    (2) Retain any content submitted to the AI detection tool for longer than is necessary to comply with this section.

    (3) Retain any personal provenance data from content submitted to the AI detection tool by a user.

22757.3. (a) A covered provider shall offer the user the option to include a manifest disclosure in image, video, or audio content, or content that is any combination thereof, created or altered by the covered provider's GenAI system that meets all of the following criteria:

    (1) The disclosure identifies content as AI-generated content.

    (2) The disclosure is clear, conspicuous, appropriate for the medium of the content, and understandable to a reasonable person.

    (3) The disclosure is permanent or extraordinarily difficult to remove, to the extent it is technically feasible.

(b) A covered provider shall include a latent disclosure in AI-generated image, video, or audio content, or content that is any combination thereof, created by the covered provider's GenAI system that meets all of the following criteria:

(1) To the extent that it is technically feasible and reasonable, the disclosure conveys all of the following information, either directly or through a link to a permanent internet website:

(A) The name of the covered provider.

(B) The name and version number of the GenAI system that created or altered the content.

(C) The time and date of the content's creation or alteration.

(D) A unique identifier.

(2) The disclosure is detectable by the covered provider's AI detection tool.

(3) The disclosure is consistent with widely accepted industry standards.

(4) The disclosure is permanent or extraordinarily difficult to remove, to the extent it is technically feasible.

(c)

(1) If a covered provider licenses its GenAI system to a third party, the covered provider shall require by contract that the licensee maintain the system's capability to include a disclosure required by subdivision (b) in content the system creates or alters.

(2) If a covered provider knows that a third-party licensee modified a licensed GenAI system such that it is no longer capable of including a disclosure required by subdivision (b) in content the system creates or alters, the covered provider shall revoke the license within 96 hours of discovering the licensee's action.

(3) A third-party licensee shall cease using a licensed GenAI system after the license for the system has been revoked by the covered provider pursuant to paragraph (2).

22757.4. (a) (1) A covered provider that violates this chapter shall be liable for a civil penalty in the amount of five thousand dollars ($5,000) per violation to be collected in a civil action filed by the Attorney General, a city attorney, or a county counsel.

(2) A prevailing plaintiff in an action brought pursuant to this subdivision shall be entitled to all reasonable attorney's costs and fees.

(b) Each day that a covered provider is in violation of this chapter shall be deemed a discrete violation.

(c) For a violation by a third-party licensee of paragraph (3) of subdivision (c) of Section 22757.3, the Attorney General, a county counsel, or a city attorney may bring a civil action for both of the following:

(1) Injunctive relief.

(2) Reasonable attorney's fees and costs.

22757.5. This chapter does not apply to any product, service, internet website, or application that provides exclusively non-user-generated video game, television, streaming, movie, or interactive experiences.

22757.6. This chapter shall become operative on January 1, 2026.

AI Basic Act
Basic Act on the Development of Artificial Intelligence and the Establishment of a Trust-Based Framework
(Republic of Korea 2024)
(machine translation)

Chapter 1 General Provisions

Article 1 (Purpose) This Act aims to protect the rights and dignity of the people and improve their quality of life and strengthen national competitiveness by stipulating the basic matters necessary for the sound development of artificial intelligence and the creation of a foundation of trust.

Article 2 (Definitions) The terms used in this Act shall have the following meanings.

1. "Artificial intelligence" means the electronic implementation of the intellectual abilities of humans, such as learning, reasoning, perception, judgment, and understanding of language.

2. "Artificial intelligence system" means an artificial intelligence-based system that has various levels of autonomy and adaptability and infers results such as predictions, recommendations, and decisions that affect real and virtual environments for a given goal.

3. "Artificial Intelligence Technology" refers to the hardware and software technology or the technology for its application necessary to implement artificial intelligence.

4. "High-impact Artificial Intelligence" refers to an artificial intelligence system that may have a significant impact on human life, physical safety, and fundamental rights or pose a risk, and is used in any of the following areas.

> A. Supply of energy under Article 2(1) of the 「Energy Act」

> B. Production process of drinking water under Article 3(1) of the Drinking Water Management Act

> C. Provision of healthcare and establishment and operation of a healthcare utilization system under Article 3(1) of the Framework Act on Health and Medical Care

D. Development and use of medical devices under Article 2(1) of the Medical Devices Act and digital medical devices under Article 2(2) of the Digital Health Products Act

D. Safe management and operation of nuclear facilities as defined in Article 2(1)(1) of the Act on Protection of Nuclear Facilities and Nuclear Emergency Preparedness and Response and nuclear materials as defined in Article 2(1)(2) of the Act on Protection of Nuclear Facilities and Nuclear Emergency Preparedness and Response.

E. Analysis and use of biometric information (personal information related to physical, physiological, and behavioral characteristics that can identify an individual, such as facial features, fingerprints, iris, and palm veins) for criminal investigations and arrests.

D. Judgments or evaluations that have a significant impact on the rights and obligations of individuals, such as hiring and loan screening.

E. The major operation and management of transportation means, transportation facilities, and transportation systems under Article 2, Paragraphs 1 through 3 of the Traffic Safety Act.

[sic]. Decision-making by the state, local governments, or public institutions (hereinafter referred to as "state agencies") that affect the public, such as verifying and determining qualifications necessary for the provision of public services or collecting fees, in accordance with Article 4 of the Act on the Operation of Public Institutions.

[sic]. Student evaluation in early childhood education, elementary education, and secondary education in accordance with Article 9(1) of the Framework Act on Education.

[sic]. Other areas that have a significant impact on the protection of human life, body safety, and fundamental rights, as determined by Presidential Decree.

5. "Generative AI" refers to an AI system that imitates the structure and characteristics of input data (data as defined in Article 2(1) of the Basic Act on the Promotion of the Data Industry and the Promotion of Data Use, hereinafter the same) to generate text, sound, images, videos, and other various results.

6. "Artificial Intelligence Industry" means the industry that develops, manufactures, produces, or distributes products (hereinafter referred to as "Artificial Intelligence

Products") using artificial intelligence or artificial intelligence technology, or provides services (hereinafter referred to as "Artificial Intelligence Services") related to such products.

7. "Artificial Intelligence Business Operator" means a corporation, organization, individual, or national institution that is engaged in a business related to the Artificial Intelligence Industry and falls under any of the following subparagraphs.

> A. Artificial Intelligence Development Business: A business that develops and provides artificial intelligence.

> B. Artificial Intelligence User Business: A business that provides artificial intelligence products or services using artificial intelligence provided by the business in Item A.

8. "User" refers to a person who receives an artificial intelligence product or service.

9. "Affected Person" refers to a person whose life, physical safety, and fundamental rights are significantly affected by an artificial intelligence product or service.

10. "Artificial Intelligence Society" refers to a society that creates value and leads development in all fields, including industry, economy, society, culture, and administration, through artificial intelligence.

11. "Artificial Intelligence Ethics" refers to the ethical standards that members of society must abide by in all areas, including the development, provision, and use of artificial intelligence, in order to realize a safe and reliable artificial intelligence society that can protect the rights and interests of the people and their lives and property, based on respect for human dignity.

Article 3 (Basic Principles and Responsibilities of the State, etc.)

1. Artificial intelligence technology and the artificial intelligence industry should be developed in a way that improves the quality of life of the people by enhancing safety and reliability.

2. Affected persons should be provided with clear and meaningful explanations of the main criteria and principles used to derive the final results of artificial intelligence to the extent technically and reasonably possible.

3. The state and local governments shall respect the creative spirit of AI businesses and strive to create a safe environment for the use of AI.

4. The state and local governments shall take measures to enable all citizens to adapt to the changes in all areas, including society, economy, culture, and daily life, brought about by AI.

Article 4 (Scope of Application)

1. This Act shall apply to acts committed outside of Korea that affect the domestic market or users.

2. This Act shall not apply to artificial intelligence developed and used for the purpose of national defense or national security, as defined by Presidential Decree.

Article 5 (Relation to Other Laws)

1. Except as otherwise provided by other laws, artificial intelligence, artificial intelligence technology, artificial intelligence industry, and artificial intelligence society (hereinafter referred to as "artificial intelligence, etc.") shall be governed by this Act.

2. When enacting or amending other laws related to artificial intelligence, etc., the purpose of this Act shall be fulfilled.

Chapter 2 Promotion System for Sound Development of Artificial Intelligence and Creation of a Trust-Based Environment

Article 6 (Establishment of the Basic Plan for Artificial Intelligence)

1. The Minister of Science and ICT shall, after hearing the opinions of the heads of relevant central administrative agencies and local governments, establish, change, and implement the Basic Plan for Artificial Intelligence (hereinafter referred to as the "Basic Plan") for the promotion of artificial intelligence technology and the artificial intelligence industry and the strengthening of national competitiveness every three years through deliberation and resolution by the National Artificial Intelligence Committee pursuant to Article 7. However, this shall not apply to changes to minor matters specified by Presidential Decree in the master plan.

2. The master plan shall include the following items.

    1. Matters concerning the basic direction and strategy of policies related to artificial intelligence, etc.

    2. Matters concerning the training of professionals for the systematic development of the artificial intelligence industry and the creation of a

foundation for the promotion of the development and use of artificial intelligence

3. Matters concerning laws, systems, and culture for the realization of a healthy artificial intelligence society, including the spread of artificial intelligence ethics.

4. Matters concerning securing financial resources and investment directions for the development of artificial intelligence technology and the promotion of the artificial intelligence industry.

5. Matters concerning the creation of a foundation of trust, including the securing of fairness, transparency, accountability, and safety of artificial intelligence.

6. Matters concerning the direction of the development of artificial intelligence technology and the changes and responses in various areas of society, including education, labor, economy, and culture, that result from such development.

7. Other matters deemed necessary by the Minister of Science and ICT for the promotion of artificial intelligence technology and the artificial intelligence industry and international cooperation, etc. to strengthen national competitiveness.

3. When establishing the master plan, the Minister of Science and ICT shall take into account the comprehensive plan under Article 6(1) of the Framework Act on Intelligent Informatization and the implementation plan under Article 7(1) of the same Act.

4. The Minister of Science and ICT may request the heads of relevant central administrative agencies, local governments, and public institutions (as defined in Article 2(16) of the Framework Act on Intelligent Informatization, hereinafter the same) to submit the data necessary for the establishment of the master plan. In this case, the heads of the institutions that have been requested to submit the data shall comply with the request unless there are special circumstances.

5. The master plan shall be regarded as a sector-specific promotion plan for the field of artificial intelligence and the artificial intelligence industry pursuant to Article 13(1) of the Framework Act on Intelligent Informatization.

6. The heads of central administrative agencies and local governments shall take the master plan into account when establishing and implementing major policies under their jurisdiction.

7. The matters necessary for the establishment, revision, and implementation of the master plan shall be determined by Presidential Decree.

Article 7 (National Artificial Intelligence Committee)

1. The National Artificial Intelligence Committee (hereinafter referred to as the "Committee") shall be established under the President to deliberate and resolve matters related to major policies for the development of artificial intelligence and the creation of a foundation for trust.

2. The Commission shall consist of no more than 45 members, including one chairperson and one vice-chairperson. In this case, the number of members under Paragraph 4, Item 4 must be the majority of the total number of members, and the Commission may not be composed of only members of a particular gender.

3. The chairperson of the Commission shall be the President, and the vice-chairperson shall be a person nominated by the President from among those who fall under Paragraph 4, Item 4.

4. The members of the committee shall be the following persons.

> 1. The head of the relevant central administrative agency as prescribed by Presidential Decree.
>
> 2. The deputy head in charge of artificial intelligence at the National Security Office.
>
> 3. The senior secretary in charge of assisting with artificial intelligence at the Office of the President.
>
> 4. A person commissioned by the President who has extensive expertise and experience in artificial intelligence.

5. The chairperson of the committee shall represent the committee and oversee the affairs of the committee.

6. The chairperson of the committee may, if necessary, have the vice-chairperson act on his/her behalf.

7. The term of office of a committee member under Paragraph 4, Item 4 shall be two years, but may be renewed once.

8. The committee shall have one secretary, who shall be a member under Paragraph 4, Item 3.

9. Members of the Committee shall not disclose to others any secrets they learn in the course of their duties or use them for purposes other than those for which they were intended. However, this shall not apply if there are special provisions in other laws.

10. The Chairperson of the Committee shall convene meetings of the Committee and preside over them.

11. Meetings of the Committee shall be convened with the attendance of a majority of the members, and resolutions shall be passed with the approval of a majority of the members present.

12. A support team shall be established to assist the committee in its work and operations.

13. The committee shall exist for five years from the date of enforcement of this Act.

14. Other matters necessary for the establishment and operation of the committee and the support team under Paragraph 12 shall be prescribed by Presidential Decree.

Article 8 (Functions of the Committee)

1. The committee shall deliberate and make decisions on the following matters.

> 1. Matters concerning the establishment, revision, and implementation inspection and analysis of the master plan
>
> 2. Matters concerning related policies such as artificial intelligence
>
> 3. Matters concerning the establishment of research and development strategies related to artificial intelligence, etc.
>
> 4. Matters concerning the establishment of investment strategies related to artificial intelligence, etc.

5. Matters concerning the discovery and improvement of regulations that hinder the development and competitiveness of the artificial intelligence industry

6. Matters concerning the expansion of infrastructure, including AI data centers (data centers under Article 40(1) of the Framework Act on Intelligent Informatization, hereinafter the same)

7. Matters concerning the promotion of the use of AI in industrial sectors such as manufacturing and service industries and the public sector

8. Matters concerning international cooperation on artificial intelligence, including the establishment of international norms for artificial intelligence.

9. Matters concerning the expression of recommendations or opinions under Paragraph 2.

10. Matters concerning the regulation of high-impact artificial intelligence.

11. Matters concerning the social changes and policy responses related to high-impact artificial intelligence.

12. Matters stipulated as matters for deliberation by the Committee under this Act or other laws.

13. Other matters deemed necessary by the Chairperson of the Committee to be brought to the Committee's meeting.

2. The Committee may make recommendations or express opinions to heads of state agencies and AI businesses regarding the proper use of AI, the practice of AI ethics, and the safety and reliability of AI technology.

3. When the Committee makes recommendations or expresses opinions to the heads of state agencies, etc. under Paragraph 2 regarding the improvement of laws and regulations or systems or the establishment of implementation plans, the heads of state agencies, etc. shall establish improvement plans and implementation plans for laws and regulations, etc.

Article 9 (Disqualification, Avoidance, and Refusal of Members)

1. A member of the Committee shall be disqualified from deliberation and decision-making on the agenda in the following cases to ensure fairness of the work.

1. If the member or the corporation or organization to which the member belongs has a direct interest in the matter.

2. If the family of the member (family as defined in Article 779 of the Civil Act) has an interest in the matter.

2. The party to the agenda under review (if the party is a corporation, organization, etc., its officers and employees are included) may file an application with the committee to avoid the committee if there are circumstances that make it difficult to expect fair performance of duties by the committee members, and the committee shall decide to avoid the committee by resolution if the application to avoid is deemed to be valid.

3. A committee member shall avoid deliberation of the agenda concerned by himself/herself if the reason for the application falls under Paragraphs 1 or 2.

Article 10 (Subcommittees, etc.)

1. The Committee may establish subcommittees as necessary to perform the Committee's work in specialized areas.

2. The Committee may establish a special committee as necessary to discuss specific issues related to artificial intelligence, etc.

3. The Committee may establish an advisory group composed of relevant experts, etc. to professionally review matters related to artificial intelligence, etc.

4. Other matters necessary for the composition and operation of subcommittees, special committees, and advisory groups shall be determined by Presidential Decree.

Article 11 (Artificial Intelligence Policy Center)

1. The Minister of Science and ICT may designate an Artificial Intelligence Policy Center (hereinafter referred to as the "Center") to comprehensively perform the tasks necessary for the development of policies related to artificial intelligence and the establishment and diffusion of international norms.

2. The Center shall carry out the following projects.

1. Support for the specialized technology necessary for the establishment and implementation of the master plan

2. Support for the specialized technology for the development of measures related to artificial intelligence and the planning and implementation of related projects

3. Investigation and analysis of the impact of the spread of the use of artificial intelligence on society, the economy, culture, and the daily lives of the people

4. Analysis of trends to support the development of policies related to artificial intelligence and artificial intelligence technology, social and cultural changes, and research on laws and systems

5. Other projects designated as the Center's work by other laws and regulations or entrusted to the Center

6. Other projects entrusted by the heads of national institutions, etc.

3. Other matters necessary for the designation of the Center shall be determined by Presidential Decree.

Article 12 (Artificial Intelligence Safety Research Institute)

1. The Minister of Science and ICT may operate the Artificial Intelligence Safety Research Institute (hereinafter referred to as the "Safety Research Institute") to professionally and efficiently perform tasks to secure a state (hereinafter referred to as "Artificial Intelligence Safety") to protect the lives, bodies, and property of the people from risks that may arise in connection with artificial intelligence and to maintain the foundation of trust in the artificial intelligence society.

2. The Safety Research Institute shall carry out the following projects.

1. Definition and analysis of risks related to AI safety

2. Research on AI safety policies

3. Research on AI safety evaluation standards and methods

4. Research on AI safety technologies and standardization

5. International exchange and cooperation related to AI safety

6. Support for ensuring the safety of AI systems in accordance with Article 32

7. Other projects related to AI safety as stipulated by Presidential Decree

3. The government may provide or support the necessary expenses for the operation of the Safety Research Institute and the promotion of its projects within the budget.

4. Other matters necessary for the operation of the Safety Research Institute shall be determined by Presidential Decree.

Chapter 3 Development of AI Technology and Promotion of the Industry

Section 1 Establishment of the AI Industry Foundation

Article 13 (Development of AI Technology and Support for Safe Use)

1. The government may support the following projects to promote the development of AI technology.

> 1. Research on the trends and levels of artificial intelligence technology and related systems at home and abroad

> 2. Research and development, testing and evaluation of artificial intelligence technology, or utilization of developed technology

> 3. Support for the practical application and commercialization of technology, including the diffusion of artificial intelligence technology, cooperation and transfer of artificial intelligence technology, etc.

> 4. Smooth distribution of information for the implementation of artificial intelligence technology and industry-academia cooperation

> 5. Other businesses prescribed by Presidential Decree in relation to the development and research and investigation of artificial intelligence technology

2. The government may support the following projects for the safe and convenient use of artificial intelligence technology.

> 1. Research and development projects that implement the matters of Article 60(1) of the Framework Act on Intelligent Informatization with artificial intelligence technology.

> 2. Support for technology research to implement emergency stop functions in artificial intelligence products or services in accordance with Article 60(3)

of the Framework Act on Intelligent Informatization and projects to spread such technology.

3. Research and development and dissemination of design standards and technologies suitable for the protection of privacy, etc. in accordance with Article 61(2) of the Framework Act on Intelligent Informatization in the development of artificial intelligence technology

4. Research and development projects for conducting and applying social impact assessments in accordance with Article 56(1) of the Framework Act on Intelligent Informatization of Artificial Intelligence Technology.

5. Research and development and dissemination projects for technologies or standards that enable artificial intelligence to be developed and used in a way that respects human dignity and fundamental rights.

6. Education and promotion projects for improving awareness, creating correct usage methods and a safe environment for the safe development and use of artificial intelligence.

7. Other projects necessary to protect the basic rights of the people, their bodies and property in the development and use of artificial intelligence.

3. The government shall disclose and disseminate the results of the projects under Paragraph 2 so that anyone can easily use them. In this case, if necessary to protect the developer of the technology, the government may set a protection period to receive a technology usage fee or protect it in other ways.

Article 14 (Standardization of Artificial Intelligence Technology)

1. The government may promote the following projects for the standardization of artificial intelligence technology, learning data pursuant to Article 15(1), and the safety and reliability of artificial intelligence.

1. Establishment, revision, and abolition of standards related to artificial intelligence technology and their dissemination

2. Investigation, research, and development of domestic and international standards related to artificial intelligence technology

3. Other standardization projects related to artificial intelligence technology

2. The government may notify the standards established in accordance with Paragraph 1, Item 1 and recommend that relevant businesses comply with them.

3. The government may provide necessary support for standardization projects related to artificial intelligence technology promoted by the private sector.

4. The government shall maintain and strengthen a cooperative system with international standardization organizations or international standardization bodies related to artificial intelligence technology standards.

5. Other necessary matters related to the promotion and support of standardization projects under Paragraphs 1 and 3 shall be determined by Presidential Decree.

Article 15 (Establishment of policies related to data for artificial intelligence learning, etc.)

1. The Minister of Science and ICT shall, in consultation with the heads of relevant central administrative agencies, promote policies necessary to promote the production, collection, management, distribution, and use of data used for the development and use of artificial intelligence (hereinafter referred to as "learning data").

2. The government may select projects to be supported and provide support within the budget to efficiently implement measures related to the production, collection, management, distribution, and use of learning data.

3. The government may implement a project to create and provide various learning data (hereinafter referred to as the "learning data construction project") to promote the production, collection, management, distribution, and use of learning data.

4. The Minister of Science and ICT shall establish and manage a system (hereinafter referred to as the "integrated provision system") that can provide and manage learning data in an integrated manner for the efficient implementation of the learning data construction project, and shall provide it for private use.

5. The Minister of Science and ICT may collect fees from those who use the integrated provision system.

6. Other matters necessary for the selection and support of projects eligible for support under Paragraph 2, the implementation of the project to build learning data, the construction and management of the integrated provision system, and the collection of costs under Paragraph 5 shall be determined by Presidential Decree.

Section 2 Development of AI technology and promotion of the AI industry

Article 16 (Support for the Introduction and Use of Artificial Intelligence Technology)

1. The state and local governments may provide the following support if necessary to promote the introduction of and spread the use of artificial intelligence technology among companies and public institutions.

> 1. Support for the development of artificial intelligence technology, artificial intelligence products, or artificial intelligence services and the dissemination of research and development results

> 2. Consulting support for companies and public institutions that want to introduce and use artificial intelligence technology

> 3. Support for education on the introduction and use of artificial intelligence technology for employees of small and medium-sized enterprises (SMEs) as defined in Article 2(1) of the 「Small and Medium Business Act」, venture companies as defined in Article 2(1) of the 「Special Act on the Promotion of Venture Businesses」, and small business owners as defined in Article 2(1) of the 「Small Business Act」 (hereinafter referred to as "SMEs, etc.")

> 4. Support for funds used to introduce and utilize artificial intelligence technology in small and medium-sized enterprises, etc.

> 5. Other matters prescribed by Presidential Decree to promote the introduction and utilization of artificial intelligence technology in companies and public institutions.

2. The matters necessary for the support under Paragraph (1) shall be prescribed by Presidential Decree.

Article 17 (Special Support for Small and Medium-sized Enterprises, etc.)

1. When implementing various support measures related to artificial intelligence technology and the artificial intelligence industry under this Act, priority shall be given to small and medium-sized enterprises, etc.

2. The government shall make efforts to promote participation of small and medium-sized enterprises in the AI industry and reflect such matters in the master plan.

3. The Minister of Science and ICT may support the implementation of measures pursuant to Article 34 and the impact assessment pursuant to Article 35 of the Act for small and medium-sized enterprises to ensure the safety and reliability of AI.

Article 18 (Promotion of Startups)

1. The government may promote the following projects to promote startups in the AI industry.

> 1. Business related to discovering and fostering and supporting founders in the AI industry

> 2. Business related to education and training to promote entrepreneurship in the AI industry

> 3. Commercialization support for excellent AI technologies of professionals in accordance with Article 21

> 4. Financial support for the valuation of AI technologies and startup funds

> 5. Provision of AI-related research and technology development results

> 6. Fostering of institutions and organizations that support startups in the AI industry

> 7. Other businesses necessary to promote the creation of businesses in the field of the artificial intelligence industry.

2. Local governments may contribute to or invest in public organizations, such as public institutions, that support the creation of businesses in the field of the artificial intelligence industry.

Article 19 (Promotion of the Convergence of Artificial Intelligence)

1. The government shall establish and implement necessary measures to promote the convergence of the artificial intelligence industry and other industries and to promote the use of artificial intelligence in all fields.

2. The government may, if necessary, give priority to research and development projects related to AI-integrated products and services in national research and development projects in accordance with the National Research and Development Innovation Act in order to support the development of AI-integrated products and services.

---

3. The government shall actively support the smooth implementation of temporary permits under Article 37 of the Special Act on the Promotion of Information and Communication and the Special Act on the Activation of Convergence and regulatory exceptions for demonstration under Article 38-2 of the same Act for AI convergence products and services developed in accordance with Paragraph 2.

Article 20 (Improvement of System, etc.)

1. The government shall endeavor to improve the relevant system, including the revision of laws and regulations, for the development of the artificial intelligence industry and the creation of a foundation of trust.

2. The government may provide administrative and financial support necessary for the study of relevant laws and systems and the collection of opinions from various sectors of society in order to promote system improvement pursuant to Paragraph (1).

Article 21 (Securing Professional Manpower)

1. The Minister of Science and ICT shall foster and support professional manpower related to artificial intelligence and artificial intelligence technology in accordance with the measures under Article 23(1) of the Framework Act on Intelligent Informatization for the development of artificial intelligence technology and the advancement of the artificial intelligence industry.

2. The government may implement the following measures to secure overseas professional manpower related to artificial intelligence and artificial intelligence technology.

> 1. Research and analysis on experts from overseas universities, research institutes, and companies related to artificial intelligence and artificial intelligence technology
>
> 2. Establishment of an international network to attract overseas experts
>
> 3. Support for overseas experts to work in Korea
>
> 4. Support for overseas expansion of Korean artificial intelligence research institutes and attracting overseas artificial intelligence research institutes to Korea

5. Support for attracting international organizations and international events related to artificial intelligence and artificial intelligence technology to Korea

6. Other matters necessary to secure overseas experts

Article 22 (Support for International Cooperation and Entry into Overseas Markets)

1. The government shall identify international trends related to artificial intelligence and promote international cooperation.

2. The government may provide the following support to individuals, companies, or organizations engaged in the artificial intelligence industry to strengthen the competitiveness of the artificial intelligence industry and promote entry into overseas markets.

1. International exchange of information, technology, and personnel related to the artificial intelligence industry

2. Collection, analysis, and provision of information on overseas expansion related to the artificial intelligence industry

3. Joint research and development of AI technology, AI products, or AI services between countries and international standardization

4. Attracting foreign capital investment in the AI industry

5. Public relations and overseas marketing, such as participation in overseas professional conferences and exhibitions related to AI, etc.

6. Establishment of sales, distribution, and cooperation systems necessary for the export of AI products or AI services

7. Identification of international trends in AI ethics and international cooperation

8. Other matters necessary to strengthen the competitiveness of the AI industry and promote entry into overseas markets.

3. The government may entrust or delegate the support under each subparagraph of Paragraph 2 to a public institution or other organization in accordance with Presidential Decree to efficiently carry out the support, and may subsidize the costs required for this.

Article 23 (Designation of AI Complex, etc.)

1. The state and local governments may promote the functional, physical, and regional integration of companies, institutions, and organizations that conduct research and development of AI and AI technologies to promote the AI industry and strengthen the competitiveness of AI development and utilization.

2. The state and local governments may designate an AI cluster (hereinafter referred to as the "AI cluster") and provide administrative, financial, and technical support as prescribed by Presidential Decree if necessary for the integration pursuant to Paragraph 1.

3. The state and local governments may cancel the designation of the AI cluster if any of the following items apply. However, the designation shall be canceled if any of the items in Paragraph 1 apply.

> 1. When the designation is obtained by falsehood or other improper means.
>
> 2. When the head of the state or local government that designated the AI cluster recognizes that it is difficult to achieve the purpose of designating the AI cluster.

4. The government may establish or designate a dedicated agency that provides comprehensive support for related tasks in order to effectively establish the integration in accordance with Paragraph 1.

5. The government may provide or subsidize all or part of the costs necessary for the operation of the dedicated institution under Paragraph 4 and for the performance of its duties.

6. Other matters necessary for the designation and cancellation of designation of the AI cluster, and the establishment or designation of the dedicated institution under Paragraph 4 shall be prescribed by Presidential Decree.

Article 24 (Creation of an AI Demonstration Base, etc.)

1. The state and local governments may build and operate facilities, equipment, and facilities necessary for testing and evaluation (hereinafter referred to as "demonstration base, etc.") to support the demonstration, performance testing, and inspection/certification (hereinafter referred to as "demonstration testing, etc.") of technologies developed or transferred by AI businesses in accordance with Article 30.

2. The state and local governments may open the demonstration bases held by the institutions specified by Presidential Decree to AI businesses to promote demonstration tests, etc.

3. Other matters necessary for the establishment, operation, and opening of the demonstration bases, etc. shall be determined by Presidential Decree.

Article 25 (Promotion of Measures Related to Artificial Intelligence Data Centers, etc.)

1. The government shall promote measures necessary to promote the construction and operation of data centers (hereinafter referred to as "artificial intelligence data centers") used for the development and utilization of artificial intelligence.

2. The government may carry out the following tasks to implement the measures under Paragraph 1.

   1. Administrative and financial support for the establishment and operation of artificial intelligence data centers.

   2. Support for the use of artificial intelligence data centers by small and medium-sized enterprises, research institutes, etc.

   3. Support for the balanced development of regional infrastructure facilities related to artificial intelligence, such as artificial intelligence data centers.

Article 26 (Establishment of the Korea Association for Artificial Intelligence Promotion)

1. Any person engaged in research and work related to artificial intelligence, etc. may establish or be designated as the Korea Association for Artificial Intelligence Promotion (hereinafter referred to as the "Association") with the approval of the Minister of Science and ICT in accordance with the Presidential Decree for the promotion of the development and use of artificial intelligence, the promotion of the artificial intelligence industry and artificial intelligence technology, and the education and promotion of artificial intelligence, etc.

2. The Association shall be a corporation.

3. The Association shall perform the following tasks.

   1. Promotion and diffusion of the use of artificial intelligence technology, artificial intelligence products, or artificial intelligence services.

2. Investigation of the current status and related statistics of artificial intelligence, etc.

3. Installation and operation of shared use facilities for artificial intelligence businesses and training for the development of specialized human resources, etc.

4. Support for the overseas expansion of artificial intelligence businesses and artificial intelligence-related specialized human resources.

5. Education and promotion for the development and use of safe and reliable artificial intelligence.

6. Businesses entrusted to the association under this Act or other laws

7. Other businesses necessary to achieve the association's founding purpose as stipulated in the articles of incorporation.

4. The state and local governments may, if necessary, provide funds necessary for the association's business execution or assist with operating expenses within the budget to develop the AI industry and build a foundation of trust.

5. The qualifications of the members of the association, the officers, the association's affairs, etc. shall be stipulated in the articles of association, and other matters to be included in the articles of association shall be stipulated by Presidential Decree.

6. The Minister of Science and ICT shall announce the fact of granting approval under Paragraph 1.

7. With regard to the association, the provisions of the Civil Code on corporations shall apply mutatis mutandis, except as provided for in this Act.

Chapter 4 Ethics and Reliability of Artificial Intelligence

Article 27 (Ethical Principles for Artificial Intelligence, etc.)

1. The government may enact and publish the Ethical Principles for Artificial Intelligence (hereinafter referred to as the "Ethical Principles") as prescribed by Presidential Decree to spread the ethics of artificial intelligence, which shall include the following items.

1. Matters concerning safety and reliability to ensure that the development and use of artificial intelligence does not harm human life, body, mental health, etc.

2. Matters concerning accessibility that allows everyone to freely and conveniently use products and services that apply artificial intelligence technology.

3. Matters concerning the development and use of artificial intelligence to contribute to the prosperity and well-being of people.

2. The Minister of Science and ICT shall collect opinions from various sectors of society to establish a plan of action to ensure that the ethical principles are implemented by all persons involved in the development and use of artificial intelligence, and shall disclose, promote, and educate on the plan.

3. When the head of a central administrative agency or local government enacts or revises the AI Ethics Standards (regulations, standards, guidelines, and other rules regarding AI ethics, regardless of their name or form), the Minister of Science and ICT may make recommendations or express opinions regarding the connection and consistency of the Ethics Principles and the implementation measures under Paragraph 2.

Article 28 (Establishment of the Private Voluntary Artificial Intelligence Ethics Committee, etc.)

1. The following organizations or groups may establish a Private Voluntary Artificial Intelligence Ethics Committee (hereinafter referred to as the "Private Voluntary Committee") to comply with the ethical principles.

1. Educational institutions or research institutes to which a person conducting research and development on artificial intelligence technology belongs

2. Artificial intelligence businesses

3. Other artificial intelligence technology-related organizations as specified by Presidential Decree

2. The private self-regulatory committee shall autonomously perform the following tasks.

1. Verification of compliance with ethical principles in the research, development, and utilization of artificial intelligence technology

2. Investigation and research on the safety and human rights violations of research, development, and utilization of artificial intelligence technology

3. Investigation and supervision of the procedures and results of research, development, and utilization of artificial intelligence technology

4. Training of researchers and employees of the relevant institution or organization on ethical principles

5. Preparation of artificial intelligence ethics guidelines for each field suitable for research, development, and utilization of artificial intelligence technology

6. Other tasks necessary for the implementation of ethical principles

3. The private self-regulatory committee shall be composed and operated in accordance with the relevant regulations of the relevant institution or organization. However, the composition of the committee shall not be limited to a specific gender, and shall include persons with experience and knowledge to evaluate the social and ethical feasibility of the committee, as well as persons who are not employed by the institution or organization.

4. The Minister of Science and ICT may prepare and distribute standard guidelines, etc. for the fair and neutral composition and operation of the private self-governing committee.

Article 29 (Preparation of Measures to Create a Foundation of Trust for Artificial Intelligence)

The government shall prepare the following measures to minimize the potential risks of artificial intelligence to the lives of the people and to create a foundation of trust for the safe use of artificial intelligence.

1. Creation of a safe and reliable environment for the use of artificial intelligence

2. Forecasts and predictions regarding the impact of the use of artificial intelligence on people's daily lives, and the maintenance of related laws and systems.

3. Support for the development and dissemination of safety and authentication technologies to ensure the safety and reliability of artificial intelligence.

4. Education and promotion for the realization of a safe and reliable artificial intelligence society and the practice of artificial intelligence ethics.

5. Support for the establishment and implementation of voluntary codes related to the safety and reliability of artificial intelligence businesses.

6. Support and spread of private activities such as autonomous cooperation for the promotion of the safety and reliability of artificial intelligence by artificial intelligence businesses, users, and other organizations related to artificial intelligence (hereinafter referred to as "organizations, etc."), and the establishment of ethical guidelines.

7. Other matters prescribed by Presidential Decree to ensure the safety and reliability of artificial intelligence.

Article 30 (Support for AI safety and reliability testing and certification, etc.)

1. The Minister of Science and ICT may promote the following projects to support the verification and certification activities (hereinafter referred to as "testing and certification, etc.") that organizations voluntarily promote to ensure the safety and reliability of AI.

1. Dissemination of guidelines on the development of AI

2. Support for research on testing and certification, etc.

3. Support for the construction and operation of equipment and systems used for inspection and certification, etc.

4. Support for the training of professionals required for inspection and certification, etc.

5. Other matters prescribed by Presidential Decree to support inspection and certification, etc.

2. The Minister of Science and ICT may provide relevant information or provide administrative and financial support to SMEs that wish to receive inspection and certification, etc. as prescribed by Presidential Decree.

3. When an artificial intelligence business provides high-impact artificial intelligence, it shall make efforts to receive inspection, certification, etc. in advance.

4. When a national institution, etc. intends to use high-impact artificial intelligence, it shall give priority consideration to products or services based on artificial intelligence that has received inspection, certification, etc.

Article 31 (Obligation to Ensure Transparency of Artificial Intelligence)

1. If an artificial intelligence business intends to provide products or services using high-impact artificial intelligence or generative artificial intelligence, it shall notify the user in advance that the product or service is operated based on the artificial intelligence.

2. If an artificial intelligence business provides products or services using generative artificial intelligence or generative artificial intelligence, it shall indicate that the results were generated by generative artificial intelligence.

3. If an artificial intelligence business provides virtual sound, image, or video results that are difficult to distinguish from the real world using an artificial intelligence system, the business must notify or display the results in a way that allows the user to clearly recognize that the results were generated by the artificial intelligence system. In this case, if the results are artistic or creative expressions or part of them, the results may be notified or displayed in a way that does not hinder their exhibition or enjoyment.

4. Other necessary matters concerning the method of prior notice under Paragraph (1), indication under Paragraph (2), notice or indication under Paragraph (3), and exceptions thereto shall be prescribed by Presidential Decree.

Article 32 (Obligation to Ensure the Safety of Artificial Intelligence)

1. An artificial intelligence business operator shall implement the following matters to ensure the safety of an artificial intelligence system whose cumulative amount of computation used for learning is equal to or greater than the standards prescribed by Presidential Decree.

> 1. Identification, evaluation, and mitigation of risks throughout the AI life cycle

> 2. Establishment of a risk management system to monitor and respond to AI-related safety accidents

2. The AI business operator shall submit the results of the implementation of the matters under each subparagraph of Paragraph 1 to the Minister of Science and ICT.

3. The Minister of Science and ICT shall set and notify the specific implementation methods of the matters under each subparagraph of Paragraph 1 and the matters necessary for the submission of the results under Paragraph 2.

Article 33 (Confirmation of High-Impact Artificial Intelligence)

1. When providing artificial intelligence or products or services using it, an artificial intelligence business operator shall review in advance whether the artificial intelligence falls under high-impact artificial intelligence, and if necessary, may request the Minister of Science and ICT to confirm whether it falls under high-impact artificial intelligence.

2. The Minister of Science and ICT shall, upon a request under Paragraph ①, check whether the AI is high-impact, and if necessary, establish a specialized committee to receive relevant advice.

3. The Minister of Science and ICT may establish and disseminate guidelines on the criteria and examples of high-impact AI.

4. Other necessary matters regarding the verification procedure under Paragraph 1 shall be determined by Presidential Decree.

Article 34 (Responsibilities of Businesses Related to High-Impact Artificial Intelligence)

1. When providing high-impact artificial intelligence or products and services using high-impact artificial intelligence, artificial intelligence businesses shall implement measures that include the following items to ensure the safety and reliability of high-impact artificial intelligence, as prescribed by Presidential Decree.

> 1. Establishment and operation of risk management plans

> 2. Establishment and implementation of a plan for explaining the final results derived by artificial intelligence within the technically possible range, the main criteria used to derive the final results of artificial intelligence, and an overview of the training data used in the development and use of artificial intelligence.

> 3. Establishment and operation of a user protection plan.

4. Management and supervision of people with high-impact artificial intelligence.

5. Creation and storage of documents that can confirm the contents of measures to ensure safety and reliability.

6. Other matters deliberated and resolved by the Committee to ensure the safety and reliability of high-impact artificial intelligence.

2. The Minister of Science and ICT may specify the specific details of the measures under each subparagraph of Paragraph 1 and notify them, and recommend that artificial intelligence businesses comply with them.

3. If an artificial intelligence business has implemented measures equivalent to those under each subparagraph of Paragraph 1 in accordance with other laws and regulations, it shall be deemed to have implemented the measures under Paragraph 1.

Article 35 (High-impact AI Impact Assessment)

1. When an AI business provides products or services using high-impact AI, it shall make efforts to assess the impact on the basic rights of people in advance (hereinafter referred to as "impact assessment").

2. When a national institution or other entity intends to use products or services using high-impact AI, it shall give priority to products or services that have undergone impact assessment.

3. Other necessary matters regarding the specific contents and methods of impact assessment shall be determined by Presidential Decree.

Article 36 (Designation of Domestic Agent)

1. An artificial intelligence business operator without an address or business office in Korea whose number of users, sales, etc. meets the standards set by Presidential Decree shall designate a person to represent the following matters (hereinafter referred to as the "Domestic Agent") in writing and report it to the Minister of Science and ICT.

1. Submission of the results of the performance pursuant to Article 32(2)

2. Request for confirmation of whether the AI falls under the category of high-impact AI pursuant to Article 33(1)

3. Support necessary for the implementation of safety and reliability assurance measures pursuant to each subparagraph of Article 34(1) (including inspection of the up-to-dateness and accuracy of documents pursuant to subparagraph 5 of the same paragraph)

2. The domestic representative shall be a person with an address or business office in Korea.

3. If a domestic agent violates this Act in relation to each of the items in Paragraph 1, the artificial intelligence business operator that designated the domestic agent shall be deemed to have committed the act.

Chapter 5 Supplementary Provisions

Article 37 (Expansion of Financial Resources for the Promotion of the Artificial Intelligence Industry, etc.)

1. The state shall establish a plan to continuously and stably expand the financial resources necessary to effectively implement the basic plan and measures under this Act.

2. The Minister of Science and ICT may recommend that public institutions provide necessary support for projects related to the promotion of the AI industry, if necessary.

3. The state and local governments shall take necessary measures to enable the private sector, including companies, to actively invest in projects related to the promotion of the AI industry.

4. The state and local governments shall strive to efficiently execute investment funds by comprehensively considering the development stage of the AI industry.

Article 38 (Survey, Statistics and Indicators)

1. The Minister of Science and ICT shall, in consultation with the head of the National Statistical Office, prepare, manage and publish domestic and international surveys, statistics and indicators on artificial intelligence in connection with the statistics under Article 26-2 of the Framework Act on Science, Technology and Information and Communication in order to plan, establish and promote basic plans, artificial intelligence-related policies and projects.

2. The Minister of Science and ICT may request the heads of relevant central administrative agencies, local governments, and public institutions to cooperate in

the preparation of statistics and indicators pursuant to Paragraph (1), including the submission of data. In this case, the head of the institution that has been requested to cooperate shall comply with the request unless there are special circumstances.

3. Other matters necessary for the investigation of the actual conditions pursuant to Paragraph (1), the preparation, management, and publication of statistics and indicators shall be prescribed by Presidential Decree.

Article 39 (Delegation of Authority and Entrustment of Work)

1. The Minister of Science and ICT or the head of a relevant central administrative agency may delegate some of the authority under this Act to the head of an affiliated institution or a special mayor, metropolitan mayor, special autonomous mayor, provincial governor, or special autonomous provincial governor (hereinafter referred to as "city/provincial governor" in this Article) as prescribed by Presidential Decree. In this case, the governor of a city or province may re-delegate some of the delegated authority to the mayor (including the administrative mayor under Article 11(2) of the Special Act on the Establishment of Jeju Special Self-Governing Province and the Creation of an International Free City), the county mayor, or the head of a district office (referring to the head of a district office in a self-governing district).

2. The government may entrust the following tasks to relevant institutions or organizations as specified by Presidential Decree.

> 1. Support for projects related to the development and use of artificial intelligence technology pursuant to Article 13.

> 2. Selection and support of projects for the production, collection, management, distribution, and utilization of learning data pursuant to Article 15(2) and (3), and promotion of learning data construction projects.

> 3. Construction, operation, and management of integrated provision systems.

> 4. Matters deemed necessary by the Minister of Science and ICT for the promotion of entrepreneurship under Article 18.

> 5. Support related to inspection and certification under Article 30(2).

> 6. Preparation of surveys, statistics, and indicators under Article 38.

> 7. Other affairs prescribed by Presidential Decree for the promotion of the AI industry and the spread of AI ethics.

Article 40 (Fact-finding, etc.)

1. The Minister of Science and ICT may request an AI business operator to submit relevant materials or have an official conduct the necessary investigation in any of the following cases.

> 1. When the Minister finds that there is a violation of Article 31(2) or (3), Article 32(1) or (2), or Article 34(1) or suspects that there is a violation

> 2. In the event that a report is received or a complaint is filed for a violation of Article 31(2) or (3), Article 32(1) or (2), or Article 34(1)

2. The Minister of Science and ICT may, if necessary for the investigation under Paragraph 1, have a government official enter the office or business site of the AI business operator to inspect the books, documents, and other materials or objects. In this case, the matters prescribed by this Act regarding the content, method, and procedure of the investigation shall be governed by the Administrative Investigation Basic Act, except as otherwise provided by this Act.

3. If the Minister of Science and ICT finds that the AI business operator has violated this Act as a result of the investigation under Paragraphs (1) and (2), the Minister may order the AI business operator to take necessary measures to stop or correct the violation.

Article 41 (Application of penalties to public officials)

1. A non-public official member of the committee shall be regarded as a public official when applying the penalties under Articles 129 to 132 of the Criminal Act.

2. Employees of an institution or organization engaged in the work entrusted to it pursuant to Article 39(2) shall be deemed public officials when applying the penalties under Articles 127 and 129 to 132 of the Criminal Act.

Chapter 6 Penalties

Article 42 (Penalties)

Any person who discloses to another person any confidential information obtained in the course of his or her duties in violation of Article 7(9) or uses such information for purposes other than those for which it was obtained shall be punished by imprisonment for not more than three years or a fine of not more than 30 million won.

Article 43 (Fines)

1. Any person falling under any of the following subparagraphs shall be fined not more than 30 million won.

> 1. A person who has not complied with Article 31(1) by not giving notice

> 2. A person who has not appointed a domestic agent in violation of Article 36(1)

> 3. A person who has not complied with a suspension or corrective order under Article 40(3)

2. The Minister of Science and ICT shall impose and collect the administrative fine under Paragraph 1 in accordance with the Presidential Decree.

Supplementary Provisions <Act No. 20676, January 21, 2025>

Article 1 (Effective Date) This Act shall enter into force on the day one year after its promulgation. However, the provisions regarding digital medical devices under Article 2(4) letter (d) shall enter into force on January 24, 2026.

Article 2 (Preparatory Actions for the Enforcement of this Act) The appointment of committee members, the formation of subcommittees, special committees, advisory groups, and support groups, etc. necessary for the enforcement of this Act may be carried out prior to the enforcement of this Act.

Article 3 (Special Provisions on Dedicated Agencies) At the time of the enforcement of this Act, among the agencies that are operating with the support of the government budget to effectively establish the integration under Article 23(1), those that meet the requirements set forth in Presidential Decree, such as organization and personnel, shall be deemed to have been designated as dedicated agencies under this Act, notwithstanding Article 23(4).

U.S. Executive Orders

White House, Executive Order 13859,
Maintaining American Leadership in Artificial Intelligence (2019)

By the authority vested in me as President by the Constitution and the laws of the United States of America, it is hereby ordered as follows:

<u>Section 1</u>. <u>Policy and Principles</u>. Artificial Intelligence (AI) promises to drive growth of the United States economy, enhance our economic and national security, and improve our quality of life. The United States is the world leader in AI research and development (R&D) and deployment. Continued American leadership in AI is of paramount importance to maintaining the economic and national security of the United States and to shaping the global evolution of AI in a manner consistent with our Nation's values, policies, and priorities. The Federal Government plays an important role in facilitating AI R&D, promoting the trust of the American people in the development and deployment of AI-related technologies, training a workforce capable of using AI in their occupations, and protecting the American AI technology base from attempted acquisition by strategic competitors and adversarial nations. Maintaining American leadership in AI requires a concerted effort to promote advancements in technology and innovation, while protecting American technology, economic and national security, civil liberties, privacy, and American values and enhancing international and industry collaboration with foreign partners and allies. It is the policy of the United States Government to sustain and enhance the scientific, technological, and economic leadership position of the United States in AI R&D and deployment through a coordinated Federal Government strategy, the American AI Initiative (Initiative), guided by five principles:

(a) The United States must drive technological breakthroughs in AI across the Federal Government, industry, and academia in order to promote scientific discovery, economic competitiveness, and national security.

(b) The United States must drive development of appropriate technical standards and reduce barriers to the safe testing and deployment of AI technologies in order to enable the creation of new AI-related industries and the adoption of AI by today's industries.

(c) The United States must train current and future generations of American workers with the skills to develop and apply AI technologies to prepare them for today's economy and jobs of the future.

(d) The United States must foster public trust and confidence in AI technologies and protect civil liberties, privacy, and American values in their application in order to fully realize the potential of AI technologies for the American people.

(e) The United States must promote an international environment that supports American AI research and innovation and opens markets for American AI industries, while protecting our technological advantage in AI and protecting our critical AI technologies from acquisition by strategic competitors and adversarial nations.

<u>Sec</u>. <u>2</u>. <u>Objectives</u>. Artificial Intelligence will affect the missions of nearly all executive departments and agencies (agencies). Agencies determined to be implementing agencies pursuant to section 3 of this order shall pursue six strategic objectives in furtherance of both promoting and protecting American advancements in AI:

(a) Promote sustained investment in AI R&D in collaboration with industry, academia, international partners and allies, and other non-Federal entities to generate technological breakthroughs in AI and related technologies and to rapidly transition those breakthroughs into capabilities that contribute to our economic and national security.

(b) Enhance access to high-quality and fully traceable Federal data, models, and computing resources to increase the value of such resources for AI R&D, while maintaining safety, security, privacy, and confidentiality protections consistent with applicable laws and policies.

(c) Reduce barriers to the use of AI technologies to promote their innovative application while protecting American technology, economic and national security, civil liberties, privacy, and values.

(d) Ensure that technical standards minimize vulnerability to attacks from malicious actors and reflect Federal priorities for innovation, public trust, and public confidence in systems that use AI technologies; and develop international standards to promote and protect those priorities.

(e) Train the next generation of American AI researchers and users through apprenticeships; skills programs; and education in science, technology, engineering, and mathematics (STEM), with an emphasis on computer science, to ensure that

American workers, including Federal workers, are capable of taking full advantage of the opportunities of AI.

(f) Develop and implement an action plan, in accordance with the National Security Presidential Memorandum of February 11, 2019 (Protecting the United States Advantage in Artificial Intelligence and Related Critical Technologies) (the NSPM) to protect the advantage of the United States in AI and technology critical to United States economic and national security interests against strategic competitors and foreign adversaries.

Sec. 3. Roles and Responsibilities. The Initiative shall be coordinated through the National Science and Technology Council (NSTC) Select Committee on Artificial Intelligence (Select Committee). Actions shall be implemented by agencies that conduct foundational AI R&D, develop and deploy applications of AI technologies, provide educational grants, and regulate and provide guidance for applications of AI technologies, as determined by the co-chairs of the NSTC Select Committee (implementing agencies).

Sec. 4. Federal Investment in AI Research and Development.

(a) Heads of implementing agencies that also perform or fund R&D (AI R&D agencies), shall consider AI as an agency R&D priority, as appropriate to their respective agencies' missions, consistent with applicable law and in accordance with the Office of Management and Budget (OMB) and the Office of Science and Technology Policy (OSTP) R&D priorities memoranda. Heads of such agencies shall take this priority into account when developing budget proposals and planning for the use of funds in Fiscal Year 2020 and in future years. Heads of these agencies shall also consider appropriate administrative actions to increase focus on AI for 2019.

(b) Heads of AI R&D agencies shall budget an amount for AI R&D that is appropriate for this prioritization.

    (i) Following the submission of the President's Budget request to the Congress, heads of such agencies shall communicate plans for achieving this prioritization to the OMB Director and the OSTP Director each fiscal year through the Networking and Information Technology Research and Development (NITRD) Program.

    (ii) Within 90 days of the enactment of appropriations for their respective agencies, heads of such agencies shall identify each year, consistent with applicable law, the programs to which the AI R&D priority will apply and estimate the total amount of such funds that will be spent on each such program. This information shall be communicated to the OMB Director and OSTP Director each fiscal year through the NITRD Program.

(c) To the extent appropriate and consistent with applicable law, heads of AI R&D agencies shall explore opportunities for collaboration with non-Federal entities, including: the private sector; academia; non-profit organizations; State, local, tribal, and territorial governments; and foreign partners and allies, so all collaborators can benefit from each other's investment and expertise in AI R&D.

<u>Sec. 5.</u> <u>Data and Computing Resources for AI Research and Development.</u>

(a) Heads of all agencies shall review their Federal data and models to identify opportunities to increase access and use by the greater non-Federal AI research community in a manner that benefits that community, while protecting safety, security, privacy, and confidentiality. Specifically, agencies shall improve data and model inventory documentation to enable discovery and usability, and shall prioritize improvements to access and quality of AI data and models based on the AI research community's user feedback.

(i)   Within 90 days of the date of this order, the OMB Director shall publish a notice in the *Federal Register* inviting the public to identify additional requests for access or quality improvements for Federal data and models that would improve AI R&D and testing. Additionally, within 90 days of the date of this order, OMB, in conjunction with the Select Committee, shall investigate barriers to access or quality limitations of Federal data and models that impede AI R&D and testing. Collectively, these actions by OMB will help to identify datasets that will facilitate non-Federal AI R&D and testing.

(ii)   Within 120 days of the date of this order, OMB, including through its interagency councils and the Select Committee, shall update implementation guidance for Enterprise Data Inventories and Source Code Inventories to support discovery and usability in AI R&D.

(iii) Within 180 days of the date of this order, and in accordance with the implementation of the Cross-Agency Priority Goal: Leveraging Federal Data as a Strategic Asset, from the March 2018 President's Management Agenda, agencies shall consider methods of improving the quality, usability, and appropriate access to priority data identified by the AI research community. Agencies shall also identify any associated resource implications.

(iv)   In identifying data and models for consideration for increased public access, agencies, in coordination with the Senior Agency Officials for Privacy established pursuant to Executive Order 13719 of February 9, 2016 (Establishment of the Federal Privacy Council), the heads of Federal statistical entities, Federal program managers, and other relevant personnel shall identify any barriers to, or requirements associated with, increased access to and use of such data and models, including:

(A)   privacy and civil liberty protections for individuals who may be affected by increased access and use, as well as confidentiality protections for individuals and other data providers;

(B)   safety and security concerns, including those related to the association or compilation of data and models;

(C)   data documentation and formatting, including the need for interoperable and machine-readable data formats;

(D)   changes necessary to ensure appropriate data and system governance; and

(E)   any other relevant considerations.

(v)   In accordance with the President's Management Agenda and the Cross-Agency Priority Goal: Leveraging Data as a Strategic Asset, agencies shall identify opportunities to use new technologies and best practices to increase access to and usability of open data and models, and explore appropriate controls on access to sensitive or restricted data and models, consistent with applicable laws and policies, privacy and confidentiality protections, and civil liberty protections.

(b)   The Secretaries of Defense, Commerce, Health and Human Services, and Energy, the Administrator of the National Aeronautics and Space Administration, and the Director of the National Science Foundation shall, to the extent appropriate and consistent with applicable law, prioritize the allocation of high-performance computing resources for AI-related applications through:

(i)   increased assignment of discretionary allocation of resources and resource reserves; or

(ii) any other appropriate mechanisms.

———

(c) Within 180 days of the date of this order, the Select Committee, in coordination with the General Services Administration (GSA), shall submit a report to the President making recommendations on better enabling the use of cloud computing resources for federally funded AI R&D.

(d) The Select Committee shall provide technical expertise to the American Technology Council on matters regarding AI and the modernization of Federal technology, data, and the delivery of digital services, as appropriate.

Sec. 6. Guidance for Regulation of AI Applications.

(a) Within 180 days of the date of this order, the OMB Director, in coordination with the OSTP Director, the Director of the Domestic Policy Council, and the Director of the National Economic Council, and in consultation with any other relevant agencies and key stakeholders as the OMB Director shall determine, shall issue a memorandum to the heads of all agencies that shall:

(i) inform the development of regulatory and non-regulatory approaches by such agencies regarding technologies and industrial sectors that are either empowered or enabled by AI, and that advance American innovation while upholding civil liberties, privacy, and American values; and

(ii) consider ways to reduce barriers to the use of AI technologies in order to promote their innovative application while protecting civil liberties, privacy, American values, and United States economic and national security.

(b) To help ensure public trust in the development and implementation of AI applications, OMB shall issue a draft version of the memorandum for public comment before it is finalized.

(c) Within 180 days of the date of the memorandum described in subsection (a) of this section, the heads of implementing agencies that also have regulatory authorities shall review their authorities relevant to applications of AI and shall submit to OMB plans to achieve consistency with the memorandum.

(d) Within 180 days of the date of this order, the Secretary of Commerce, through the Director of the National Institute of Standards and Technology (NIST), shall issue a plan for Federal engagement in the development of technical standards and related tools in support of reliable, robust, and trustworthy systems that use AI technologies. NIST shall lead the development of this plan with participation from relevant agencies as the Secretary of Commerce shall determine.

(i) Consistent with OMB Circular A-119, this plan shall include:

(A) Federal priority needs for standardization of AI systems development and deployment;

(B) identification of standards development entities in which Federal agencies should seek membership with the goal of establishing or supporting United States technical leadership roles; and

(C) opportunities for and challenges to United States leadership in standardization related to AI technologies.

(ii) This plan shall be developed in consultation with the Select Committee, as needed, and in consultation with the private sector, academia, non-governmental entities, and other stakeholders, as appropriate.

Sec. 7. AI and the American Workforce.

(a) Heads of implementing agencies that also provide educational grants shall, to the extent consistent with applicable law, consider AI as a priority area within existing Federal fellowship and service programs.

(i) Eligible programs for prioritization shall give preference to American citizens, to the extent permitted by law, and shall include:

(A) high school, undergraduate, and graduate fellowship; alternative education; and training programs;

(B) programs to recognize and fund early-career university faculty who conduct AI R&D, including through Presidential awards and recognitions;

(C) scholarship for service programs;

(D) direct commissioning programs of the United States Armed Forces; and

(E) programs that support the development of instructional programs and curricula that encourage the integration of AI technologies into courses in order to facilitate personalized and adaptive learning experiences for formal and informal education and training.

(ii) Agencies shall annually communicate plans for achieving this prioritization to the co-chairs of the Select Committee.

(b) Within 90 days of the date of this order, the Select Committee shall provide recommendations to the NSTC Committee on STEM Education regarding AI-related educational and workforce development considerations that focus on American citizens.

(c) The Select Committee shall provide technical expertise to the National Council for the American Worker on matters regarding AI and the American workforce, as appropriate.

Sec. 8. <u>Action Plan for Protection of the United States Advantage in AI Technologies.</u>

(a) As directed by the NSPM, the Assistant to the President for National Security Affairs, in coordination with the OSTP Director and the recipients of the NSPM, shall organize the development of an action plan to protect the United States advantage in AI and AI technology critical to United States economic and national security interests against strategic competitors and adversarial nations.

(b) The action plan shall be provided to the President within 120 days of the date of this order, and may be classified in full or in part, as appropriate.

(c) Upon approval by the President, the action plan shall be implemented by all agencies who are recipients of the NSPM, for all AI-related activities, including those conducted pursuant to this order.

Sec. 9. <u>Definitions.</u> As used in this order:

(a) the term "artificial intelligence" means the full extent of Federal investments in AI, to include: R&D of core AI techniques and technologies; AI prototype systems; application and adaptation of AI techniques; architectural and systems support for AI; and cyberinfrastructure, data sets, and standards for AI; and

(b) the term "open data" shall, in accordance with OMB Circular A-130 and memorandum M-13-13, mean "publicly available data structured in a way that enables the data to be fully discoverable and usable by end users."

Sec. 10. <u>General Provisions.</u>

(a) Nothing in this order shall be construed to impair or otherwise affect:

(i) the authority granted by law to an executive department or agency, or the head thereof; or

(ii) the functions of the Director of OMB relating to budgetary, administrative, or legislative proposals.

(b) This order shall be implemented consistent with applicable law and subject to the availability of appropriations.

(c) This order is not intended to, and does not, create any right or benefit, substantive or procedural, enforceable at law or in equity by any party against the United States, its departments, agencies, or entities, its officers, employees, or agents, or any other person.

DONALD J. TRUMP

THE WHITE HOUSE,
February 11, 2019

.

White House, Executive Order 13960, Promoting the Use of Trustworthy Artificial Intelligence in the Federal Government (2020)

By the authority vested in me as President by the Constitution and the laws of the United States of America, it is hereby ordered as follows:

Section 1. *Purpose.* Artificial intelligence (AI) promises to drive the growth of the United States economy and improve the quality of life of all Americans. In alignment with Executive Order 13859 of February 11, 2019 (Maintaining American Leadership in Artificial Intelligence), executive departments and agencies (agencies) have recognized the power of AI to improve their operations, processes, and procedures; meet strategic goals; reduce costs; enhance oversight of the use of taxpayer funds; increase efficiency and mission effectiveness; improve quality of services; improve safety; train workforces; and support decision making by the Federal workforce, among other positive developments. Given the broad applicability of AI, nearly every agency and those served by those agencies can benefit from the appropriate use of AI.

Agencies are already leading the way in the use of AI by applying it to accelerate regulatory reform; review Federal solicitations for regulatory compliance; combat fraud, waste, and abuse committed against taxpayers; identify information security threats and assess trends in related illicit activities; enhance the security and interoperability of Federal Government information systems; facilitate review of large datasets; streamline processes for grant applications; model weather patterns; facilitate predictive maintenance; and much more.

Agencies are encouraged to continue to use AI, when appropriate, to benefit the American people. The ongoing adoption and acceptance of AI will depend significantly on public trust. Agencies must therefore design, develop, acquire, and use AI in a manner that fosters public trust and confidence while protecting privacy, civil rights, civil liberties, and American values, consistent with applicable law and the goals of Executive Order 13859.

Certain agencies have already adopted guidelines and principles for the use of AI for national security or defense purposes, such as the Department of Defense's *Ethical Principles for Artificial Intelligence* (February 24, 2020), and the Office of the Director of National Intelligence's *Principles of Artificial Intelligence Ethics for the Intelligence Community* (July 23, 2020) and its *Artificial Intelligence Ethics Framework for the Intelligence Community* (July 23, 2020). Such guidelines and principles ensure that the use of AI in those contexts will benefit the American people and be worthy of their trust.

Section 3 of this order establishes additional principles (Principles) for the use of AI in the Federal Government for purposes other than national security and defense, to similarly ensure that such uses are consistent with our Nation's values and are beneficial to the public. This order further establishes a process for implementing these Principles through common policy guidance across agencies.

Sec. 2 . *Policy.* (a) It is the policy of the United States to promote the innovation and use of AI, where appropriate, to improve Government operations and services in a manner that fosters public trust, builds confidence in AI, protects our Nation's values, and remains consistent with all applicable laws, including those related to privacy, civil rights, and civil liberties.

(b) It is the policy of the United States that responsible agencies, as defined in section 8 of this order, shall, when considering the design, development, acquisition, and use of AI in Government, be guided by the common set of Principles set forth in section 3 of this order, which are designed to foster public trust and confidence in the use of AI, protect our Nation's values, and ensure that the use of AI remains consistent with all applicable laws, including those related to privacy, civil rights, and civil liberties.

(c) It is the policy of the United States that the Principles for the use of AI in Government shall be governed by common policy guidance issued by the Office of Management and Budget (OMB) as outlined in section 4 of this order, consistent with applicable law.

Sec. 3. *Principles for Use of AI in Government.* When designing, developing, acquiring, and using AI in the Federal Government, agencies shall adhere to the following Principles:

(a) Lawful and respectful of our Nation's values. Agencies shall design, develop, acquire, and use AI in a manner that exhibits due respect for our Nation's values and is consistent with the Constitution and all other applicable laws and policies, including those addressing privacy, civil rights, and civil liberties.

(b) Purposeful and performance-driven. Agencies shall seek opportunities for designing, developing, acquiring, and using AI, where the benefits of doing so significantly outweigh the risks, and the risks can be assessed and managed.

(c) Accurate, reliable, and effective. Agencies shall ensure that their application of AI is consistent with the use cases for which that AI was trained, and such use is accurate, reliable, and effective.

(d) Safe, secure, and resilient. Agencies shall ensure the safety, security, and resiliency of their AI applications, including resilience when confronted with systematic vulnerabilities, adversarial manipulation, and other malicious exploitation.

(e) Understandable. Agencies shall ensure that the operations and outcomes of their AI applications are sufficiently understandable by subject matter experts, users, and others, as appropriate.

(f) Responsible and traceable. Agencies shall ensure that human roles and responsibilities are clearly defined, understood, and appropriately assigned for the design, development, acquisition, and use of AI. Agencies shall ensure that AI is used in a manner consistent with these Principles and the purposes for which each use of AI is intended. The design, development, acquisition, and use of AI, as well as relevant inputs and outputs of particular AI applications, should be well documented and traceable, as appropriate and to the extent practicable.

(g) Regularly monitored. Agencies shall ensure that their AI applications are regularly tested against these Principles. Mechanisms should be maintained to supersede, disengage, or deactivate existing applications of AI that demonstrate performance or outcomes that are inconsistent with their intended use or this order.

(h) Transparent. Agencies shall be transparent in disclosing relevant information regarding their use of AI to appropriate stakeholders, including the Congress and the public, to the extent practicable and in accordance with applicable laws and policies, including with respect to the protection of privacy and of sensitive law enforcement, national security, and other protected information.

(i) Accountable. Agencies shall be accountable for implementing and enforcing appropriate safeguards for the proper use and functioning of their applications of AI, and shall monitor, audit, and document compliance with those safeguards. Agencies shall provide appropriate training to all agency personnel responsible for the design, development, acquisition, and use of AI.

Sec. 4. *Implementation of Principles.* (a) Existing OMB policies currently address many aspects of information and information technology design, development, acquisition, and use that apply, but are not unique, to AI. To the extent they are consistent with the Principles set forth in this order and applicable law, these existing policies shall continue to apply to relevant aspects of AI use in Government.

(b) Within 180 days of the date of this order, the Director of OMB (Director), in coordination with key stakeholders identified by the Director, shall publicly post a roadmap for the policy guidance that OMB intends to create or revise to better support the use of AI, consistent with this order. This roadmap shall include, where appropriate, a schedule for engaging with the public and timelines for finalizing relevant policy guidance. In addressing novel aspects of the use of AI in Government, OMB shall consider updates to the breadth of its policy guidance, including OMB Circulars and Management Memoranda.

(c) Agencies shall continue to use voluntary consensus standards developed with industry participation, where available, when such use would not be inconsistent with applicable law or otherwise impracticable. Such standards shall also be taken into consideration by OMB when revising or developing AI guidance.

Sec. 5. *Agency Inventory of AI Use Cases.* (a) Within 60 days of the date of this order, the Federal Chief Information Officers Council (CIO Council), in coordination with other interagency bodies as it deems appropriate, shall identify, provide guidance on, and make publicly available the criteria, format, and mechanisms for agency inventories of non-classified and non-sensitive use cases of AI by agencies.

(b) Within 180 days of the CIO Council's completion of the directive in section 5(a) of this order, and annually thereafter, each agency shall prepare an inventory of its non-classified and non-sensitive use cases of AI, within the scope defined by section 9 of this order, including current and planned uses, consistent with the agency's mission.

(c) As part of their respective inventories of AI use cases, agencies shall identify, review, and assess existing AI deployed and operating in support of agency missions for any inconsistencies with this order.

(i) Within 120 days of completing their respective inventories, agencies shall develop plans either to achieve consistency with this order for each AI application or to retire AI applications found to be developed or used in a manner that is not consistent with this order. These plans must be approved by the agency-designated responsible official(s), as described in section 8 of this order, within this same 120-day time period.

(ii) In coordination with the Agency Data Governance Body and relevant officials from agencies not represented within that body, agencies shall strive to implement the approved plans within 180 days of plan approval, subject to existing resource levels.

(d) Within 60 days of the completion of their respective inventories of use cases of AI, agencies shall share their inventories with other agencies, to the extent practicable and consistent with applicable law and policy, including those concerning protection of privacy and of sensitive law enforcement, national security, and other protected information. This sharing shall be coordinated through the CIO and Chief Data Officer Councils, as well as other interagency bodies, as appropriate, to improve interagency coordination and information sharing for common use cases.

(e) Within 120 days of the completion of their inventories, agencies shall make their inventories available to the public, to the extent practicable and in accordance with applicable law and policy, including those concerning the protection of privacy and of sensitive law enforcement, national security, and other protected information.

Sec. 6. *Interagency Coordination.* Agencies are expected to participate in interagency bodies for the purpose of advancing the implementation of the Principles and the use of AI consistent with this order. Within 45 days of this order, the CIO Council shall publish a list of recommended interagency bodies and forums in which agencies may elect to participate, as appropriate and consistent with their respective authorities and missions.

Sec. 7. *AI Implementation Expertise.* (a) Within 90 days of the date of this order, the Presidential Innovation Fellows (PIF) program, administered by the General Services Administration (GSA) in collaboration with other agencies, shall identify priority areas of expertise and establish an AI track to attract experts from industry and academia to undertake a period of work at an agency. These PIF experts will work within agencies to further the design, development, acquisition, and use of AI in Government, consistent with this order.

(b) Within 45 days of the date of this order, the Office of Personnel Management (OPM), in coordination with GSA and relevant agencies, shall create an inventory of Federal Government rotational programs and determine how these programs can be used to expand the number of employees with AI expertise at the agencies.

(c) Within 180 days of the creation of the inventory of Government rotational programs described in section 7(b) of this order, OPM shall issue a report with recommendations for how the programs in the inventory can be best used to expand the number of employees with AI expertise at the agencies. This report shall be shared with the interagency coordination bodies identified pursuant to section 6 of this order, enabling agencies to better use these programs for the use of AI, consistent with this order.

Sec. 8. *Responsible Agencies and Officials.* (a) For purposes of this order, the term "agency" refers to all agencies described in section 3502, subsection (1), of title 44, United States Code, except for the agencies described in section 3502, subsection (5), of title 44.

(b) This order applies to agencies that have use cases for AI that fall within the scope defined in section 9 of this order, and excludes the Department of Defense and those agencies and agency components with functions that lie wholly within the Intelligence Community. The term "Intelligence Community" has the meaning given the term in section 3003 of title 50, United States Code.

(c) Within 30 days of the date of this order, each agency shall specify the responsible official(s) at that agency who will coordinate implementation of the Principles set forth in section 3 of this order with the Agency Data Governance Body and other relevant officials and will collaborate with the interagency coordination bodies identified pursuant to section 6 of this order.

Sec. 9. *Scope of Application.* (a) This order uses the definition of AI set forth in section 238(g) of the National Defense Authorization Act for Fiscal Year 2019 as a reference point. As Federal Government use of AI matures and evolves, OMB guidance developed or revised pursuant to section 4 of this order shall include such definitions as are necessary to ensure the application of the Principles in this order to appropriate use cases.

(b) Except for the exclusions set forth in section 9(d) of this order, or provided for by applicable law, the Principles and implementation guidance in this order shall apply to AI designed, developed, acquired, or used specifically to advance the execution of agencies' missions, enhance decision making, or provide the public with a specified benefit.

(c) This order applies to both existing and new uses of AI; both stand-alone AI and AI embedded within other systems or applications; AI developed both by the agency or by third parties on behalf of agencies for the fulfilment of specific agency missions, including relevant data inputs used to train AI and outputs used in support of decision making; and agencies' procurement of AI applications.

(d) This order does not apply to:

(i) AI used in defense or national security systems (as defined in 44 U.S.C. 3552(b)(6) or as determined by the agency), in whole or in part, although agencies shall adhere to other applicable guidelines and principles for defense and national security purposes, such as those adopted by the Department of Defense and the Office of the Director of National Intelligence;

(ii) AI embedded within common commercial products, such as word processors or map navigation systems, while noting that Government use of such products must nevertheless comply with applicable law and policy to assure the protection of safety, security, privacy, civil rights, civil liberties, and American values; and

(iii) AI research and development (R&D) activities, although the Principles and OMB implementation guidance should inform any R&D directed at potential future applications of AI in the Federal Government.

Sec. 10. *General Provisions.* (a) Nothing in this order shall be construed to impair or otherwise affect:

(i) the authority granted by law to an executive department or agency, or the head thereof; or

(ii) the functions of the Director relating to budgetary, administrative, or legislative proposals.

(b) This order shall be implemented consistent with applicable law and subject to the availability of appropriations.

(c) This order is not intended to, and does not, create any right or benefit, substantive or procedural, enforceable at law or in equity by any party against the United States, its departments, agencies, or entities, its officers, employees, or agents, or any other person.

DONALD J. TRUMP

THE WHITE HOUSE,
December 3, 2020

White House, Executive Order 14110, The Safe, Secure, and Trustworthy Development and Use of Artificial Intelligence (2023)

By the authority vested in me as President by the Constitution and the laws of the United States of America, it is hereby ordered as follows:

Section 1. *Purpose*. Artificial intelligence (AI) holds extraordinary potential for both promise and peril. Responsible AI use has the potential to help solve urgent challenges while making our world more prosperous, productive, innovative, and secure. At the same time, irresponsible use could exacerbate societal harms such as fraud, discrimination, bias, and disinformation; displace and disempower workers; stifle competition; and pose risks to national security. Harnessing AI for good and realizing its myriad benefits requires mitigating its substantial risks. This endeavor demands a society-wide effort that includes government, the private sector, academia, and civil society.

My Administration places the highest urgency on governing the development and use of AI safely and responsibly, and is therefore advancing a coordinated, Federal Government-wide approach to doing so. The rapid speed at which AI capabilities are advancing compels the United States to lead in this moment for the sake of our security, economy, and society.

In the end, AI reflects the principles of the people who build it, the people who use it, and the data upon which it is built. I firmly believe that the power of our ideals; the foundations of our society; and the creativity, diversity, and decency of our people are the reasons that America thrived in past eras of rapid change. They are the reasons we will succeed again in this moment. We are more than capable of harnessing AI for justice, security, and opportunity for all.

Sec. 2. *Policy and Principles*. It is the policy of my Administration to advance and govern the development and use of AI in accordance with eight guiding principles and priorities. When undertaking the actions set forth in this order, executive departments and agencies (agencies) shall, as appropriate and consistent with applicable law, adhere to these principles, while, as feasible, taking into account the views of other agencies, industry, members of academia, civil society, labor unions, international allies and partners, and other relevant organizations:

(a) Artificial Intelligence must be safe and secure. Meeting this goal requires robust, reliable, repeatable, and standardized evaluations of AI systems, as well as policies, institutions, and, as appropriate, other mechanisms to test, understand, and mitigate risks from these systems before they are put to use. It also requires addressing AI systems' most pressing security risks—including with respect to biotechnology, cybersecurity, critical infrastructure, and other national security dangers—while navigating AI's opacity and complexity. Testing and evaluations, including post-deployment performance monitoring, will help ensure that AI systems function as intended, are resilient against misuse or dangerous modifications, are ethically developed and operated in a secure manner, and are compliant with applicable Federal laws and policies. Finally, my Administration will help develop effective labeling and content provenance mechanisms, so that Americans are able to determine when content is generated using AI and when it is not. These actions will provide a vital foundation for an approach that addresses AI's risks without unduly reducing its benefits.

(b) Promoting responsible innovation, competition, and collaboration will allow the United States to lead in AI and unlock the technology's potential to solve some of society's most difficult challenges. This effort requires investments in AI-related education, training, development, research, and capacity, while simultaneously tackling novel intellectual property (IP) questions and other problems to protect inventors and creators. Across the Federal Government, my Administration will support programs to provide Americans the skills they need for the age of AI and attract the world's AI talent to our shores—not just to study, but to stay—so that the companies and technologies of the future are made in America. The Federal Government will promote a fair, open, and competitive ecosystem and marketplace for AI and related technologies so that small developers and entrepreneurs can continue to drive innovation. Doing so requires stopping unlawful collusion and addressing risks from dominant firms' use of key assets such as semiconductors, computing power, cloud storage, and data to disadvantage competitors, and it requires supporting a marketplace that harnesses the benefits of AI to provide new opportunities for small businesses, workers, and entrepreneurs.

(c) The responsible development and use of AI require a commitment to supporting American workers. As AI creates new jobs and industries, all workers need a seat at the table, including through collective bargaining, to ensure that they benefit from these opportunities. My Administration will seek to adapt job training and education to support a diverse workforce and help provide access to opportunities that AI creates. In the workplace itself, AI should not be deployed in ways that undermine rights, worsen job quality, encourage undue worker surveillance, lessen market competition, introduce new health and safety risks, or cause harmful labor-force disruptions. The critical next steps in AI development should be built on the views of workers, labor unions, educators, and employers to support responsible uses of AI that improve workers' lives, positively augment human work, and help all people safely enjoy the gains and opportunities from technological innovation.

———

376

(d) Artificial Intelligence policies must be consistent with my Administration's dedication to advancing equity and civil rights. My Administration cannot—and will not—tolerate the use of AI to disadvantage those who are already too often denied equal opportunity and justice. From hiring to housing to healthcare, we have seen what happens when AI use deepens discrimination and bias, rather than improving quality of life. Artificial Intelligence systems deployed irresponsibly have reproduced and intensified existing inequities, caused new types of harmful discrimination, and exacerbated online and physical harms. My Administration will build on the important steps that have already been taken—such as issuing the Blueprint for an AI Bill of Rights, the AI Risk Management Framework, and Executive Order 14091 of February 16, 2023 (Further Advancing Racial Equity and Support for Underserved Communities Through the Federal Government)—in seeking to ensure that AI complies with all Federal laws and to promote robust technical evaluations, careful oversight, engagement with affected communities, and rigorous regulation. It is necessary to hold those developing and deploying AI accountable to standards that protect against unlawful discrimination and abuse, including in the justice system and the Federal Government. Only then can Americans trust AI to advance civil rights, civil liberties, equity, and justice for all.

(e) The interests of Americans who increasingly use, interact with, or purchase AI and AI-enabled products in their daily lives must be protected. Use of new technologies, such as AI, does not excuse organizations from their legal obligations, and hard-won consumer protections are more important than ever in moments of technological change. The Federal Government will enforce existing consumer protection laws and principles and enact appropriate safeguards against fraud, unintended bias, discrimination, infringements on privacy, and other harms from AI. Such protections are especially important in critical fields like healthcare, financial services, education, housing, law, and transportation, where mistakes by or misuse of AI could harm patients, cost consumers or small businesses, or jeopardize safety or rights. At the same time, my Administration will promote responsible uses of AI that protect consumers, raise the quality of goods and services, lower their prices, or expand selection and availability.

(f) Americans' privacy and civil liberties must be protected as AI continues advancing. Artificial Intelligence is making it easier to extract, re-identify, link, infer, and act on sensitive information about people's identities, locations, habits, and desires. Artificial Intelligence's capabilities in these areas can increase the risk that personal data could be exploited and exposed. To combat this risk, the Federal Government will ensure that the collection, use, and retention of data is lawful, is secure, and mitigates privacy and confidentiality risks. Agencies shall use available policy and technical tools, including privacy-enhancing technologies (PETs) where appropriate, to protect privacy and to combat the broader legal and societal risks—

including the chilling of First Amendment rights—that result from the improper collection and use of people's data.

(g) It is important to manage the risks from the Federal Government's own use of AI and increase its internal capacity to regulate, govern, and support responsible use of AI to deliver better results for Americans. These efforts start with people, our Nation's greatest asset. My Administration will take steps to attract, retain, and develop public service-oriented AI professionals, including from underserved communities, across disciplines—including technology, policy, managerial, procurement, regulatory, ethical, governance, and legal fields—and ease AI professionals' path into the Federal Government to help harness and govern AI. The Federal Government will work to ensure that all members of its workforce receive adequate training to understand the benefits, risks, and limitations of AI for their job functions, and to modernize Federal Government information technology infrastructure, remove bureaucratic obstacles, and ensure that safe and rights-respecting AI is adopted, deployed, and used.

(h) The Federal Government should lead the way to global societal, economic, and technological progress, as the United States has in previous eras of disruptive innovation and change. This leadership is not measured solely by the technological advancements our country makes. Effective leadership also means pioneering those systems and safeguards needed to deploy technology responsibly—and building and promoting those safeguards with the rest of the world. My Administration will engage with international allies and partners in developing a framework to manage AI's risks, unlock AI's potential for good, and promote common approaches to shared challenges. The Federal Government will seek to promote responsible AI safety and security principles and actions with other nations, including our competitors, while leading key global conversations and collaborations to ensure that AI benefits the whole world, rather than exacerbating inequities, threatening human rights, and causing other harms.

Sec. 3. *Definitions*. For purposes of this order:

(a) The term "agency" means each agency described in 44 U.S.C. 3502(1), except for the independent regulatory agencies described in 44 U.S.C. 3502(5).

(b) The term "artificial intelligence" or "AI" has the meaning set forth in 15 U.S.C. 9401(3): a machine-based system that can, for a given set of human-defined objectives, make predictions, recommendations, or decisions influencing real or virtual environments. Artificial intelligence systems use machine- and human-based inputs to perceive real and virtual environments; abstract such perceptions into models through analysis in an automated manner; and use model inference to formulate options for information or action.

(c) The term "AI model" means a component of an information system that implements AI technology and uses computational, statistical, or machine-learning techniques to produce outputs from a given set of inputs.

(d) The term "AI red-teaming" means a structured testing effort to find flaws and vulnerabilities in an AI system, often in a controlled environment and in collaboration with developers of AI. Artificial Intelligence red-teaming is most often performed by dedicated "red teams" that adopt adversarial methods to identify flaws and vulnerabilities, such as harmful or discriminatory outputs from an AI system, unforeseen or undesirable system behaviors, limitations, or potential risks associated with the misuse of the system.

(e) The term "AI system" means any data system, software, hardware, application, tool, or utility that operates in whole or in part using AI.

(f) The term "commercially available information" means any information or data about an individual or group of individuals, including an individual's or group of individuals' device or location, that is made available or obtainable and sold, leased, or licensed to the general public or to governmental or non-governmental entities.

(g) The term "crime forecasting" means the use of analytical techniques to attempt to predict future crimes or crime-related information. It can include machine-generated predictions that use algorithms to analyze large volumes of data, as well as other forecasts that are generated without machines and based on statistics, such as historical crime statistics.

(h) The term "critical and emerging technologies" means those technologies listed in the February 2022 Critical and Emerging Technologies List Update issued by the National Science and Technology Council (NSTC), as amended by subsequent updates to the list issued by the NSTC.

(i) The term "critical infrastructure" has the meaning set forth in section 1016(e) of the USA PATRIOT Act of 2001, 42 U.S.C. 5195c(e).

(j) The term "differential-privacy guarantee" means protections that allow information about a group to be shared while provably limiting the improper access, use, or disclosure of personal information about particular entities.

(k) The term "dual-use foundation model" means an AI model that is trained on broad data; generally uses self-supervision; contains at least tens of billions of parameters; is applicable across a wide range of contexts; and that exhibits, or could be easily modified to exhibit, high levels of performance at tasks that pose a serious risk to security, national economic security, national public health or safety, or any combination of those matters, such as by:

(i) substantially lowering the barrier of entry for non-experts to design, synthesize, acquire, or use chemical, biological, radiological, or nuclear (CBRN) weapons;

(ii) enabling powerful offensive cyber operations through automated vulnerability discovery and exploitation against a wide range of potential targets of cyber attacks; or

(iii) permitting the evasion of human control or oversight through means of deception or obfuscation.

Models meet this definition even if they are provided to end users with technical safeguards that attempt to prevent users from taking advantage of the relevant unsafe capabilities.

(l) The term "Federal law enforcement agency" has the meaning set forth in section 21(a) of Executive Order 14074 of May 25, 2022 (Advancing Effective, Accountable Policing and Criminal Justice Practices To Enhance Public Trust and Public Safety).

(m) The term "floating-point operation" means any mathematical operation or assignment involving floating-point numbers, which are a subset of the real numbers typically represented on computers by an integer of fixed precision scaled by an integer exponent of a fixed base.

(n) The term "foreign person" has the meaning set forth in section 5(c) of Executive Order 13984 of January 19, 2021 (Taking Additional Steps To Address the National Emergency With Respect to Significant Malicious Cyber-Enabled Activities).

(o) The terms "foreign reseller" and "foreign reseller of United States Infrastructure as a Service Products" mean a foreign person who has established an Infrastructure as a Service Account to provide Infrastructure as a Service Products subsequently, in whole or in part, to a third party.

(p) The term "generative AI" means the class of AI models that emulate the structure and characteristics of input data in order to generate derived synthetic content. This can include images, videos, audio, text, and other digital content.

(q) The terms "Infrastructure as a Service Product," "United States Infrastructure as a Service Product," "United States Infrastructure as a Service Provider," and "Infrastructure as a Service Account" each have the respective meanings given to those terms in section 5 of Executive Order 13984.

(r) The term "integer operation" means any mathematical operation or assignment involving only integers, or whole numbers expressed without a decimal point.

(s) The term "Intelligence Community" has the meaning given to that term in section 3.5(h) of Executive Order 12333 of December 4, 1981 (United States Intelligence Activities), as amended.

(t) The term "machine learning" means a set of techniques that can be used to train AI algorithms to improve performance at a task based on data.

(u) The term "model weight" means a numerical parameter within an AI model that helps determine the model's outputs in response to inputs.

(v) The term "national security system" has the meaning set forth in 44 U.S.C. 3552(b)(6).

(w) The term "omics" means biomolecules, including nucleic acids, proteins, and metabolites, that make up a cell or cellular system.

(x) The term "Open RAN" means the Open Radio Access Network approach to telecommunications-network standardization adopted by the O-RAN Alliance, Third Generation Partnership Project, or any similar set of published open standards for multi-vendor network equipment interoperability.

(y) The term "personally identifiable information" has the meaning set forth in Office of Management and Budget (OMB) Circular No. A-130.

(z) The term "privacy-enhancing technology" means any software or hardware solution, technical process, technique, or other technological means of mitigating privacy risks arising from data processing, including by enhancing predictability, manageability, disassociability, storage, security, and confidentiality. These technological means may include secure multiparty computation, homomorphic encryption, zero-knowledge proofs, federated learning, secure enclaves, differential privacy, and synthetic-data-generation tools. This is also sometimes referred to as "privacy-preserving technology."

(aa) The term "privacy impact assessment" has the meaning set forth in OMB Circular No. A-130.

(bb) The term "Sector Risk Management Agency" has the meaning set forth in 6 U.S.C. 650(23).

(cc) The term "self-healing network" means a telecommunications network that automatically diagnoses and addresses network issues to permit self-restoration.

(dd) The term "synthetic biology" means a field of science that involves redesigning organisms, or the biomolecules of organisms, at the genetic level to give them new characteristics. Synthetic nucleic acids are a type of biomolecule redesigned through synthetic-biology methods.

(ee) The term "synthetic content" means information, such as images, videos, audio clips, and text, that has been significantly modified or generated by algorithms, including by AI.

(ff) The term "testbed" means a facility or mechanism equipped for conducting rigorous, transparent, and replicable testing of tools and technologies, including AI and PETs, to help evaluate the functionality, usability, and performance of those tools or technologies.

(gg) The term "watermarking" means the act of embedding information, which is typically difficult to remove, into outputs created by AI—including into outputs such as photos, videos, audio clips, or text—for the purposes of verifying the authenticity of the output or the identity or characteristics of its provenance, modifications, or conveyance.

Sec. 4. *Ensuring the Safety and Security of AI Technology.*

4.1. *Developing Guidelines, Standards, and Best Practices for AI Safety and Security.* (a) Within 270 days of the date of this order, to help ensure the development of safe, secure, and trustworthy AI systems, the Secretary of Commerce, acting through the Director of the National Institute of Standards and Technology (NIST), in coordination with the Secretary of Energy, the Secretary of Homeland Security, and the heads of other relevant agencies as the Secretary of Commerce may deem appropriate, shall:

(i) Establish guidelines and best practices, with the aim of promoting consensus industry standards, for developing and deploying safe, secure, and trustworthy AI systems, including:

(A) developing a companion resource to the AI Risk Management Framework, NIST AI 100-1, for generative AI;

382

(B) developing a companion resource to the Secure Software Development Framework to incorporate secure development practices for generative AI and for dual-use foundation models; and

(C) launching an initiative to create guidance and benchmarks for evaluating and auditing AI capabilities, with a focus on capabilities through which AI could cause harm, such as in the areas of cybersecurity and biosecurity.

(ii) Establish appropriate guidelines (except for AI used as a component of a national security system), including appropriate procedures and processes, to enable developers of AI, especially of dual-use foundation models, to conduct AI red-teaming tests to enable deployment of safe, secure, and trustworthy systems. These efforts shall include:

(A) coordinating or developing guidelines related to assessing and managing the safety, security, and trustworthiness of dual-use foundation models; and

(B) in coordination with the Secretary of Energy and the Director of the National Science Foundation (NSF), developing and helping to ensure the availability of testing environments, such as testbeds, to support the development of safe, secure, and trustworthy AI technologies, as well as to support the design, development, and deployment of associated PETs, consistent with section 9(b) of this order.

(b) Within 270 days of the date of this order, to understand and mitigate AI security risks, the Secretary of Energy, in coordination with the heads of other Sector Risk Management Agencies (SRMAs) as the Secretary of Energy may deem appropriate, shall develop and, to the extent permitted by law and available appropriations, implement a plan for developing the Department of Energy's AI model evaluation tools and AI testbeds. The Secretary shall undertake this work using existing solutions where possible, and shall develop these tools and AI testbeds to be capable of assessing near-term extrapolations of AI systems' capabilities. At a minimum, the Secretary shall develop tools to evaluate AI capabilities to generate outputs that may represent nuclear, non-proliferation, biological, chemical, critical infrastructure, and energy-security threats or hazards. The Secretary shall do this work solely for the purposes of guarding against these threats, and shall also develop model guardrails that reduce such risks. The Secretary shall, as appropriate, consult with private AI laboratories, academia, civil society, and third-party evaluators, and shall use existing solutions.

4.2. *Ensuring Safe and Reliable AI.* (a) Within 90 days of the date of this order, to ensure and verify the continuous availability of safe, reliable, and effective AI in accordance with the Defense Production Act, as amended, 50 U.S.C. 4501 ET SEQ., including for the national defense and the protection of critical infrastructure, the Secretary of Commerce shall require:

(i) Companies developing or demonstrating an intent to develop potential dual-use foundation models to provide the Federal Government, on an ongoing basis, with information, reports, or records regarding the following:

(A) any ongoing or planned activities related to training, developing, or producing dual-use foundation models, including the physical and cybersecurity protections taken to assure the integrity of that training process against sophisticated threats;

(B) the ownership and possession of the model weights of any dual-use foundation models, and the physical and cybersecurity measures taken to protect those model weights; and

(C) the results of any developed dual-use foundation model's performance in relevant AI red-team testing based on guidance developed by NIST pursuant to subsection 4.1(a)(ii) of this section, and a description of any associated measures the company has taken to meet safety objectives, such as mitigations to improve performance on these red-team tests and strengthen overall model security. Prior to the development of guidance on red-team testing standards by NIST pursuant to subsection 4.1(a)(ii) of this section, this description shall include the results of any red-team testing that the company has conducted relating to lowering the barrier to entry for the development, acquisition, and use of biological weapons by non-state actors; the discovery of software vulnerabilities and development of associated exploits; the use of software or tools to influence real or virtual events; the possibility for self-replication or propagation; and associated measures to meet safety objectives; and

(ii) Companies, individuals, or other organizations or entities that acquire, develop, or possess a potential large-scale computing cluster to report any such acquisition, development, or possession, including the existence and location of these clusters and the amount of total computing power available in each cluster.

(b) The Secretary of Commerce, in consultation with the Secretary of State, the Secretary of Defense, the Secretary of Energy, and the Director of National Intelligence, shall define, and thereafter update as needed on a regular basis, the set of technical conditions for models and computing clusters that would be subject to the reporting requirements of subsection 4.2(a) of this section. Until such technical conditions are defined, the Secretary shall require compliance with these reporting requirements for:

(i) any model that was trained using a quantity of computing power greater than 10^{26} integer or floating-point operations, or using primarily biological sequence data and using a quantity of computing power greater than 10^{23} integer or floating-point operations; and

(ii) any computing cluster that has a set of machines physically co-located in a single datacenter, transitively connected by data center networking of over 100 Gbit/s, and having a theoretical maximum computing capacity of 10^{20} integer or floating-point operations per second for training AI.

(c) Because I find that additional steps must be taken to deal with the national emergency related to significant malicious cyber-enabled activities declared in Executive Order 13694 of April 1, 2015 (Blocking the Property of Certain Persons Engaging in Significant Malicious Cyber-Enabled Activities), as amended by Executive Order 13757 of December 28, 2016 (Taking Additional Steps to Address the National Emergency With Respect to Significant Malicious Cyber-Enabled Activities), and further amended by Executive Order 13984, to address the use of United States Infrastructure as a Service (IaaS) Products by foreign malicious cyber actors, including to impose additional record-keeping obligations with respect to foreign transactions and to assist in the investigation of transactions involving foreign malicious cyber actors, I hereby direct the Secretary of Commerce, within 90 days of the date of this order, to:

(i) Propose regulations that require United States IaaS Providers to submit a report to the Secretary of Commerce when a foreign person transacts with that United States IaaS Provider to train a large AI model with potential capabilities that could be used in malicious cyber-enabled activity (a "training run"). Such reports shall include, at a minimum, the identity of the foreign person and the existence of any training run of an AI model meeting the criteria set forth in this section, or other criteria defined by the Secretary in regulations, as well as any additional information identified by the Secretary.

(ii) Include a requirement in the regulations proposed pursuant to subsection 4.2(c)(i) of this section that United States IaaS Providers prohibit any foreign reseller of their United States IaaS Product from providing those products unless such foreign reseller submits to the United States IaaS Provider a report, which the United States IaaS Provider must provide to the Secretary of Commerce, detailing each instance in which a foreign person transacts with the foreign reseller to use the United States IaaS Product to conduct a training run described in subsection 4.2(c)(i) of this section. Such reports shall include, at a minimum, the information specified in subsection 4.2(c)(i) of this section as well as any additional information identified by the Secretary.

(iii) Determine the set of technical conditions for a large AI model to have potential capabilities that could be used in malicious cyber-enabled activity, and revise that determination as necessary and appropriate. Until the Secretary makes such a determination, a model shall be considered to have potential capabilities that could be used in malicious cyber-enabled activity if it requires a quantity of computing power greater than 10^{26} integer or floating-point operations and is trained on a computing cluster that has a set of machines physically co-located in a single datacenter, transitively connected by data center networking of over 100 Gbit/s, and having a theoretical maximum compute capacity of 10^{20} integer or floating-point operations per second for training AI.

(d) Within 180 days of the date of this order, pursuant to the finding set forth in subsection 4.2(c) of this section, the Secretary of Commerce shall propose regulations that require United States IaaS Providers to ensure that foreign resellers of United States IaaS Products verify the identity of any foreign person that obtains an IaaS account (account) from the foreign reseller. These regulations shall, at a minimum:

(i) Set forth the minimum standards that a United States IaaS Provider must require of foreign resellers of its United States IaaS Products to verify the identity of a foreign person who opens an account or maintains an existing account with a foreign reseller, including:

(A) the types of documentation and procedures that foreign resellers of United States IaaS Products must require to verify the identity of any foreign person acting as a lessee or sub-lessee of these products or services;

(B) records that foreign resellers of United States IaaS Products must securely maintain regarding a foreign person that obtains an account, including information establishing:

(1) the identity of such foreign person, including name and address;

(2) the means and source of payment (including any associated financial institution and other identifiers such as credit card number, account number, customer identifier, transaction identifiers, or virtual currency wallet or wallet address identifier);

(3) the electronic mail address and telephonic contact information used to verify a foreign person's identity; and

(4) the internet Protocol addresses used for access or administration and the date and time of each such access or administrative action related to ongoing verification of such foreign person's ownership of such an account; and

(C) methods that foreign resellers of United States IaaS Products must implement to limit all third-party access to the information described in this subsection, except insofar as such access is otherwise consistent with this order and allowed under applicable law;

(ii) Take into consideration the types of accounts maintained by foreign resellers of United States IaaS Products, methods of opening an account, and types of identifying information available to accomplish the objectives of identifying foreign malicious cyber actors using any such products and avoiding the imposition of an undue burden on such resellers; and

(iii) Provide that the Secretary of Commerce, in accordance with such standards and procedures as the Secretary may delineate and in consultation with the Secretary of Defense, the Attorney General, the Secretary of Homeland Security, and the Director of National Intelligence, may exempt a United States IaaS Provider with respect to any specific foreign reseller of their United States IaaS Products, or with respect to any specific type of account or lessee, from the requirements of any regulation issued pursuant to this subsection. Such standards and procedures may include a finding by the Secretary that such foreign reseller, account, or lessee complies with security best practices to otherwise deter abuse of United States IaaS Products.

(e) The Secretary of Commerce is hereby authorized to take such actions, including the promulgation of rules and regulations, and to employ all powers granted to the President by the International Emergency Economic Powers Act, 50 U.S.C. 1701 ET SEQ., as may be necessary to carry out the purposes of subsections 4.2(c) and (d) of this section. Such actions may include a requirement that United States IaaS Providers require foreign resellers of United States IaaS Products to provide United States IaaS Providers verifications relative to those subsections.

4.3. *Managing AI in Critical Infrastructure and in Cybersecurity*. (a) To ensure the protection of critical infrastructure, the following actions shall be taken:

(i) Within 90 days of the date of this order, and at least annually thereafter, the head of each agency with relevant regulatory authority over critical infrastructure and the heads of relevant SRMAs, in coordination with the Director of the Cybersecurity and Infrastructure Security Agency within the Department of Homeland Security for consideration of cross-sector risks, shall evaluate and provide to the Secretary of

Homeland Security an assessment of potential risks related to the use of AI in critical infrastructure sectors involved, including ways in which deploying AI may make critical infrastructure systems more vulnerable to critical failures, physical attacks, and cyber attacks, and shall consider ways to mitigate these vulnerabilities. Independent regulatory agencies are encouraged, as they deem appropriate, to contribute to sector-specific risk assessments.

(ii) Within 150 days of the date of this order, the Secretary of the Treasury shall issue a public report on best practices for financial institutions to manage AI-specific cybersecurity risks.

(iii) Within 180 days of the date of this order, the Secretary of Homeland Security, in coordination with the Secretary of Commerce and with SRMAs and other regulators as determined by the Secretary of Homeland Security, shall incorporate as appropriate the AI Risk Management Framework, NIST AI 100-1, as well as other appropriate security guidance, into relevant safety and security guidelines for use by critical infrastructure owners and operators.

(iv) Within 240 days of the completion of the guidelines described in subsection 4.3(a)(iii) of this section, the Assistant to the President for National Security Affairs and the Director of OMB, in consultation with the Secretary of Homeland Security, shall coordinate work by the heads of agencies with authority over critical infrastructure to develop and take steps for the Federal Government to mandate such guidelines, or appropriate portions thereof, through regulatory or other appropriate action. Independent regulatory agencies are encouraged, as they deem appropriate, to consider whether to mandate guidance through regulatory action in their areas of authority and responsibility.

(v) The Secretary of Homeland Security shall establish an Artificial Intelligence Safety and Security Board as an advisory committee pursuant to section 871 of the Homeland Security Act of 2002 (Public Law 107-296). The Advisory Committee shall include AI experts from the private sector, academia, and government, as appropriate, and provide to the Secretary of Homeland Security and the Federal Government's critical infrastructure community advice, information, or recommendations for improving security, resilience, and incident response related to AI usage in critical infrastructure.

(b) To capitalize on AI's potential to improve United States cyber defenses:

(i) The Secretary of Defense shall carry out the actions described in subsections 4.3(b)(ii) and (iii) of this section for national security systems, and the Secretary of Homeland Security shall carry out these actions for non-national security systems. Each shall do so in consultation with the heads of other relevant agencies as the

Secretary of Defense and the Secretary of Homeland Security may deem appropriate.

(ii) As set forth in subsection 4.3(b)(i) of this section, within 180 days of the date of this order, the Secretary of Defense and the Secretary of Homeland Security shall, consistent with applicable law, each develop plans for, conduct, and complete an operational pilot project to identify, develop, test, evaluate, and deploy AI capabilities, such as large-language models, to aid in the discovery and remediation of vulnerabilities in critical United States Government software, systems, and networks.

(iii) As set forth in subsection 4.3(b)(i) of this section, within 270 days of the date of this order, the Secretary of Defense and the Secretary of Homeland Security shall each provide a report to the Assistant to the President for National Security Affairs on the results of actions taken pursuant to the plans and operational pilot projects required by subsection 4.3(b)(ii) of this section, including a description of any vulnerabilities found and fixed through the development and deployment of AI capabilities and any lessons learned on how to identify, develop, test, evaluate, and deploy AI capabilities effectively for cyber defense.

4.4. *Reducing Risks at the Intersection of AI and CBRN Threats.* (a) To better understand and mitigate the risk of AI being misused to assist in the development or use of CBRN threats—with a particular focus on biological weapons—the following actions shall be taken:

(i) Within 180 days of the date of this order, the Secretary of Homeland Security, in consultation with the Secretary of Energy and the Director of the Office of Science and Technology Policy (OSTP), shall evaluate the potential for AI to be misused to enable the development or production of CBRN threats, while also considering the benefits and application of AI to counter these threats, including, as appropriate, the results of work conducted under section 8(b) of this order. The Secretary of Homeland Security shall:

(A) consult with experts in AI and CBRN issues from the Department of Energy, private AI laboratories, academia, and third-party model evaluators, as appropriate, to evaluate AI model capabilities to present CBRN threats—for the sole purpose of guarding against those threats—as well as options for minimizing the risks of AI model misuse to generate or exacerbate those threats; and

(B) submit a report to the President that describes the progress of these efforts, including an assessment of the types of AI models that may present CBRN risks to the United States, and that makes recommendations for regulating or overseeing the training, deployment, publication, or use of these models, including requirements for safety evaluations and guardrails for mitigating potential threats to national security.

(ii) Within 120 days of the date of this order, the Secretary of Defense, in consultation with the Assistant to the President for National Security Affairs and the Director of OSTP, shall enter into a contract with the National Academies of Sciences, Engineering, and Medicine to conduct—and submit to the Secretary of Defense, the Assistant to the President for National Security Affairs, the Director of the Office of Pandemic Preparedness and Response Policy, the Director of OSTP, and the Chair of the Chief Data Officer Council—a study that:

(A) assesses the ways in which AI can increase biosecurity risks, including risks from generative AI models trained on biological data, and makes recommendations on how to mitigate these risks;

(B) considers the national security implications of the use of data and datasets, especially those associated with pathogens and omics studies, that the United States Government hosts, generates, funds the creation of, or otherwise owns, for the training of generative AI models, and makes recommendations on how to mitigate the risks related to the use of these data and datasets;

(C) assesses the ways in which AI applied to biology can be used to reduce biosecurity risks, including recommendations on opportunities to coordinate data and high-performance computing resources; and

(D) considers additional concerns and opportunities at the intersection of AI and synthetic biology that the Secretary of Defense deems appropriate.

(b) To reduce the risk of misuse of synthetic nucleic acids, which could be substantially increased by AI's capabilities in this area, and improve biosecurity measures for the nucleic acid synthesis industry, the following actions shall be taken:

(i) Within 180 days of the date of this order, the Director of OSTP, in consultation with the Secretary of State, the Secretary of Defense, the Attorney General, the Secretary of Commerce, the Secretary of Health and Human Services (HHS), the Secretary of Energy, the Secretary of Homeland Security, the Director of National Intelligence, and the heads of other relevant agencies as the Director of OSTP may deem appropriate, shall establish a framework, incorporating, as appropriate, existing United States Government guidance, to encourage providers of synthetic nucleic acid sequences to implement comprehensive, scalable, and verifiable synthetic nucleic acid procurement screening mechanisms, including standards and recommended incentives. As part of this framework, the Director of OSTP shall:

(A) establish criteria and mechanisms for ongoing identification of biological sequences that could be used in a manner that would pose a risk to the national security of the United States; and

(B) determine standardized methodologies and tools for conducting and verifying the performance of sequence synthesis procurement screening, including customer screening approaches to support due diligence with respect to managing security risks posed by purchasers of biological sequences identified in subsection 4.4(b)(i)(A) of this section, and processes for the reporting of concerning activity to enforcement entities.

(ii) Within 180 days of the date of this order, the Secretary of Commerce, acting through the Director of NIST, in coordination with the Director of OSTP, and in consultation with the Secretary of State, the Secretary of HHS, and the heads of other relevant agencies as the Secretary of Commerce may deem appropriate, shall initiate an effort to engage with industry and relevant stakeholders, informed by the framework developed under subsection 4.4(b)(i) of this section, to develop and refine for possible use by synthetic nucleic acid sequence providers:

(A) specifications for effective nucleic acid synthesis procurement screening;

(B) best practices, including security and access controls, for managing sequence-of-concern databases to support such screening;

(C) technical implementation guides for effective screening; and

(D) conformity-assessment best practices and mechanisms.

(iii) Within 180 days of the establishment of the framework pursuant to subsection 4.4(b)(i) of this section, all agencies that fund life-sciences research shall, as appropriate and consistent with applicable law, establish that, as a requirement of funding, synthetic nucleic acid procurement is conducted through providers or manufacturers that adhere to the framework, such as through an attestation from the provider or manufacturer. The Assistant to the President for National Security Affairs and the Director of OSTP shall coordinate the process of reviewing such funding requirements to facilitate consistency in implementation of the framework across funding agencies.

(iv) In order to facilitate effective implementation of the measures described in subsections 4.4(b)(i)-(iii) of this section, the Secretary of Homeland Security, in consultation with the heads of other relevant agencies as the Secretary of Homeland Security may deem appropriate, shall:

(A) within 180 days of the establishment of the framework pursuant to subsection 4.4(b)(i) of this section, develop a framework to conduct structured evaluation and stress testing of nucleic acid synthesis procurement screening, including the systems developed in accordance with subsections 4.4(b)(i)-(ii) of this section and implemented by providers of synthetic nucleic acid sequences; and

(B) following development of the framework pursuant to subsection 4.4(b)(iv)(A) of this section, submit an annual report to the Assistant to the President for National Security Affairs, the Director of the Office of Pandemic Preparedness and Response Policy, and the Director of OSTP on any results of the activities conducted pursuant to subsection 4.4(b)(iv)(A) of this section, including recommendations, if any, on how to strengthen nucleic acid synthesis procurement screening, including customer screening systems.

4.5. REDUCING THE RISKS POSED BY SYNTHETIC CONTENT. To foster capabilities for identifying and labeling synthetic content produced by AI systems, and to establish the authenticity and provenance of digital content, both synthetic and not synthetic, produced by the Federal Government or on its behalf:

(a) Within 240 days of the date of this order, the Secretary of Commerce, in consultation with the heads of other relevant agencies as the Secretary of Commerce may deem appropriate, shall submit a report to the Director of OMB and the Assistant to the President for National Security Affairs identifying the existing standards, tools, methods, and practices, as well as the potential development of further science-backed standards and techniques, for:

(i) authenticating content and tracking its provenance;

(ii) labeling synthetic content, such as using watermarking;

(iii) detecting synthetic content;

(iv) preventing generative AI from producing child sexual abuse material or producing non-consensual intimate imagery of real individuals (to include intimate digital depictions of the body or body parts of an identifiable individual);

(v) testing software used for the above purposes; and

(vi) auditing and maintaining synthetic content.

(b) Within 180 days of submitting the report required under subsection 4.5(a) of this section, and updated periodically thereafter, the Secretary of Commerce, in coordination with the Director of OMB, shall develop guidance regarding the existing tools and practices for digital content authentication and synthetic content detection measures. The guidance shall include measures for the purposes listed in subsection 4.5(a) of this section.

(c) Within 180 days of the development of the guidance required under subsection 4.5(b) of this section, and updated periodically thereafter, the Director of OMB, in consultation with the Secretary of State; the Secretary of Defense; the Attorney General; the Secretary of Commerce, acting through the Director of NIST; the Secretary of Homeland Security; the Director of National Intelligence; and the heads of other agencies that the Director of OMB deems appropriate, shall—for the purpose of strengthening public confidence in the integrity of official United States Government digital content—issue guidance to agencies for labeling and authenticating such content that they produce or publish.

(d) The Federal Acquisition Regulatory Council shall, as appropriate and consistent with applicable law, consider amending the Federal Acquisition Regulation to take into account the guidance established under subsection 4.5 of this section.

4.6. *Soliciting Input on Dual-Use Foundation Models with Widely Available Model Weights*. When the weights for a dual-use foundation model are widely available— such as when they are publicly posted on the internet—there can be substantial benefits to innovation, but also substantial security risks, such as the removal of safeguards within the model. To address the risks and potential benefits of dual-use foundation models with widely available weights, within 270 days of the date of this order, the Secretary of Commerce, acting through the Assistant Secretary of Commerce for Communications and Information, and in consultation with the Secretary of State, shall:

(a) solicit input from the private sector, academia, civil society, and other stakeholders through a public consultation process on potential risks, benefits, other implications, and appropriate policy and regulatory approaches related to dual-use foundation models for which the model weights are widely available, including:

(i) risks associated with actors fine-tuning dual-use foundation models for which the model weights are widely available or removing those models' safeguards;

(ii) benefits to AI innovation and research, including research into AI safety and risk management, of dual-use foundation models for which the model weights are widely available; and

(iii) potential voluntary, regulatory, and international mechanisms to manage the risks and maximize the benefits of dual-use foundation models for which the model weights are widely available; and

(b) based on input from the process described in subsection 4.6(a) of this section, and in consultation with the heads of other relevant agencies as the Secretary of Commerce deems appropriate, submit a report to the President on the potential benefits, risks, and implications of dual-use foundation models for which the model weights are widely available, as well as policy and regulatory recommendations pertaining to those models.

4.7. *Promoting Safe Release and Preventing the Malicious Use of Federal Data for AI Training.* To improve public data access and manage security risks, and consistent with the objectives of the Open, Public, Electronic, and Necessary Government Data Act (title II of Public Law 115-435) to expand public access to Federal data assets in a machine-readable format while also taking into account security considerations, including the risk that information in an individual data asset in isolation does not pose a security risk but, when combined with other available information, may pose such a risk:

(a) within 270 days of the date of this order, the Chief Data Officer Council, in consultation with the Secretary of Defense, the Secretary of Commerce, the Secretary of Energy, the Secretary of Homeland Security, and the Director of National Intelligence, shall develop initial guidelines for performing security reviews, including reviews to identify and manage the potential security risks of releasing Federal data that could aid in the development of CBRN weapons as well as the development of autonomous offensive cyber capabilities, while also providing public access to Federal Government data in line with the goals stated in the Open, Public, Electronic, and Necessary Government Data Act (title II of Public Law 115-435); and

(b) within 180 days of the development of the initial guidelines required by subsection 4.7(a) of this section, agencies shall conduct a security review of all data assets in the comprehensive data inventory required under 44 U.S.C. 3511(a)(1) and (2)(B) and shall take steps, as appropriate and consistent with applicable law, to address the highest-priority potential security risks that releasing that data could raise with respect to CBRN weapons, such as the ways in which that data could be used to train AI systems.

4.8. *Directing the Development of a National Security Memorandum.* To develop a coordinated executive branch approach to managing AI's security risks, the Assistant to the President for National Security Affairs and the Assistant to the

President and Deputy Chief of Staff for Policy shall oversee an interagency process with the purpose of, within 270 days of the date of this order, developing and submitting a proposed National Security Memorandum on AI to the President. The memorandum shall address the governance of AI used as a component of a national security system or for military and intelligence purposes. The memorandum shall take into account current efforts to govern the development and use of AI for national security systems. The memorandum shall outline actions for the Department of Defense, the Department of State, other relevant agencies, and the Intelligence Community to address the national security risks and potential benefits posed by AI. In particular, the memorandum shall:

(a) provide guidance to the Department of Defense, other relevant agencies, and the Intelligence Community on the continued adoption of AI capabilities to advance the United States national security mission, including through directing specific AI assurance and risk-management practices for national security uses of AI that may affect the rights or safety of United States persons and, in appropriate contexts, non-United States persons; and

(b) direct continued actions, as appropriate and consistent with applicable law, to address the potential use of AI systems by adversaries and other foreign actors in ways that threaten the capabilities or objectives of the Department of Defense or the Intelligence Community, or that otherwise pose risks to the security of the United States or its allies and partners.

Sec. 5. *Promoting Innovation and Competition.*

5.1. Attracting AI Talent to the United States. (a) Within 90 days of the date of this order, to attract and retain talent in AI and other critical and emerging technologies in the United States economy, the Secretary of State and the Secretary of Homeland Security shall take appropriate steps to:

(i) streamline processing times of visa petitions and applications, including by ensuring timely availability of visa appointments, for noncitizens who seek to travel to the United States to work on, study, or conduct research in AI or other critical and emerging technologies; and

(ii) facilitate continued availability of visa appointments in sufficient volume for applicants with expertise in AI or other critical and emerging technologies.

(b) Within 120 days of the date of this order, the Secretary of State shall:

(i) consider initiating a rulemaking to establish new criteria to designate countries and skills on the Department of State's Exchange Visitor Skills List as it relates to the 2-year foreign residence requirement for certain J-1 nonimmigrants, including those skills that are critical to the United States;

(ii) consider publishing updates to the 2009 Revised Exchange Visitor Skills List (74 FR 20108); and

(iii) consider implementing a domestic visa renewal program under 22 CFR 41.111(b) to facilitate the ability of qualified applicants, including highly skilled talent in AI and critical and emerging technologies, to continue their work in the United States without unnecessary interruption.

(c) Within 180 days of the date of this order, the Secretary of State shall:

(i) consider initiating a rulemaking to expand the categories of nonimmigrants who qualify for the domestic visa renewal program covered under 22 CFR 41.111(b) to include academic J-1 research scholars and F-1 students in science, technology, engineering, and mathematics (STEM); and

(ii) establish, to the extent permitted by law and available appropriations, a program to identify and attract top talent in AI and other critical and emerging technologies at universities, research institutions, and the private sector overseas, and to establish and increase connections with that talent to educate them on opportunities and resources for research and employment in the United States, including overseas educational components to inform top STEM talent of nonimmigrant and immigrant visa options and potential expedited adjudication of their visa petitions and applications.

(d) Within 180 days of the date of this order, the Secretary of Homeland Security shall:

(i) review and initiate any policy changes the Secretary determines necessary and appropriate to clarify and modernize immigration pathways for experts in AI and other critical and emerging technologies, including O-1A and EB-1 noncitizens of extraordinary ability; EB-2 advanced-degree holders and noncitizens of exceptional ability; and startup founders in AI and other critical and emerging technologies using the International Entrepreneur Rule; and

(ii) continue its rulemaking process to modernize the H-1B program and enhance its integrity and usage, including by experts in AI and other critical and emerging technologies, and consider initiating a rulemaking to enhance the process for noncitizens, including experts in AI and other critical and emerging technologies and their spouses, dependents, and children, to adjust their status to lawful permanent resident.

(e) Within 45 days of the date of this order, for purposes of considering updates to the "Schedule A" list of occupations, 20 CFR 656.5, the Secretary of Labor shall publish a request for information (RFI) to solicit public input, including from industry and worker-advocate communities, identifying AI and other STEM-related occupations, as well as additional occupations across the economy, for which there is an insufficient number of ready, willing, able, and qualified United States workers.

(f) The Secretary of State and the Secretary of Homeland Security shall, consistent with applicable law and implementing regulations, use their discretionary authorities to support and attract foreign nationals with special skills in AI and other critical and emerging technologies seeking to work, study, or conduct research in the United States.

(g) Within 120 days of the date of this order, the Secretary of Homeland Security, in consultation with the Secretary of State, the Secretary of Commerce, and the Director of OSTP, shall develop and publish informational resources to better attract and retain experts in AI and other critical and emerging technologies, including:

(i) a clear and comprehensive guide for experts in AI and other critical and emerging technologies to understand their options for working in the United States, to be published in multiple relevant languages on AI.gov; and

(ii) a public report with relevant data on applications, petitions, approvals, and other key indicators of how experts in AI and other critical and emerging technologies have utilized the immigration system through the end of Fiscal Year 2023.

5.2. *Promoting Innovation.* (a) To develop and strengthen public-private partnerships for advancing innovation, commercialization, and risk-mitigation methods for AI, and to help promote safe, responsible, fair, privacy-protecting, and trustworthy AI systems, the Director of NSF shall take the following steps:

(i) Within 90 days of the date of this order, in coordination with the heads of agencies that the Director of NSF deems appropriate, launch a pilot program implementing the National AI Research Resource (NAIRR), consistent with past recommendations of the NAIRR Task Force. The program shall pursue the infrastructure, governance mechanisms, and user interfaces to pilot an initial integration of distributed computational, data, model, and training resources to be made available to the research community in support of AI-related research and development. The Director of NSF shall identify Federal and private sector computational, data, software, and training resources appropriate for inclusion in the NAIRR pilot program. To assist with such work, within 45 days of the date of this order, the heads of agencies whom the Director of NSF identifies for coordination pursuant to this subsection shall each submit to the Director of NSF a report identifying the agency resources that could be developed and integrated into such a pilot program. These reports shall include a description of such resources, including their current status and availability; their format, structure, or technical specifications; associated agency expertise that will be provided; and the benefits and risks associated with their inclusion in the NAIRR pilot program. The heads of independent regulatory agencies are encouraged to take similar steps, as they deem appropriate.

(ii) Within 150 days of the date of this order, fund and launch at least one NSF Regional Innovation Engine that prioritizes AI-related work, such as AI-related research, societal, or workforce needs.

(iii) Within 540 days of the date of this order, establish at least four new National AI Research Institutes, in addition to the 25 currently funded as of the date of this order.

(b) Within 120 days of the date of this order, to support activities involving high-performance and data-intensive computing, the Secretary of Energy, in coordination with the Director of NSF, shall, in a manner consistent with applicable law and available appropriations, establish a pilot program to enhance existing successful training programs for scientists, with the goal of training 500 new researchers by 2025 capable of meeting the rising demand for AI talent.

(c) To promote innovation and clarify issues related to AI and inventorship of patentable subject matter, the Under Secretary of Commerce for Intellectual Property and Director of the United States Patent and Trademark Office (USPTO Director) shall:

(i) within 120 days of the date of this order, publish guidance to USPTO patent examiners and applicants addressing inventorship and the use of AI, including generative AI, in the inventive process, including illustrative examples in which AI systems play different roles in inventive processes and how, in each example, inventorship issues ought to be analyzed;

(ii) subsequently, within 270 days of the date of this order, issue additional guidance to USPTO patent examiners and applicants to address other considerations at the intersection of AI and IP, which could include, as the USPTO Director deems necessary, updated guidance on patent eligibility to address innovation in AI and critical and emerging technologies; and

(iii) within 270 days of the date of this order or 180 days after the United States Copyright Office of the Library of Congress publishes its forthcoming AI study that will address copyright issues raised by AI, whichever comes later, consult with the Director of the United States Copyright Office and issue recommendations to the President on potential executive actions relating to copyright and AI. The recommendations shall address any copyright and related issues discussed in the United States Copyright Office's study, including the scope of protection for works produced using AI and the treatment of copyrighted works in AI training.

(d) Within 180 days of the date of this order, to assist developers of AI in combatting AI-related IP risks, the Secretary of Homeland Security, acting through the Director of the National Intellectual Property Rights Coordination Center, and in consultation with the Attorney General, shall develop a training, analysis, and evaluation program to mitigate AI-related IP risks. Such a program shall:

(i) include appropriate personnel dedicated to collecting and analyzing reports of AI-related IP theft, investigating such incidents with implications for national security, and, where appropriate and consistent with applicable law, pursuing related enforcement actions;

(ii) implement a policy of sharing information and coordinating on such work, as appropriate and consistent with applicable law, with the Federal Bureau of Investigation; United States Customs and Border Protection; other agencies; State and local agencies; and appropriate international organizations, including through work-sharing agreements;

(iii) develop guidance and other appropriate resources to assist private sector actors with mitigating the risks of AI-related IP theft;

(iv) share information and best practices with AI developers and law enforcement personnel to identify incidents, inform stakeholders of current legal requirements, and evaluate AI systems for IP law violations, as well as develop mitigation strategies and resources; and

(v) assist the Intellectual Property Enforcement Coordinator in updating the Intellectual Property Enforcement Coordinator Joint Strategic Plan on Intellectual Property Enforcement to address AI-related issues.

(e) To advance responsible AI innovation by a wide range of healthcare technology developers that promotes the welfare of patients and workers in the healthcare sector, the Secretary of HHS shall identify and, as appropriate and consistent with applicable law and the activities directed in section 8 of this order, prioritize grantmaking and other awards, as well as undertake related efforts, to support responsible AI development and use, including:

(i) collaborating with appropriate private sector actors through HHS programs that may support the advancement of AI-enabled tools that develop personalized immune-response profiles for patients, consistent with section 4 of this order;

(ii) prioritizing the allocation of 2024 Leading Edge Acceleration Project cooperative agreement awards to initiatives that explore ways to improve healthcare-data quality to support the responsible development of AI tools for clinical care, real-world-evidence programs, population health, public health, and related research; and

(iii) accelerating grants awarded through the National Institutes of Health Artificial Intelligence/Machine Learning Consortium to Advance Health Equity and Researcher Diversity (AIM-AHEAD) program and showcasing current AIM-AHEAD activities in underserved communities.

(f) To advance the development of AI systems that improve the quality of veterans' healthcare, and in order to support small businesses' innovative capacity, the Secretary of Veterans Affairs shall:

(i) within 365 days of the date of this order, host two 3-month nationwide AI Tech Sprint competitions; and

(ii) as part of the AI Tech Sprint competitions and in collaboration with appropriate partners, provide participants access to technical assistance, mentorship opportunities, individualized expert feedback on products under development, potential contract opportunities, and other programming and resources.

(g) Within 180 days of the date of this order, to support the goal of strengthening our Nation's resilience against climate change impacts and building an equitable clean energy economy for the future, the Secretary of Energy, in consultation with the Chair of the Federal Energy Regulatory Commission, the Director of OSTP, the Chair of the Council on Environmental Quality, the Assistant to the President and National Climate Advisor, and the heads of other relevant agencies as the Secretary of Energy may deem appropriate, shall:

(i) issue a public report describing the potential for AI to improve planning, permitting, investment, and operations for electric grid infrastructure and to enable the provision of clean, affordable, reliable, resilient, and secure electric power to all Americans;

(ii) develop tools that facilitate building foundation models useful for basic and applied science, including models that streamline permitting and environmental reviews while improving environmental and social outcomes;

(iii) collaborate, as appropriate, with private sector organizations and members of academia to support development of AI tools to mitigate climate change risks;

(iv) take steps to expand partnerships with industry, academia, other agencies, and international allies and partners to utilize the Department of Energy's computing capabilities and AI testbeds to build foundation models that support new applications in science and energy, and for national security, including partnerships that increase community preparedness for climate-related risks, enable clean-energy deployment (including addressing delays in permitting reviews), and enhance grid reliability and resilience; and

(v) establish an office to coordinate development of AI and other critical and emerging technologies across Department of Energy programs and the 17 National Laboratories.

(h) Within 180 days of the date of this order, to understand AI's implications for scientific research, the President's Council of Advisors on Science and Technology shall submit to the President and make publicly available a report on the potential role of AI, especially given recent developments in AI, in research aimed at tackling major societal and global challenges. The report shall include a discussion of issues that may hinder the effective use of AI in research and practices needed to ensure that AI is used responsibly for research.

5.3. *Promoting Competition.* (a) The head of each agency developing policies and regulations related to AI shall use their authorities, as appropriate and consistent with applicable law, to promote competition in AI and related technologies, as well as in other markets. Such actions include addressing risks arising from concentrated control of key inputs, taking steps to stop unlawful collusion and prevent dominant firms from disadvantaging competitors, and working to provide new opportunities for small businesses and entrepreneurs. In particular, the Federal Trade Commission is encouraged to consider, as it deems appropriate, whether to exercise the Commission's existing authorities, including its rulemaking authority under the Federal Trade Commission Act, 15 U.S.C. 41 ET SEQ., to ensure fair competition in the AI marketplace and to ensure that consumers and workers are protected from harms that may be enabled by the use of AI.

(b) To promote competition and innovation in the semiconductor industry, recognizing that semiconductors power AI technologies and that their availability is critical to AI competition, the Secretary of Commerce shall, in implementing division A of Public Law 117-167, known as the Creating Helpful Incentives to Produce Semiconductors (CHIPS) Act of 2022, promote competition by:

(i) implementing a flexible membership structure for the National Semiconductor Technology Center that attracts all parts of the semiconductor and microelectronics ecosystem, including startups and small firms;

(ii) implementing mentorship programs to increase interest and participation in the semiconductor industry, including from workers in underserved communities;

(iii) increasing, where appropriate and to the extent permitted by law, the availability of resources to startups and small businesses, including:

(A) funding for physical assets, such as specialty equipment or facilities, to which startups and small businesses may not otherwise have access;

(B) datasets—potentially including test and performance data—collected, aggregated, or shared by CHIPS research and development programs;

(C) workforce development programs;

(D) design and process technology, as well as IP, as appropriate; and

(E) other resources, including technical and intellectual property assistance, that could accelerate commercialization of new technologies by startups and small businesses, as appropriate; and

(iv) considering the inclusion, to the maximum extent possible, and as consistent with applicable law, of competition-increasing measures in notices of funding availability for commercial research-and-development facilities focused on semiconductors, including measures that increase access to facility capacity for startups or small firms developing semiconductors used to power AI technologies.

(c) To support small businesses innovating and commercializing AI, as well as in responsibly adopting and deploying AI, the Administrator of the Small Business Administration shall:

(i) prioritize the allocation of Regional Innovation Cluster program funding for clusters that support planning activities related to the establishment of one or more Small Business AI Innovation and Commercialization Institutes that provide support, technical assistance, and other resources to small businesses seeking to innovate, commercialize, scale, or otherwise advance the development of AI;

(ii) prioritize the allocation of up to $2 million in Growth Accelerator Fund Competition bonus prize funds for accelerators that support the incorporation or expansion of AI-related curricula, training, and technical assistance, or other AI-related resources within their programming; and

(iii) assess the extent to which the eligibility criteria of existing programs, including the State Trade Expansion Program, Technical and Business Assistance funding, and capital-access programs—such as the 7(a) loan program, 504 loan program, and Small Business Investment Company (SBIC) program—support appropriate expenses by small businesses related to the adoption of AI and, if feasible and appropriate, revise eligibility criteria to improve support for these expenses.

(d) The Administrator of the Small Business Administration, in coordination with resource partners, shall conduct outreach regarding, and raise awareness of, opportunities for small businesses to use capital-access programs described in subsection 5.3(c) of this section for eligible AI-related purposes, and for eligible investment funds with AI-related expertise—particularly those seeking to serve or with experience serving underserved communities—to apply for an SBIC license.

Sec. 6. *Supporting Workers.* (a) To advance the Government's understanding of AI's implications for workers, the following actions shall be taken within 180 days of the date of this order:

(i) The Chairman of the Council of Economic Advisers shall prepare and submit a report to the President on the labor-market effects of AI.

(ii) To evaluate necessary steps for the Federal Government to address AI-related workforce disruptions, the Secretary of Labor shall submit to the President a report analyzing the abilities of agencies to support workers displaced by the adoption of AI and other technological advancements. The report shall, at a minimum:

(A) assess how current or formerly operational Federal programs designed to assist workers facing job disruptions—including unemployment insurance and programs authorized by the Workforce Innovation and Opportunity Act (Public Law 113-128)—could be used to respond to possible future AI-related disruptions; and

(B) identify options, including potential legislative measures, to strengthen or develop additional Federal support for workers displaced by AI and, in consultation with the Secretary of Commerce and the Secretary of Education, strengthen and expand education and training opportunities that provide individuals pathways to occupations related to AI.

(b) To help ensure that AI deployed in the workplace advances employees' well-being:

(i) The Secretary of Labor shall, within 180 days of the date of this order and in consultation with other agencies and with outside entities, including labor unions and workers, as the Secretary of Labor deems appropriate, develop and publish principles and best practices for employers that could be used to mitigate AI's potential harms to employees' well-being and maximize its potential benefits. The principles and best practices shall include specific steps for employers to take with regard to AI, and shall cover, at a minimum:

(A) job-displacement risks and career opportunities related to AI, including effects on job skills and evaluation of applicants and workers;

(B) labor standards and job quality, including issues related to the equity, protected-activity, compensation, health, and safety implications of AI in the workplace; and

(C) implications for workers of employers' AI-related collection and use of data about them, including transparency, engagement, management, and activity protected under worker-protection laws.

(ii) After principles and best practices are developed pursuant to subsection (b)(i) of this section, the heads of agencies shall consider, in consultation with the Secretary of Labor, encouraging the adoption of these guidelines in their programs to the extent appropriate for each program and consistent with applicable law.

(iii) To support employees whose work is monitored or augmented by AI in being compensated appropriately for all of their work time, the Secretary of Labor shall issue guidance to make clear that employers that deploy AI to monitor or augment employees' work must continue to comply with protections that ensure that workers are compensated for their hours worked, as defined under the Fair Labor Standards Act of 1938, 29 U.S.C. 201 ET SEQ., and other legal requirements.

(c) To foster a diverse AI-ready workforce, the Director of NSF shall prioritize available resources to support AI-related education and AI-related workforce development through existing programs. The Director shall additionally consult with agencies, as appropriate, to identify further opportunities for agencies to allocate resources for those purposes. The actions by the Director shall use appropriate fellowship programs and awards for these purposes.

Sec. 7. Advancing Equity and Civil Rights.

7.1. Strengthening AI and Civil Rights in the Criminal Justice System. (a) To address unlawful discrimination and other harms that may be exacerbated by AI, the Attorney General shall:

(i) consistent with Executive Order 12250 of November 2, 1980 (Leadership and Coordination of Nondiscrimination Laws), Executive Order 14091, and 28 CFR 0.50-51, coordinate with and support agencies in their implementation and enforcement of existing Federal laws to address civil rights and civil liberties violations and discrimination related to AI;

(ii) direct the Assistant Attorney General in charge of the Civil Rights Division to convene, within 90 days of the date of this order, a meeting of the heads of Federal civil rights offices—for which meeting the heads of civil rights offices within independent regulatory agencies will be encouraged to join—to discuss comprehensive use of their respective authorities and offices to: prevent and address discrimination in the use of automated systems, including algorithmic discrimination; increase coordination between the Department of Justice's Civil Rights Division and Federal civil rights offices concerning issues related to AI and algorithmic discrimination; improve external stakeholder engagement to promote public awareness of potential discriminatory uses and effects of AI; and develop, as appropriate, additional training, technical assistance, guidance, or other resources; and

(iii) consider providing, as appropriate and consistent with applicable law, guidance, technical assistance, and training to State, local, Tribal, and territorial investigators and prosecutors on best practices for investigating and prosecuting civil rights violations and discrimination related to automated systems, including AI.

(b) To promote the equitable treatment of individuals and adhere to the Federal Government's fundamental obligation to ensure fair and impartial justice for all, with respect to the use of AI in the criminal justice system, the Attorney General shall, in consultation with the Secretary of Homeland Security and the Director of OSTP:

(i) within 365 days of the date of this order, submit to the President a report that addresses the use of AI in the criminal justice system, including any use in:

(A) sentencing;

(B) parole, supervised release, and probation;

(C) bail, pretrial release, and pretrial detention;

(D) risk assessments, including pretrial, earned time, and early release or transfer to home-confinement determinations;

(E) police surveillance;

(F) crime forecasting and predictive policing, including the ingestion of historical crime data into AI systems to predict high-density "hot spots";

(G) prison-management tools; and

(H) forensic analysis;

(ii) within the report set forth in subsection 7.1(b)(i) of this section:

(A) identify areas where AI can enhance law enforcement efficiency and accuracy, consistent with protections for privacy, civil rights, and civil liberties; and

(B) recommend best practices for law enforcement agencies, including safeguards and appropriate use limits for AI, to address the concerns set forth in section 13(e)(i) of Executive Order 14074 as well as the best practices and the guidelines set forth in section 13(e)(iii) of Executive Order 14074; and

(iii) supplement the report set forth in subsection 7.1(b)(i) of this section as appropriate with recommendations to the President, including with respect to requests for necessary legislation.

(c) To advance the presence of relevant technical experts and expertise (such as machine-learning engineers, software and infrastructure engineering, data privacy experts, data scientists, and user experience researchers) among law enforcement professionals:

(i) The interagency working group created pursuant to section 3 of Executive Order 14074 shall, within 180 days of the date of this order, identify and share best practices for recruiting and hiring law enforcement professionals who have the technical skills mentioned in subsection 7.1(c) of this section, and for training law enforcement professionals about responsible application of AI.

(ii) Within 270 days of the date of this order, the Attorney General shall, in consultation with the Secretary of Homeland Security, consider those best practices and the guidance developed under section 3(d) of Executive Order 14074 and, if necessary, develop additional general recommendations for State, local, Tribal, and territorial law enforcement agencies and criminal justice agencies seeking to recruit, hire, train, promote, and retain highly qualified and service-oriented officers and staff with relevant technical knowledge. In considering this guidance, the Attorney General shall consult with State, local, Tribal, and territorial law enforcement agencies, as appropriate.

(iii) Within 365 days of the date of this order, the Attorney General shall review the work conducted pursuant to section 2(b) of Executive Order 14074 and, if appropriate, reassess the existing capacity to investigate law enforcement deprivation of rights under color of law resulting from the use of AI, including through improving and increasing training of Federal law enforcement officers, their supervisors, and Federal prosecutors on how to investigate and prosecute cases related to AI involving the deprivation of rights under color of law pursuant to 18 U.S.C. 242.

7.2. *Protecting Civil Rights Related to Government Benefits and Programs.* (a) To advance equity and civil rights, consistent with the directives of Executive Order 14091, and in addition to complying with the guidance on Federal Government use of AI issued pursuant to section 10.1(b) of this order, agencies shall use their respective civil rights and civil liberties offices and authorities—as appropriate and consistent with applicable law—to prevent and address unlawful discrimination and other harms that result from uses of AI in Federal Government programs and benefits administration. This directive does not apply to agencies' civil or criminal enforcement authorities. Agencies shall consider opportunities to ensure that their respective civil rights and civil liberties offices are appropriately consulted on agency decisions regarding the design, development, acquisition, and use of AI in Federal Government programs and benefits administration. To further these objectives, agencies shall also consider opportunities to increase coordination, communication, and engagement about AI as appropriate with community-based organizations; civil-rights and civil-liberties organizations; academic institutions; industry; State, local, Tribal, and territorial governments; and other stakeholders.

(b) To promote equitable administration of public benefits:

(i) The Secretary of HHS shall, within 180 days of the date of this order and in consultation with relevant agencies, publish a plan, informed by the guidance issued pursuant to section 10.1(b) of this order, addressing the use of automated or algorithmic systems in the implementation by States and localities of public benefits and services administered by the Secretary, such as to promote: assessment of access to benefits by qualified recipients; notice to recipients about the presence of such systems; regular evaluation to detect unjust denials; processes to retain appropriate levels of discretion of expert agency staff; processes to appeal denials to human reviewers; and analysis of whether algorithmic systems in use by benefit programs achieve equitable and just outcomes.

(ii) The Secretary of Agriculture shall, within 180 days of the date of this order and as informed by the guidance issued pursuant to section 10.1(b) of this order, issue guidance to State, local, Tribal, and territorial public-benefits administrators on the use of automated or algorithmic systems in implementing benefits or in providing customer support for benefit programs administered by the Secretary, to ensure that programs using those systems:

(A) maximize program access for eligible recipients;

(B) employ automated or algorithmic systems in a manner consistent with any requirements for using merit systems personnel in public-benefits programs;

(C) identify instances in which reliance on automated or algorithmic systems would require notification by the State, local, Tribal, or territorial government to the Secretary;

(D) identify instances when applicants and participants can appeal benefit determinations to a human reviewer for reconsideration and can receive other customer support from a human being;

(E) enable auditing and, if necessary, remediation of the logic used to arrive at an individual decision or determination to facilitate the evaluation of appeals; and

(F) enable the analysis of whether algorithmic systems in use by benefit programs achieve equitable outcomes.

7.3. *Strengthening AI and Civil Rights in the Broader Economy.* (a) Within 365 days of the date of this order, to prevent unlawful discrimination from AI used for hiring, the Secretary of Labor shall publish guidance for Federal contractors regarding nondiscrimination in hiring involving AI and other technology-based hiring systems.

(b) To address discrimination and biases against protected groups in housing markets and consumer financial markets, the Director of the Federal Housing Finance Agency and the Director of the Consumer Financial Protection Bureau are encouraged to consider using their authorities, as they deem appropriate, to require their respective regulated entities, where possible, to use appropriate methodologies including AI tools to ensure compliance with Federal law and:

(i) evaluate their underwriting models for bias or disparities affecting protected groups; and

(ii) evaluate automated collateral-valuation and appraisal processes in ways that minimize bias.

(c) Within 180 days of the date of this order, to combat unlawful discrimination enabled by automated or algorithmic tools used to make decisions about access to housing and in other real estate-related transactions, the Secretary of Housing and Urban Development shall, and the Director of the Consumer Financial Protection Bureau is encouraged to, issue additional guidance:

(i) addressing the use of tenant screening systems in ways that may violate the Fair Housing Act (Public Law 90-284), the Fair Credit Reporting Act (Public Law 91-508), or other relevant Federal laws, including how the use of data, such as criminal records, eviction records, and credit information, can lead to discriminatory outcomes in violation of Federal law; and

(ii) addressing how the Fair Housing Act, the Consumer Financial Protection Act of 2010 (title X of Public Law 111-203), or the Equal Credit Opportunity Act (Public Law 93-495) apply to the advertising of housing, credit, and other real estate-related transactions through digital platforms, including those that use algorithms to facilitate advertising delivery, as well as on best practices to avoid violations of Federal law.

(d) To help ensure that people with disabilities benefit from AI's promise while being protected from its risks, including unequal treatment from the use of biometric data like gaze direction, eye tracking, gait analysis, and hand motions, the Architectural and Transportation Barriers Compliance Board is encouraged, as it deems appropriate, to solicit public participation and conduct community engagement; to issue technical assistance and recommendations on the risks and benefits of AI in using biometric data as an input; and to provide people with disabilities access to information and communication technology and transportation services.

Sec. 8. Protecting Consumers, Patients, Passengers, and Students. (a) Independent regulatory agencies are encouraged, as they deem appropriate, to consider using their full range of authorities to protect American consumers from fraud, discrimination, and threats to privacy and to address other risks that may arise from the use of AI, including risks to financial stability, and to consider rulemaking, as well as emphasizing or clarifying where existing regulations and guidance apply to AI, including clarifying the responsibility of regulated entities to conduct due diligence on and monitor any third-party AI services they use, and emphasizing or clarifying requirements and expectations related to the transparency of AI models and regulated entities' ability to explain their use of AI models.

(b) To help ensure the safe, responsible deployment and use of AI in the healthcare, public-health, and human-services sectors:

(i) Within 90 days of the date of this order, the Secretary of HHS shall, in consultation with the Secretary of Defense and the Secretary of Veterans Affairs, establish an HHS AI Task Force that shall, within 365 days of its creation, develop a strategic plan that includes policies and frameworks—possibly including regulatory action, as appropriate—on responsible deployment and use of AI and AI-enabled technologies in the health and human services sector (including research and discovery, drug and device safety, healthcare delivery and financing, and public health), and identify appropriate guidance and resources to promote that deployment, including in the following areas:

(A) development, maintenance, and use of predictive and generative AI-enabled technologies in healthcare delivery and financing—including quality measurement, performance improvement, program integrity, benefits administration, and patient experience—taking into account considerations such as appropriate human oversight of the application of AI-generated output;

(B) long-term safety and real-world performance monitoring of AI-enabled technologies in the health and human services sector, including clinically relevant or significant modifications and performance across population groups, with a means to communicate product updates to regulators, developers, and users;

(C) incorporation of equity principles in AI-enabled technologies used in the health and human services sector, using disaggregated data on affected populations and representative population data sets when developing new models, monitoring algorithmic performance against discrimination and bias in existing models, and helping to identify and mitigate discrimination and bias in current systems;

(D) incorporation of safety, privacy, and security standards into the software-development lifecycle for protection of personally identifiable information, including measures to address AI-enhanced cybersecurity threats in the health and human services sector;

(E) development, maintenance, and availability of documentation to help users determine appropriate and safe uses of AI in local settings in the health and human services sector;

(F) work to be done with State, local, Tribal, and territorial health and human services agencies to advance positive use cases and best practices for use of AI in local settings; and

(G) identification of uses of AI to promote workplace efficiency and satisfaction in the health and human services sector, including reducing administrative burdens.

(ii) Within 180 days of the date of this order, the Secretary of HHS shall direct HHS components, as the Secretary of HHS deems appropriate, to develop a strategy, in consultation with relevant agencies, to determine whether AI-enabled technologies in the health and human services sector maintain appropriate levels of quality, including, as appropriate, in the areas described in subsection (b)(i) of this section. This work shall include the development of AI assurance policy—to evaluate important aspects of the performance of AI-enabled healthcare tools—and infrastructure needs for enabling pre-market assessment and post-market oversight of AI-enabled healthcare-technology algorithmic system performance against real-world data.

(iii) Within 180 days of the date of this order, the Secretary of HHS shall, in consultation with relevant agencies as the Secretary of HHS deems appropriate, consider appropriate actions to advance the prompt understanding of, and compliance with, Federal nondiscrimination laws by health and human services providers that receive Federal financial assistance, as well as how those laws relate to AI. Such actions may include:

(A) convening and providing technical assistance to health and human services providers and payers about their obligations under Federal nondiscrimination and privacy laws as they relate to AI and the potential consequences of noncompliance; and

(B) issuing guidance, or taking other action as appropriate, in response to any complaints or other reports of noncompliance with Federal nondiscrimination and privacy laws as they relate to AI.

(iv) Within 365 days of the date of this order, the Secretary of HHS shall, in consultation with the Secretary of Defense and the Secretary of Veterans Affairs, establish an AI safety program that, in partnership with voluntary federally listed Patient Safety Organizations:

(A) establishes a common framework for approaches to identifying and capturing clinical errors resulting from AI deployed in healthcare settings as well as specifications for a central tracking repository for associated incidents that cause harm, including through bias or discrimination, to patients, caregivers, or other parties;

(B) analyzes captured data and generated evidence to develop, wherever appropriate, recommendations, best practices, or other informal guidelines aimed at avoiding these harms; and

(C) disseminates those recommendations, best practices, or other informal guidance to appropriate stakeholders, including healthcare providers.

(v) Within 365 days of the date of this order, the Secretary of HHS shall develop a strategy for regulating the use of AI or AI-enabled tools in drug-development processes. The strategy shall, at a minimum:

(A) define the objectives, goals, and high-level principles required for appropriate regulation throughout each phase of drug development;

(B) identify areas where future rulemaking, guidance, or additional statutory authority may be necessary to implement such a regulatory system;

(C) identify the existing budget, resources, personnel, and potential for new public/private partnerships necessary for such a regulatory system; and

(D) consider risks identified by the actions undertaken to implement section 4 of this order.

(c) To promote the safe and responsible development and use of AI in the transportation sector, in consultation with relevant agencies:

(i) Within 30 days of the date of this order, the Secretary of Transportation shall direct the Nontraditional and Emerging Transportation Technology (NETT) Council to assess the need for information, technical assistance, and guidance regarding the use of AI in transportation. The Secretary of Transportation shall further direct the NETT Council, as part of any such efforts, to:

(A) support existing and future initiatives to pilot transportation-related applications of AI, as they align with policy priorities articulated in the Department of Transportation's (DOT) Innovation Principles, including, as appropriate, through technical assistance and connecting stakeholders;

(B) evaluate the outcomes of such pilot programs in order to assess when DOT, or other Federal or State agencies, have sufficient information to take regulatory actions, as appropriate, and recommend appropriate actions when that information is available; and

(C) establish a new DOT Cross-Modal Executive Working Group, which will consist of members from different divisions of DOT and coordinate applicable work among these divisions, to solicit and use relevant input from appropriate stakeholders.

(ii) Within 90 days of the date of this order, the Secretary of Transportation shall direct appropriate Federal Advisory Committees of the DOT to provide advice on the safe and responsible use of AI in transportation. The committees shall include the Advanced Aviation Advisory Committee, the Transforming Transportation Advisory Committee, and the Intelligent Transportation Systems Program Advisory Committee.

(iii) Within 180 days of the date of this order, the Secretary of Transportation shall direct the Advanced Research Projects Agency-Infrastructure (ARPA-I) to explore the transportation-related opportunities and challenges of AI—including regarding software-defined AI enhancements impacting autonomous mobility ecosystems. The Secretary of Transportation shall further encourage ARPA-I to prioritize the allocation of grants to those opportunities, as appropriate. The work tasked to ARPA-I shall include soliciting input on these topics through a public consultation process, such as an RFI.

(d) To help ensure the responsible development and deployment of AI in the education sector, the Secretary of Education shall, within 365 days of the date of this order, develop resources, policies, and guidance regarding AI. These resources shall address safe, responsible, and nondiscriminatory uses of AI in education, including the impact AI systems have on vulnerable and underserved communities, and shall be developed in consultation with stakeholders as appropriate. They shall also include the development of an "AI toolkit" for education leaders implementing recommendations from the Department of Education's AI and the Future of Teaching and Learning report, including appropriate human review of AI decisions, designing AI systems to enhance trust and safety and align with privacy-related laws and regulations in the educational context, and developing education-specific guardrails.

(e) The Federal Communications Commission is encouraged to consider actions related to how AI will affect communications networks and consumers, including by:

(i) examining the potential for AI to improve spectrum management, increase the efficiency of non-Federal spectrum usage, and expand opportunities for the sharing of non-Federal spectrum;

(ii) coordinating with the National Telecommunications and Information Administration to create opportunities for sharing spectrum between Federal and non-Federal spectrum operations;

(iii) providing support for efforts to improve network security, resiliency, and interoperability using next-generation technologies that incorporate AI, including self-healing networks, 6G, and Open RAN; and

(iv) encouraging, including through rulemaking, efforts to combat unwanted robocalls and robotexts that are facilitated or exacerbated by AI and to deploy AI technologies that better serve consumers by blocking unwanted robocalls and robotexts.

Sec. 9. *Protecting Privacy.* (a) To mitigate privacy risks potentially exacerbated by AI—including by AI's facilitation of the collection or use of information about individuals, or the making of inferences about individuals—the Director of OMB shall:

(i) evaluate and take steps to identify commercially available information (CAI) procured by agencies, particularly CAI that contains personally identifiable information and including CAI procured from data brokers and CAI procured and processed indirectly through vendors, in appropriate agency inventory and reporting processes (other than when it is used for the purposes of national security);

(ii) evaluate, in consultation with the Federal Privacy Council and the Interagency Council on Statistical Policy, agency standards and procedures associated with the collection, processing, maintenance, use, sharing, dissemination, and disposition of CAI that contains personally identifiable information (other than when it is used for the purposes of national security) to inform potential guidance to agencies on ways to mitigate privacy and confidentiality risks from agencies' activities related to CAI;

(iii) within 180 days of the date of this order, in consultation with the Attorney General, the Assistant to the President for Economic Policy, and the Director of OSTP, issue an RFI to inform potential revisions to guidance to agencies on implementing the privacy provisions of the E-Government Act of 2002 (Public Law 107-347). The RFI shall seek feedback regarding how privacy impact assessments may be more effective at mitigating privacy risks, including those that are further exacerbated by AI; and

(iv) take such steps as are necessary and appropriate, consistent with applicable law, to support and advance the near-term actions and long-term strategy identified through the RFI process, including issuing new or updated guidance or RFIs or consulting other agencies or the Federal Privacy Council.

(b) Within 365 days of the date of this order, to better enable agencies to use PETs to safeguard Americans' privacy from the potential threats exacerbated by AI, the Secretary of Commerce, acting through the Director of NIST, shall create guidelines for agencies to evaluate the efficacy of differential-privacy-guarantee protections, including for AI. The guidelines shall, at a minimum, describe the significant factors that bear on differential-privacy safeguards and common risks to realizing differential privacy in practice.

(c) To advance research, development, and implementation related to PETs:

(i) Within 120 days of the date of this order, the Director of NSF, in collaboration with the Secretary of Energy, shall fund the creation of a Research Coordination Network (RCN) dedicated to advancing privacy research and, in particular, the development, deployment, and scaling of PETs. The RCN shall serve to enable privacy researchers to share information, coordinate and collaborate in research, and develop standards for the privacy-research community.

(ii) Within 240 days of the date of this order, the Director of NSF shall engage with agencies to identify ongoing work and potential opportunities to incorporate PETs into their operations. The Director of NSF shall, where feasible and appropriate, prioritize research—including efforts to translate research discoveries into practical applications—that encourage the adoption of leading-edge PETs solutions for agencies' use, including through research engagement through the RCN described in subsection (c)(i) of this section.

(iii) The Director of NSF shall use the results of the United States-United Kingdom PETs Prize Challenge to inform the approaches taken, and opportunities identified, for PETs research and adoption.

Sec. 10. *Advancing Federal Government Use of AI.*

10.1. *Providing Guidance for AI Management.* (a) To coordinate the use of AI across the Federal Government, within 60 days of the date of this order and on an ongoing basis as necessary, the Director of OMB shall convene and chair an interagency council to coordinate the development and use of AI in agencies' programs and operations, other than the use of AI in national security systems. The Director of OSTP shall serve as Vice Chair for the interagency council. The interagency council's membership shall include, at minimum, the heads of the agencies identified in 31 U.S.C. 901(b), the Director of National Intelligence, and other agencies as identified by the Chair. Until agencies designate their permanent Chief AI Officers consistent with the guidance described in subsection 10.1(b) of this section, they shall be represented on the interagency council by an appropriate official at the Assistant Secretary level or equivalent, as determined by the head of each agency.

(b) To provide guidance on Federal Government use of AI, within 150 days of the date of this order and updated periodically thereafter, the Director of OMB, in coordination with the Director of OSTP, and in consultation with the interagency council established in subsection 10.1(a) of this section, shall issue guidance to agencies to strengthen the effective and appropriate use of AI, advance AI innovation, and manage risks from AI in the Federal Government. The Director of OMB's guidance shall specify, to the extent appropriate and consistent with applicable law:

(i) the requirement to designate at each agency within 60 days of the issuance of the guidance a Chief Artificial Intelligence Officer who shall hold primary responsibility in their agency, in coordination with other responsible officials, for coordinating their agency's use of AI, promoting AI innovation in their agency, managing risks from their agency's use of AI, and carrying out the responsibilities described in section 8(c) of Executive Order 13960 of December 3, 2020 (Promoting the Use of Trustworthy Artificial Intelligence in the Federal Government), and section 4(b) of Executive Order 14091;

(ii) the Chief Artificial Intelligence Officers' roles, responsibilities, seniority, position, and reporting structures;

(iii) for the agencies identified in 31 U.S.C. 901(b), the creation of internal Artificial Intelligence Governance Boards, or other appropriate mechanisms, at each agency within 60 days of the issuance of the guidance to coordinate and govern AI issues through relevant senior leaders from across the agency;

(iv) required minimum risk-management practices for Government uses of AI that impact people's rights or safety, including, where appropriate, the following practices derived from OSTP's Blueprint for an AI Bill of Rights and the NIST AI Risk Management Framework: conducting public consultation; assessing data quality; assessing and mitigating disparate impacts and algorithmic discrimination; providing notice of the use of AI; continuously monitoring and evaluating deployed AI; and granting human consideration and remedies for adverse decisions made using AI;

(v) specific Federal Government uses of AI that are presumed by default to impact rights or safety;

416

(vi) recommendations to agencies to reduce barriers to the responsible use of AI, including barriers related to information technology infrastructure, data, workforce, budgetary restrictions, and cybersecurity processes;

(vii) requirements that agencies identified in 31 U.S.C. 901(b) develop AI strategies and pursue high-impact AI use cases;

(viii) in consultation with the Secretary of Commerce, the Secretary of Homeland Security, and the heads of other appropriate agencies as determined by the Director of OMB, recommendations to agencies regarding:

(A) external testing for AI, including AI red-teaming for generative AI, to be developed in coordination with the Cybersecurity and Infrastructure Security Agency;

(B) testing and safeguards against discriminatory, misleading, inflammatory, unsafe, or deceptive outputs, as well as against producing child sexual abuse material and against producing non-consensual intimate imagery of real individuals (including intimate digital depictions of the body or body parts of an identifiable individual), for generative AI;

(C) reasonable steps to watermark or otherwise label output from generative AI;

(D) application of the mandatory minimum risk-management practices defined under subsection 10.1(b)(iv) of this section to procured AI;

(E) independent evaluation of vendors' claims concerning both the effectiveness and risk mitigation of their AI offerings;

(F) documentation and oversight of procured AI;

(G) maximizing the value to agencies when relying on contractors to use and enrich Federal Government data for the purposes of AI development and operation;

(H) provision of incentives for the continuous improvement of procured AI; and

(I) training on AI in accordance with the principles set out in this order and in other references related to AI listed herein; and

(ix) requirements for public reporting on compliance with this guidance.

(c) To track agencies' AI progress, within 60 days of the issuance of the guidance established in subsection 10.1(b) of this section and updated periodically thereafter, the Director of OMB shall develop a method for agencies to track and assess their ability to adopt AI into their programs and operations, manage its risks, and comply with Federal policy on AI. This method should draw on existing related efforts as appropriate and should address, as appropriate and consistent with applicable law, the practices, processes, and capabilities necessary for responsible AI adoption, training, and governance across, at a minimum, the areas of information technology infrastructure, data, workforce, leadership, and risk management.

(d) To assist agencies in implementing the guidance to be established in subsection 10.1(b) of this section:

(i) within 90 days of the issuance of the guidance, the Secretary of Commerce, acting through the Director of NIST, and in coordination with the Director of OMB and the Director of OSTP, shall develop guidelines, tools, and practices to support implementation of the minimum risk-management practices described in subsection 10.1(b)(iv) of this section; and

(ii) within 180 days of the issuance of the guidance, the Director of OMB shall develop an initial means to ensure that agency contracts for the acquisition of AI systems and services align with the guidance described in subsection 10.1(b) of this section and advance the other aims identified in section 7224(d)(1) of the Advancing American AI Act (Public Law 117-263, div. G, title LXXII, subtitle B).

(e) To improve transparency for agencies' use of AI, the Director of OMB shall, on an annual basis, issue instructions to agencies for the collection, reporting, and publication of agency AI use cases, pursuant to section 7225(a) of the Advancing American AI Act. Through these instructions, the Director shall, as appropriate, expand agencies' reporting on how they are managing risks from their AI use cases and update or replace the guidance originally established in section 5 of Executive Order 13960.

(f) To advance the responsible and secure use of generative AI in the Federal Government:

(i) As generative AI products become widely available and common in online platforms, agencies are discouraged from imposing broad general bans or blocks on agency use of generative AI. Agencies should instead limit access, as necessary, to specific generative AI services based on specific risk assessments; establish guidelines and limitations on the appropriate use of generative AI; and, with appropriate safeguards in place, provide their personnel and programs with access to secure and reliable generative AI capabilities, at least for the purposes of experimentation and routine tasks that carry a low risk of impacting Americans' rights. To protect Federal Government information, agencies are also encouraged to employ risk-management practices, such as training their staff on proper use, protection, dissemination, and disposition of Federal information; negotiating appropriate terms of service with vendors; implementing measures designed to ensure compliance with record-keeping, cybersecurity, confidentiality, privacy, and data protection requirements; and deploying other measures to prevent misuse of Federal Government information in generative AI.

(ii) Within 90 days of the date of this order, the Administrator of General Services, in coordination with the Director of OMB, and in consultation with the Federal Secure Cloud Advisory Committee and other relevant agencies as the Administrator of General Services may deem appropriate, shall develop and issue a framework for prioritizing critical and emerging technologies offerings in the Federal Risk and Authorization Management Program authorization process, starting with generative AI offerings that have the primary purpose of providing large language model-based chat interfaces, code-generation and debugging tools, and associated application programming interfaces, as well as prompt-based image generators. This framework shall apply for no less than 2 years from the date of its issuance. Agency Chief Information Officers, Chief Information Security Officers, and authorizing officials are also encouraged to prioritize generative AI and other critical and emerging technologies in granting authorities for agency operation of information technology systems and any other applicable release or oversight processes, using continuous authorizations and approvals wherever feasible.

(iii) Within 180 days of the date of this order, the Director of the Office of Personnel Management (OPM), in coordination with the Director of OMB, shall develop guidance on the use of generative AI for work by the Federal workforce.

(g) Within 30 days of the date of this order, to increase agency investment in AI, the Technology Modernization Board shall consider, as it deems appropriate and consistent with applicable law, prioritizing funding for AI projects for the Technology Modernization Fund for a period of at least 1 year. Agencies are encouraged to submit to the Technology Modernization Fund project funding proposals that include AI—and particularly generative AI—in service of mission delivery.

(h) Within 180 days of the date of this order, to facilitate agencies' access to commercial AI capabilities, the Administrator of General Services, in coordination with the Director of OMB, and in collaboration with the Secretary of Defense, the Secretary of Homeland Security, the Director of National Intelligence, the Administrator of the National Aeronautics and Space Administration, and the head of any other agency identified by the Administrator of General Services, shall take steps consistent with applicable law to facilitate access to Federal Government-wide acquisition solutions for specified types of AI services and products, such as through the creation of a resource guide or other tools to assist the acquisition workforce. Specified types of AI capabilities shall include generative AI and specialized computing infrastructure.

(i) The initial means, instructions, and guidance issued pursuant to subsections 10.1(a)-(h) of this section shall not apply to AI when it is used as a component of a national security system, which shall be addressed by the proposed National Security Memorandum described in subsection 4.8 of this order.

10.2. *Increasing AI Talent in Government*. (a) Within 45 days of the date of this order, to plan a national surge in AI talent in the Federal Government, the Director of OSTP and the Director of OMB, in consultation with the Assistant to the President for National Security Affairs, the Assistant to the President for Economic Policy, the Assistant to the President and Domestic Policy Advisor, and the Assistant to the President and Director of the Gender Policy Council, shall identify priority mission areas for increased Federal Government AI talent, the types of talent that are highest priority to recruit and develop to ensure adequate implementation of this order and use of relevant enforcement and regulatory authorities to address AI risks, and accelerated hiring pathways.

(b) Within 45 days of the date of this order, to coordinate rapid advances in the capacity of the Federal AI workforce, the Assistant to the President and Deputy Chief of Staff for Policy, in coordination with the Director of OSTP and the Director of OMB, and in consultation with the National Cyber Director, shall convene an AI and Technology Talent Task Force, which shall include the Director of OPM, the Director of the General Services Administration's Technology Transformation Services, a representative from the Chief Human Capital Officers Council, the Assistant to the President for Presidential Personnel, members of appropriate agency technology talent programs, a representative of the Chief Data Officer Council, and a representative of the interagency council convened under subsection 10.1(a) of this section. The Task Force's purpose shall be to accelerate and track the hiring of AI and AI-enabling talent across the Federal Government, including through the following actions:

(i) within 180 days of the date of this order, tracking and reporting progress to the President on increasing AI capacity across the Federal Government, including submitting to the President a report and recommendations for further increasing capacity;

(ii) identifying and circulating best practices for agencies to attract, hire, retain, train, and empower AI talent, including diversity, inclusion, and accessibility best practices, as well as to plan and budget adequately for AI workforce needs;

(iii) coordinating, in consultation with the Director of OPM, the use of fellowship programs and agency technology-talent programs and human-capital teams to build hiring capabilities, execute hires, and place AI talent to fill staffing gaps; and

(iv) convening a cross-agency forum for ongoing collaboration between AI professionals to share best practices and improve retention.

(c) Within 45 days of the date of this order, to advance existing Federal technology talent programs, the United States Digital Service, Presidential Innovation Fellowship, United States Digital Corps, OPM, and technology talent programs at agencies, with support from the AI and Technology Talent Task Force described in subsection 10.2(b) of this section, as appropriate and permitted by law, shall develop and begin to implement plans to support the rapid recruitment of individuals as part of a Federal Government-wide AI talent surge to accelerate the placement of key AI and AI-enabling talent in high-priority areas and to advance agencies' data and technology strategies.

(d) To meet the critical hiring need for qualified personnel to execute the initiatives in this order, and to improve Federal hiring practices for AI talent, the Director of OPM, in consultation with the Director of OMB, shall:

(i) within 60 days of the date of this order, conduct an evidence-based review on the need for hiring and workplace flexibility, including Federal Government-wide direct-hire authority for AI and related data-science and technical roles, and, where the Director of OPM finds such authority is appropriate, grant it; this review shall include the following job series at all General Schedule (GS) levels: IT Specialist (2210), Computer Scientist (1550), Computer Engineer (0854), and Program Analyst (0343) focused on AI, and any subsequently developed job series derived from these job series;

(ii) within 60 days of the date of this order, consider authorizing the use of excepted service appointments under 5 CFR 213.3102(i)(3) to address the need for hiring additional staff to implement directives of this order;

(iii) within 90 days of the date of this order, coordinate a pooled-hiring action informed by subject-matter experts and using skills-based assessments to support the recruitment of AI talent across agencies;

(iv) within 120 days of the date of this order, as appropriate and permitted by law, issue guidance for agency application of existing pay flexibilities or incentive pay programs for AI, AI-enabling, and other key technical positions to facilitate appropriate use of current pay incentives;

(v) within 180 days of the date of this order, establish guidance and policy on skills-based, Federal Government-wide hiring of AI, data, and technology talent in order to increase access to those with nontraditional academic backgrounds to Federal AI, data, and technology roles;

(vi) within 180 days of the date of this order, establish an interagency working group, staffed with both human-resources professionals and recruiting technical experts, to facilitate Federal Government-wide hiring of people with AI and other technical skills;

(vii) within 180 days of the date of this order, review existing Executive Core Qualifications (ECQs) for Senior Executive Service (SES) positions informed by data and AI literacy competencies and, within 365 days of the date of this order, implement new ECQs as appropriate in the SES assessment process;

(viii) within 180 days of the date of this order, complete a review of competencies for civil engineers (GS-0810 series) and, if applicable, other related occupations, and make recommendations for ensuring that adequate AI expertise and credentials in these occupations in the Federal Government reflect the increased use of AI in critical infrastructure; and

(ix) work with the Security, Suitability, and Credentialing Performance Accountability Council to assess mechanisms to streamline and accelerate personnel-vetting requirements, as appropriate, to support AI and fields related to other critical and emerging technologies.

(e) To expand the use of special authorities for AI hiring and retention, agencies shall use all appropriate hiring authorities, including Schedule A(r) excepted service hiring and direct-hire authority, as applicable and appropriate, to hire AI talent and AI-enabling talent rapidly. In addition to participating in OPM-led pooled hiring actions, agencies shall collaborate, where appropriate, on agency-led pooled hiring under the Competitive Service Act of 2015 (Public Law 114-137) and other shared hiring. Agencies shall also, where applicable, use existing incentives, pay-setting authorities, and other compensation flexibilities, similar to those used for cyber and information technology positions, for AI and data-science professionals, as well as plain-language job titles, to help recruit and retain these highly skilled professionals. Agencies shall ensure that AI and other related talent needs (such as technology governance and privacy) are reflected in strategic workforce planning and budget formulation.

(f) To facilitate the hiring of data scientists, the Chief Data Officer Council shall develop a position-description library for data scientists (job series 1560) and a hiring guide to support agencies in hiring data scientists.

(g) To help train the Federal workforce on AI issues, the head of each agency shall implement—or increase the availability and use of—AI training and familiarization programs for employees, managers, and leadership in technology as well as relevant policy, managerial, procurement, regulatory, ethical, governance, and legal fields. Such training programs should, for example, empower Federal employees, managers, and leaders to develop and maintain an operating knowledge of emerging AI technologies to assess opportunities to use these technologies to enhance the delivery of services to the public, and to mitigate risks associated with these technologies. Agencies that provide professional-development opportunities, grants, or funds for their staff should take appropriate steps to ensure that employees who do not serve in traditional technical roles, such as policy, managerial, procurement, or legal fields, are nonetheless eligible to receive funding for programs and courses that focus on AI, machine learning, data science, or other related subject areas.

(h) Within 180 days of the date of this order, to address gaps in AI talent for national defense, the Secretary of Defense shall submit a report to the President through the Assistant to the President for National Security Affairs that includes:

(i) recommendations to address challenges in the Department of Defense's ability to hire certain noncitizens, including at the Science and Technology Reinvention Laboratories;

(ii) recommendations to clarify and streamline processes for accessing classified information for certain noncitizens through Limited Access Authorization at Department of Defense laboratories;

(iii) recommendations for the appropriate use of enlistment authority under 10 U.S.C. 504(b)(2) for experts in AI and other critical and emerging technologies; and

(iv) recommendations for the Department of Defense and the Department of Homeland Security to work together to enhance the use of appropriate authorities for the retention of certain noncitizens of vital importance to national security by the Department of Defense and the Department of Homeland Security.

Sec. 11. *Strengthening American Leadership Abroad.* (a) To strengthen United States leadership of global efforts to unlock AI's potential and meet its challenges, the Secretary of State, in coordination with the Assistant to the President for National Security Affairs, the Assistant to the President for Economic Policy, the Director of OSTP, and the heads of other relevant agencies as appropriate, shall:

(i) lead efforts outside of military and intelligence areas to expand engagements with international allies and partners in relevant bilateral, multilateral, and multi-stakeholder fora to advance those allies' and partners' understanding of existing and planned AI-related guidance and policies of the United States, as well as to enhance international collaboration; and

(ii) lead efforts to establish a strong international framework for managing the risks and harnessing the benefits of AI, including by encouraging international allies and partners to support voluntary commitments similar to those that United States companies have made in pursuit of these objectives and coordinating the activities directed by subsections (b), (c), (d), and (e) of this section, and to develop common regulatory and other accountability principles for foreign nations, including to manage the risk that AI systems pose.

(b) To advance responsible global technical standards for AI development and use outside of military and intelligence areas, the Secretary of Commerce, in coordination with the Secretary of State and the heads of other relevant agencies as appropriate, shall lead preparations for a coordinated effort with key international allies and partners and with standards development organizations, to drive the development and implementation of AI-related consensus standards, cooperation and coordination, and information sharing. In particular, the Secretary of Commerce shall:

(i) within 270 days of the date of this order, establish a plan for global engagement on promoting and developing AI standards, with lines of effort that may include:

(A) AI nomenclature and terminology;

(B) best practices regarding data capture, processing, protection, privacy, confidentiality, handling, and analysis;

(C) trustworthiness, verification, and assurance of AI systems; and

(D) AI risk management;

(ii) within 180 days of the date the plan is established, submit a report to the President on priority actions taken pursuant to the plan; and

(iii) ensure that such efforts are guided by principles set out in the NIST AI Risk Management Framework and United States Government National Standards Strategy for Critical and Emerging Technology.

(c) Within 365 days of the date of this order, to promote safe, responsible, and rights-affirming development and deployment of AI abroad:

(i) The Secretary of State and the Administrator of the United States Agency for International Development, in coordination with the Secretary of Commerce, acting through the director of NIST, shall publish an AI in Global Development Playbook that incorporates the AI Risk Management Framework's principles, guidelines, and best practices into the social, technical, economic, governance, human rights, and security conditions of contexts beyond United States borders. As part of this work, the Secretary of State and the Administrator of the United States Agency for International Development shall draw on lessons learned from programmatic uses of AI in global development.

(ii) The Secretary of State and the Administrator of the United States Agency for International Development, in collaboration with the Secretary of Energy and the Director of NSF, shall develop a Global AI Research Agenda to guide the objectives and implementation of AI-related research in contexts beyond United States borders. The Agenda shall:

(A) include principles, guidelines, priorities, and best practices aimed at ensuring the safe, responsible, beneficial, and sustainable global development and adoption of AI; and

(B) address AI's labor-market implications across international contexts, including by recommending risk mitigations.

(d) To address cross-border and global AI risks to critical infrastructure, the Secretary of Homeland Security, in coordination with the Secretary of State, and in consultation with the heads of other relevant agencies as the Secretary of Homeland Security deems appropriate, shall lead efforts with international allies and partners to enhance cooperation to prevent, respond to, and recover from potential critical infrastructure disruptions resulting from incorporation of AI into critical infrastructure systems or malicious use of AI.

(i) Within 270 days of the date of this order, the Secretary of Homeland Security, in coordination with the Secretary of State, shall develop a plan for multilateral engagements to encourage the adoption of the AI safety and security guidelines for use by critical infrastructure owners and operators developed in section 4.3(a) of this order.

(ii) Within 180 days of establishing the plan described in subsection (d)(i) of this section, the Secretary of Homeland Security shall submit a report to the President on priority actions to mitigate cross-border risks to critical United States infrastructure.

Sec. 12. *Implementation.* (a) There is established, within the Executive Office of the President, the White House Artificial Intelligence Council (White House AI Council). The function of the White House AI Council is to coordinate the activities of agencies across the Federal Government to ensure the effective formulation, development, communication, industry engagement related to, and timely implementation of AI-related policies, including policies set forth in this order.

(b) The Assistant to the President and Deputy Chief of Staff for Policy shall serve as Chair of the White House AI Council.

(c) In addition to the Chair, the White House AI Council shall consist of the following members, or their designees:

(i) the Secretary of State;

(ii) the Secretary of the Treasury;

(iii) the Secretary of Defense;

(iv) the Attorney General;

(v) the Secretary of Agriculture;

(vi) the Secretary of Commerce;

(vii) the Secretary of Labor;

(viii) the Secretary of HHS;

(ix) the Secretary of Housing and Urban Development;

(x) the Secretary of Transportation;

(xi) the Secretary of Energy;

(xii) the Secretary of Education;

(xiii) the Secretary of Veterans Affairs;

(xiv) the Secretary of Homeland Security;

(xv) the Administrator of the Small Business Administration;

(xvi) the Administrator of the United States Agency for International Development;

(xvii) the Director of National Intelligence;

(xviii) the Director of NSF;

(xix) the Director of OMB;

(xx) the Director of OSTP;

(xxi) the Assistant to the President for National Security Affairs;

(xxii) the Assistant to the President for Economic Policy;

(xxiii) the Assistant to the President and Domestic Policy Advisor;

(xxiv) the Assistant to the President and Chief of Staff to the Vice President;

(xxv) the Assistant to the President and Director of the Gender Policy Council;

(xxvi) the Chairman of the Council of Economic Advisers;

(xxvii) the National Cyber Director;

(xxviii) the Chairman of the Joint Chiefs of Staff; and

(xxix) the heads of such other agencies, independent regulatory agencies, and executive offices as the Chair may from time to time designate or invite to participate.

(d) The Chair may create and coordinate subgroups consisting of White House AI Council members or their designees, as appropriate.

Sec. 13. *General Provisions.* (a) Nothing in this order shall be construed to impair or otherwise affect:

(i) the authority granted by law to an executive department or agency, or the head thereof; or

(ii) the functions of the Director of the Office of Management and Budget relating to budgetary, administrative, or legislative proposals.

(b) This order shall be implemented consistent with applicable law and subject to the availability of appropriations.

(c) This order is not intended to, and does not, create any right or benefit, substantive or procedural, enforceable at law or in equity by any party against the United States, its departments, agencies, or entities, its officers, employees, or agents, or any other person.

JOSEPH R. BIDEN, JR.

THE WHITE HOUSE,
October 30, 2023

National Security Memorandum on AI
Memorandum on Advancing the United States' Leadership in Artificial Intelligence; Harnessing Artificial Intelligence to Fulfill National Security Objectives; and Fostering the Safety, Security, and Trustworthiness of Artificial Intelligence
(2024)

Section 1. *Policy.* (a) This memorandum fulfills the directive set forth in subsection 4.8 of Executive Order 14110 of October 30, 2023 (Safe, Secure, and Trustworthy Development and Use of Artificial Intelligence). This memorandum provides further direction on appropriately harnessing artificial intelligence (AI) models and AI-enabled technologies in the United States Government, especially in the context of national security systems (NSS), while protecting human rights, civil rights, civil liberties, privacy, and safety in AI-enabled national security activities. A classified annex to this memorandum addresses additional sensitive national security issues, including countering adversary use of AI that poses risks to United States national security.

(b) United States national security institutions have historically triumphed during eras of technological transition. To meet changing times, they developed new capabilities, from submarines and aircraft to space systems and cyber tools. To gain a decisive edge and protect national security, they pioneered technologies such as radar, the Global Positioning System, and nuclear propulsion, and unleashed these hard-won breakthroughs on the battlefield. With each paradigm shift, they also developed new systems for tracking and countering adversaries' attempts to wield cutting-edge technology for their own advantage.

(c) AI has emerged as an era-defining technology and has demonstrated significant and growing relevance to national security. The United States must lead the world in the responsible application of AI to appropriate national security functions. AI, if used appropriately and for its intended purpose, can offer great benefits. If misused, AI could threaten United States national security, bolster authoritarianism worldwide, undermine democratic institutions and processes, facilitate human rights abuses, and weaken the rules-based international order. Harmful outcomes could occur even without malicious intent if AI systems and processes lack sufficient protections.

(d) Recent innovations have spurred not only an increase in AI use throughout society, but also a paradigm shift within the AI field — one that has occurred mostly outside of Government. This era of AI development and deployment rests atop unprecedented aggregations of specialized computational power, as well as deep scientific and engineering expertise, much of which is concentrated in the private sector. This trend is most evident with the rise of large language models, but it extends to a broader class of increasingly general-purpose and computationally intensive systems. The United States Government must urgently consider how this current AI paradigm specifically could transform the national security mission.

(e) Predicting technological change with certainty is impossible, but the foundational drivers that have underpinned recent AI progress show little sign of abating. These factors include compounding algorithmic improvements, increasingly efficient computational hardware, a growing willingness in industry to invest substantially in research and development, and the expansion of training data sets. AI under the current paradigm may continue to become more powerful and general-purpose. Developing and effectively using these systems requires an evolving array of resources, infrastructure, competencies, and workflows that in many cases differ from what was required to harness prior technologies, including previous paradigms of AI.

(f) If the United States Government does not act with responsible speed and in partnership with industry, civil society, and academia to make use of AI capabilities in service of the national security mission — and to ensure the safety, security, and trustworthiness of American AI innovation writ large — it risks losing ground to strategic competitors. Ceding the United States' technological edge would not only greatly harm American national security, but it would also undermine United States foreign policy objectives and erode safety, human rights, and democratic norms worldwide.

(g) Establishing national security leadership in AI will require making deliberate and meaningful changes to aspects of the United States Government's strategies, capabilities, infrastructure, governance, and organization. AI is likely to affect almost all domains with national security significance, and its use cannot be relegated to a single institutional silo. The increasing generality of AI means that many functions that to date have been served by individual bespoke tools may, going forward, be better fulfilled by systems that, at least in part, rely on a shared, multi-purpose AI capability. Such integration will only succeed if paired with appropriately redesigned United States Government organizational and informational infrastructure.

(h) In this effort, the United States Government must also protect human rights, civil rights, civil liberties, privacy, and safety, and lay the groundwork for a stable and responsible international AI governance landscape. Throughout its history, the United States has been a global leader in shaping the design, development, and use of new technologies not only to advance national security, but also to protect and promote democratic values. The United States Government must develop safeguards for its use of AI tools, and take an active role in steering global AI norms and institutions. The AI frontier is moving quickly, and the United States Government must stay attuned to ongoing technical developments without losing focus on its guiding principles.

(i) This memorandum aims to catalyze needed change in how the United States Government approaches AI national security policy. In line with Executive Order 14110, it directs actions to strengthen and protect the United States AI ecosystem; improve the safety, security, and trustworthiness of AI systems developed and used in the United States; enhance the United States Government's appropriate, responsible, and effective adoption of AI in service of the national security mission; and minimize the misuse of AI worldwide.

Sec. 2. *Objectives*. It is the policy of the United States Government that the following three objectives will guide its activities with respect to AI and national security.

(a) First, the United States must lead the world's development of safe, secure, and trustworthy AI. To that end, the United States Government must — in partnership with industry, civil society, and academia — promote and secure the foundational capabilities across the United States that power AI development. The United States Government cannot take the unmatched vibrancy and innovativeness of the United States AI ecosystem for granted; it must proactively strengthen it, ensuring that the United States remains the most attractive destination for global talent and home to the world's most sophisticated computational facilities. The United States Government must also provide appropriate safety and security guidance to AI developers and users, and rigorously assess and help mitigate the risks that AI systems could pose.

(b) Second, the United States Government must harness powerful AI, with appropriate safeguards, to achieve national security objectives. Emerging AI capabilities, including increasingly general-purpose models, offer profound opportunities for enhancing national security, but employing these systems effectively will require significant technical, organizational, and policy changes. The United States must understand AI's limitations as it harnesses the technology's benefits, and any use of AI must respect democratic values with regard to transparency, human rights, civil rights, civil liberties, privacy, and safety.

(c) Third, the United States Government must continue cultivating a stable and responsible framework to advance international AI governance that fosters safe, secure, and trustworthy AI development and use; manages AI risks; realizes democratic values; respects human rights, civil rights, civil liberties, and privacy; and promotes worldwide benefits from AI. It must do so in collaboration with a wide range of allies and partners. Success for the United States in the age of AI will be measured not only by the preeminence of United States technology and innovation, but also by the United States' leadership in developing effective global norms and engaging in institutions rooted in international law, human rights, civil rights, and democratic values.

Sec. 3. *Promoting and Securing the United States' Foundational AI Capabilities.* (a) To preserve and expand United States advantages in AI, it is the policy of the United States Government to promote progress, innovation, and competition in domestic AI development; protect the United States AI ecosystem against foreign intelligence threats; and manage risks to AI safety, security, and trustworthiness. Leadership in responsible AI development benefits United States national security by enabling applications directly relevant to the national security mission, unlocking economic growth, and avoiding strategic surprise. United States technological leadership also confers global benefits by enabling like-minded entities to collectively mitigate the risks of AI misuse and accidents, prevent the unchecked spread of digital authoritarianism, and prioritize vital research.

3.1. Promoting Progress, Innovation, and Competition in United States AI Development. (a) The United States' competitive edge in AI development will be at risk absent concerted United States Government efforts to promote and secure domestic AI progress, innovation, and competition. Although the United States has benefited from a head start in AI, competitors are working hard to catch up, have identified AI as a top strategic priority, and may soon devote resources to research and development that United States AI developers cannot match without appropriately supportive Government policies and action. It is therefore the policy of the United States Government to enhance innovation and competition by bolstering key drivers of AI progress, such as technical talent and computational power.

(b) It is the policy of the United States Government that advancing the lawful ability of noncitizens highly skilled in AI and related fields to enter and work in the United States constitutes a national security priority. Today, the unparalleled United States AI industry rests in substantial part on the insights of brilliant scientists, engineers, and entrepreneurs who moved to the United States in pursuit of academic, social, and economic opportunity. Preserving and expanding United States talent advantages requires developing talent at home and continuing to attract and retain top international minds.

(c) Consistent with these goals

(i) On an ongoing basis, the Department of State, the Department of Defense (DOD), and the Department of Homeland Security (DHS) shall each use all available legal authorities to assist in attracting and rapidly bringing to the United States individuals with relevant technical expertise who would improve United States competitiveness in AI and related fields, such as semiconductor design and production. These activities shall include all appropriate vetting of these individuals and shall be consistent with all appropriate risk mitigation measures. This tasking is consistent with and additive to the taskings on attracting AI talent in section 5 of Executive Order 14110.

(ii) Within 180 days of the date of this memorandum, the Chair of the Council of Economic Advisers shall prepare an analysis of the AI talent market in the United States and overseas, to the extent that reliable data is available.

(iii) Within 180 days of the date of this memorandum, the Assistant to the President for Economic Policy and Director of the National Economic Council shall coordinate an economic assessment of the relative competitive advantage of the United States private sector AI ecosystem, the key sources of the United States private sector's competitive advantage, and possible risks to that position, and shall recommend policies to mitigate them. The assessment could include areas including (1) the design, manufacture, and packaging of chips critical in AI-related activities; (2) the availability of capital; (3) the availability of workers highly skilled in AI-related fields; (4) computational resources and the associated electricity requirements; and (5) technological platforms or institutions with the requisite scale of capital and data resources for frontier AI model development, as well as possible other factors.

(iv) Within 90 days of the date of this memorandum, the Assistant to the President for National Security Affairs (APNSA) shall convene appropriate executive departments and agencies (agencies) to explore actions for prioritizing and streamlining administrative processing operations for all visa applicants working with sensitive technologies. Doing so shall assist with streamlined processing of highly skilled applicants in AI and other critical and emerging technologies. This effort shall explore options for ensuring the adequate resourcing of such operations and narrowing the criteria that trigger secure advisory opinion requests for such applicants, as consistent with national security objectives.

(d) The current paradigm of AI development depends heavily on computational resources. To retain its lead in AI, the United States must continue developing the world's most sophisticated AI semiconductors and constructing its most advanced AI-dedicated computational infrastructure.

(e) Consistent with these goals:

(i)   DOD, the Department of Energy (DOE) (including national laboratories), and the Intelligence Community (IC) shall, when planning for and constructing or renovating computational facilities, consider the applicability of large-scale AI to their mission. Where appropriate, agencies shall design and build facilities capable of harnessing frontier AI for relevant scientific research domains and intelligence analysis. Those investments shall be consistent with the Federal Mission Resilience Strategy adopted in Executive Order 13961 of December 7, 2020 (Governance and Integration of Federal Mission Resilience).

(ii)   On an ongoing basis, the National Science Foundation (NSF) shall, consistent with its authorities, use the National AI Research Resource (NAIRR) pilot project and any future NAIRR efforts to distribute computational resources, data, and other critical assets for AI development to a diverse array of actors that otherwise would lack access to such capabilities — such as universities, nonprofits, and independent researchers (including trusted international collaborators) — to ensure that AI research in the United States remains competitive and innovative. This tasking is consistent with the NAIRR pilot assigned in section 5 of Executive Order 14110.

(iii) Within 180 days of the date of this memorandum, DOE shall launch a pilot project to evaluate the performance and efficiency of federated AI and data sources for frontier AI-scale training, fine-tuning, and inference.

(iv)   The Office of the White House Chief of Staff, in coordination with DOE and other relevant agencies, shall coordinate efforts to streamline permitting, approvals, and incentives for the construction of AI-enabling infrastructure, as well as surrounding assets supporting the resilient operation of this infrastructure, such as clean energy generation, power transmission lines, and high-capacity fiber data links. These efforts shall include coordination, collaboration, consultation, and partnership with State, local, Tribal, and territorial governments, as appropriate, and shall be consistent with the United States' goals for managing climate risks.

(v)   The Department of State, DOD, DOE, the IC, and the Department of Commerce (Commerce) shall, as appropriate and consistent with applicable law, use existing authorities to make public investments and encourage private investments in strategic domestic and foreign AI technologies and adjacent fields. These agencies shall assess the need for new authorities for the purposes of facilitating public and private investment in AI and adjacent capabilities.

3.2. *Protecting United States AI from Foreign Intelligence Threats.* (a) In addition to pursuing industrial strategies that support their respective AI industries, foreign states almost certainly aim to obtain and repurpose the fruits of AI innovation in the United States to serve their national security goals. Historically, such competitors have employed techniques including research collaborations, investment schemes, insider threats, and advanced cyber espionage to collect and exploit United States scientific insights. It is the policy of the United States Government to protect United States industry, civil society, and academic AI intellectual property and related infrastructure from foreign intelligence threats to maintain a lead in foundational capabilities and, as necessary, to provide appropriate Government assistance to relevant non-government entities.

(b) Consistent with these goals:

(i) Within 90 days of the date of this memorandum, the National Security Council (NSC) staff and the Office of the Director of National Intelligence (ODNI) shall review the President's Intelligence Priorities and the National Intelligence Priorities Framework consistent with National Security Memorandum 12 of July 12, 2022 (The President's Intelligence Priorities), and make recommendations to ensure that such priorities improve identification and assessment of foreign intelligence threats to the United States AI ecosystem and closely related enabling sectors, such as those involved in semiconductor design and production.

(ii) Within 180 days of the date of this memorandum, and on an ongoing basis thereafter, ODNI, in coordination with DOD, the Department of Justice (DOJ), Commerce, DOE, DHS, and other IC elements as appropriate, shall identify critical nodes in the AI supply chain, and develop a list of the most plausible avenues through which these nodes could be disrupted or compromised by foreign actors. On an ongoing basis, these agencies shall take all steps, as appropriate and consistent with applicable law, to reduce such risks.

(c) Foreign actors may also seek to obtain United States intellectual property through gray-zone methods, such as technology transfer and data localization requirements. AI-related intellectual property often includes critical technical artifacts (CTAs) that would substantially lower the costs of recreating, attaining, or using powerful AI capabilities. The United States Government must guard against these risks.

(d) Consistent with these goals:

(i) In furtherance of Executive Order 14083 of September 15, 2022 (Ensuring Robust Consideration of Evolving National Security Risks by the Committee on

Foreign Investment in the United States), the Committee on Foreign Investment in the United States shall, as appropriate, consider whether a covered transaction involves foreign actor access to proprietary information on AI training techniques, algorithmic improvements, hardware advances, CTAs, or other proprietary insights that shed light on how to create and effectively use powerful AI systems.

3.3. *Managing Risks to AI Safety, Security, and Trustworthiness.* (a) Current and near-future AI systems could pose significant safety, security, and trustworthiness risks, including those stemming from deliberate misuse and accidents. Across many technological domains, the United States has historically led the world not only in advancing capabilities, but also in developing the tests, standards, and norms that underpin reliable and beneficial global adoption. The United States approach to AI should be no different, and proactively constructing testing infrastructure to assess and mitigate AI risks will be essential to realizing AI's positive potential and to preserving United States AI leadership.

(b) It is the policy of the United States Government to pursue new technical and policy tools that address the potential challenges posed by AI. These tools include processes for reliably testing AI models' applicability to harmful tasks and deeper partnerships with institutions in industry, academia, and civil society capable of advancing research related to AI safety, security, and trustworthiness.

(c) Commerce, acting through the AI Safety Institute (AISI) within the National Institute of Standards and Technology (NIST), shall serve as the primary United States Government point of contact with private sector AI developers to facilitate voluntary pre- and post-public deployment testing for safety, security, and trustworthiness of frontier AI models. In coordination with relevant agencies as appropriate, Commerce shall establish an enduring capability to lead voluntary unclassified pre-deployment safety testing of frontier AI models on behalf of the United States Government, including assessments of risks relating to cybersecurity, biosecurity, chemical weapons, system autonomy, and other risks as appropriate (not including nuclear risk, the assessment of which shall be led by DOE). Voluntary unclassified safety testing shall also, as appropriate, address risks to human rights, civil rights, and civil liberties, such as those related to privacy, discrimination and bias, freedom of expression, and the safety of individuals and groups. Other agencies, as identified in subsection 3.3(f) of this section, shall establish enduring capabilities to perform complementary voluntary classified testing in appropriate areas of expertise. The directives set forth in this subsection are consistent with broader taskings on AI safety in section 4 of Executive Order 14110, and provide additional clarity on agencies' respective roles and responsibilities.

(d) Nothing in this subsection shall inhibit agencies from performing their own evaluations of AI systems, including tests performed before those systems are released to the public, for the purposes of evaluating suitability for that agency's acquisition and procurement. AISI's responsibilities do not extend to the evaluation of AI systems for the potential use by the United States Government for national security purposes; those responsibilities lie with agencies considering such use, as outlined in subsection 4.2(e) of this memorandum and the associated framework described in that subsection.

(e) Consistent with these goals, Commerce, acting through AISI within NIST, shall take the following actions to aid in the evaluation of current and near-future AI systems:

(i) Within 180 days of the date of this memorandum and subject to private sector cooperation, AISI shall pursue voluntary preliminary testing of at least two frontier AI models prior to their public deployment or release to evaluate capabilities that might pose a threat to national security. This testing shall assess models' capabilities to aid offensive cyber operations, accelerate development of biological and/or chemical weapons, autonomously carry out malicious behavior, automate development and deployment of other models with such capabilities, and give rise to other risks identified by AISI. AISI shall share feedback with the APNSA, interagency counterparts as appropriate, and the respective model developers regarding the results of risks identified during such testing and any appropriate mitigations prior to deployment.

(ii) Within 180 days of the date of this memorandum, AISI shall issue guidance for AI developers on how to test, evaluate, and manage risks to safety, security, and trustworthiness arising from dual-use foundation models, building on guidelines issued pursuant to subsection 4.1(a) of Executive Order 14110. AISI shall issue guidance on topics including:

(A) How to measure capabilities that are relevant to the risk that AI models could enable the development of biological and chemical weapons or the automation of offensive cyber operations;

(B) How to address societal risks, such as the misuse of models to harass or impersonate individuals;

(C) How to develop mitigation measures to prevent malicious or improper use of models;

(D) How to test the efficacy of safety and security mitigations; and

(E) How to apply risk management practices throughout the development and deployment lifecycle (pre-development, development, and deployment/release).

(iii) Within 180 days of the date of this memorandum, AISI, in consultation with other agencies as appropriate, shall develop or recommend benchmarks or other methods for assessing AI systems' capabilities and limitations in science, mathematics, code generation, and general reasoning, as well as other categories of activity that AISI deems relevant to assessing general-purpose capabilities likely to have a bearing on national security and public safety.

(iv) In the event that AISI or another agency determines that a dual-use foundation model's capabilities could be used to harm public safety significantly, AISI shall serve as the primary point of contact through which the United States Government communicates such findings and any associated recommendations regarding risk mitigation to the developer of the model.

(v) Within 270 days of the date of this memorandum, and at least annually thereafter, AISI shall submit to the President, through the APNSA, and provide to other interagency counterparts as appropriate, at minimum one report that shall include the following:

(A) A summary of findings from AI safety assessments of frontier AI models that have been conducted by or shared with AISI;

(B) A summary of whether AISI deemed risk mitigation necessary to resolve any issues identified in the assessments, along with conclusions regarding any mitigations' efficacy; and

(C) A summary of the adequacy of the science-based tools and methods used to inform such assessments.

(f) Consistent with these goals, other agencies specified below shall take the following actions, in coordination with Commerce, acting through AISI within NIST, to provide classified sector-specific evaluations of current and near-future AI systems for cyber, nuclear, and radiological risks:

(i) All agencies that conduct or fund safety testing and evaluations of AI systems shall share the results of such evaluations with AISI within 30 days of their completion, consistent with applicable protections for classified and controlled information.

(ii) Within 120 days of the date of this memorandum, the National Security Agency (NSA), acting through its AI Security Center (AISC) and in coordination with AISI, shall develop the capability to perform rapid systematic classified testing of AI models' capacity to detect, generate, and/or exacerbate offensive cyber threats. Such tests shall assess the degree to which AI systems, if misused, could accelerate offensive cyber operations.

(iii) Within 120 days of the date of this memorandum, DOE, acting primarily through the National Nuclear Security Administration (NNSA) and in close coordination with AISI and NSA, shall seek to develop the capability to perform rapid systematic testing of AI models' capacity to generate or exacerbate nuclear and radiological risks. This initiative shall involve the development and maintenance of infrastructure capable of running classified and unclassified tests, including using restricted data and relevant classified threat information. This initiative shall also feature the creation and regular updating of automated evaluations, the development of an interface for enabling human-led red-teaming, and the establishment of technical and legal tooling necessary for facilitating the rapid and secure transfer of United States Government, open-weight, and proprietary models to these facilities. As part of this initiative:

(A) Within 180 days of the date of this memorandum, DOE shall use the capability described in subsection 3.3(f)(iii) of this section to complete initial evaluations of the radiological and nuclear knowledge, capabilities, and implications of a frontier AI model no more than 30 days after the model has been made available to NNSA at an appropriate classification level. These evaluations shall involve tests of AI systems both without significant modifications and, as appropriate, with fine-tuning or other modifications that could enhance performance.

(B) Within 270 days of the date of this memorandum, and at least annually thereafter, DOE shall submit to the President, through the APNSA, at minimum one assessment that shall include the following:

(1) A concise summary of the findings of each AI model evaluation for radiological and nuclear risk, described in subsection 3.3(f)(iii)(A) of this section, that DOE has performed in the preceding 12 months;

(2) A recommendation as to whether corrective action is necessary to resolve any issues identified in the evaluations, including but not limited to actions necessary for attaining and sustaining compliance conditions appropriate to safeguard and prevent unauthorized disclosure of restricted data or other classified information, pursuant to the Atomic Energy Act of 1954; and

(3) A concise statement regarding the adequacy of the science-based tools and methods used to inform the evaluations.

(iv) On an ongoing basis, DHS, acting through the Cybersecurity and Infrastructure Security Agency (CISA), shall continue to fulfill its responsibilities with respect to the application of AISI guidance, as identified in National Security Memorandum 22 of April 30, 2024 (Critical Infrastructure Security and Resilience), and section 4 of Executive Order 14110.

(g) Consistent with these goals, and to reduce the chemical and biological risks that could emerge from AI:

(i) The United States Government shall advance classified evaluations of advanced AI models' capacity to generate or exacerbate deliberate chemical and biological threats. As part of this initiative:

(A) Within 210 days of the date of this memorandum, DOE, DHS, and AISI, in consultation with DOD and other relevant agencies, shall coordinate to develop a roadmap for future classified evaluations of advanced AI models' capacity to generate or exacerbate deliberate chemical and biological threats, to be shared with the APNSA. This roadmap shall consider the scope, scale, and priority of classified evaluations; proper safeguards to ensure that evaluations and simulations are not misconstrued as offensive capability development; proper safeguards for testing sensitive and/or classified information; and sustainable implementation of evaluation methodologies.

(B) On an ongoing basis, DHS shall provide expertise, threat and risk information, and other technical support to assess the feasibility of proposed biological and chemical classified evaluations; interpret and contextualize evaluation results; and advise relevant agencies on potential risk mitigations.

(C) Within 270 days of the date of this memorandum, DOE shall establish a pilot project to provide expertise, infrastructure, and facilities capable of conducting classified tests in this area.

(ii) Within 240 days of the date of this memorandum, DOD, the Department of Health and Human Services (HHS), DOE (including national laboratories), DHS, NSF, and other agencies pursuing the development of AI systems substantially trained on biological and chemical data shall, as appropriate, support efforts to utilize high-performance computing resources and AI systems to enhance biosafety and biosecurity. These efforts shall include:

(A) The development of tools for screening in silico chemical and biological research and technology;

(B) The creation of algorithms for nucleic acid synthesis screening;

(C) The construction of high-assurance software foundations for novel biotechnologies;

(D) The screening of complete orders or data streams from cloud labs and biofoundries; and

(E) The development of risk mitigation strategies such as medical countermeasures.

(iii) After the publication of biological and chemical safety guidance by AISI outlined in subsection 3.3(e) of this section, all agencies that directly develop relevant dual-use foundation AI models that are made available to the public and are substantially trained on biological or chemical data shall incorporate this guidance into their agency's practices, as appropriate and feasible.

(iv) Within 180 days of the date of this memorandum, NSF, in coordination with DOD, Commerce (acting through AISI within NIST), HHS, DOE, the Office of Science and Technology Policy (OSTP), and other relevant agencies, shall seek to convene academic research institutions and scientific publishers to develop voluntary best practices and standards for publishing computational biological and chemical models, data sets, and approaches, including those that use AI and that could contribute to the production of knowledge, information, technologies, and products that could be misused to cause harm. This is in furtherance of the activities described in subsections 4.4 and 4.7 of Executive Order 14110.

(v) Within 540 days of the date of this memorandum, and informed by the United States Government Policy for Oversight of Dual Use Research of Concern and Pathogens with Enhanced Pandemic Potential, OSTP, NSC staff, and the Office of Pandemic Preparedness and Response Policy, in consultation with relevant agencies and external stakeholders as appropriate, shall develop guidance promoting the benefits of and mitigating the risks associated with in silico biological and chemical research.

(h) Agencies shall take the following actions to improve foundational understanding of AI safety, security, and trustworthiness:

(i) DOD, Commerce, DOE, DHS, ODNI, NSF, NSA, and the National Geospatial-Intelligence Agency (NGA) shall, as appropriate and consistent with applicable law, prioritize research on AI safety and trustworthiness. As appropriate and consistent with existing authorities, they shall pursue partnerships as appropriate with leading public sector, industry, civil society, academic, and other institutions with expertise in these domains, with the objective of accelerating technical and socio-technical progress in AI safety and trustworthiness. This work may include research on interpretability, formal methods, privacy enhancing technologies, techniques to address risks to civil liberties and human rights, human-AI interaction, and/or the socio-technical effects of detecting and labeling synthetic and authentic content (for example, to address the malicious use of AI to generate misleading videos or images, including those of a strategically damaging or non-consensual intimate nature, of political or public figures).

(ii) DOD, Commerce, DOE, DHS, ODNI, NSF, NSA, and NGA shall, as appropriate and consistent with applicable law, prioritize research to improve the security, robustness, and reliability of AI systems and controls. These entities shall, as appropriate and consistent with applicable law, partner with other agencies, industry, civil society, and academia. Where appropriate, DOD, DHS (acting through CISA), the Federal Bureau of Investigation, and NSA (acting through AISC) shall publish unclassified guidance concerning known AI cybersecurity vulnerabilities and threats; best practices for avoiding, detecting, and mitigating such issues during model training and deployment; and the integration of AI into other software systems. This work shall include an examination of the role of and vulnerabilities potentially caused by AI systems used in critical infrastructure.

(i) Agencies shall take actions to protect classified and controlled information, given the potential risks posed by AI:

(i) In the course of regular updates to policies and procedures, DOD, DOE, and the IC shall consider how analysis enabled by AI tools may affect decisions related to declassification of material, standards for sufficient anonymization, and similar activities, as well as the robustness of existing operational security and equity controls to protect classified or controlled information, given that AI systems have demonstrated the capacity to extract previously inaccessible insight from redacted and anonymized data.

Sec. 4. *Responsibly Harnessing AI to Achieve National Security Objectives.* (a) It is the policy of the United States Government to act decisively to enable the effective and responsible use of AI in furtherance of its national security mission. Achieving global leadership in national security applications of AI will require effective partnership with organizations outside Government, as well as significant internal transformation, including strengthening effective oversight and governance functions.

4.1. *Enabling Effective and Responsible Use of AI.* (a) It is the policy of the United States Government to adapt its partnerships, policies, and infrastructure to use AI capabilities appropriately, effectively, and responsibly. These modifications must balance each agency's unique oversight, data, and application needs with the substantial benefits associated with sharing powerful AI and computational resources across the United States Government. Modifications must also be grounded in a clear understanding of the United States Government's comparative advantages relative to industry, civil society, and academia, and must leverage offerings from external collaborators and contractors as appropriate. The United States Government must make the most of the rich United States AI ecosystem by incentivizing innovation in safe, secure, and trustworthy AI and promoting industry competition when selecting contractors, grant recipients, and research collaborators. Finally, the United States Government must address important technical and policy considerations in ways that ensure the integrity and interoperability needed to pursue its objectives while protecting human rights, civil rights, civil liberties, privacy, and safety.

(b) The United States Government needs an updated set of Government-wide procedures for attracting, hiring, developing, and retaining AI and AI-enabling talent for national security purposes.

(c) Consistent with these goals:

(i) In the course of regular legal, policy, and compliance framework reviews, the Department of State, DOD, DOJ, DOE, DHS, and IC elements shall revise, as appropriate, their hiring and retention policies and strategies to accelerate responsible AI adoption. Agencies shall account for technical talent needs required to adopt AI and integrate it into their missions and other roles necessary to use AI effectively, such as AI-related governance, ethics, and policy positions. These policies and strategies shall identify financial, organizational, and security hurdles, as well as potential mitigations consistent with applicable law. Such measures shall also include consideration of programs to attract experts with relevant technical expertise from industry, academia, and civil society — including scholarship for service programs — and similar initiatives that would expose Government employees to relevant non-government entities in ways that build technical, organizational, and cultural familiarity with the AI industry. These policies and strategies shall use all available authorities, including expedited security clearance procedures as appropriate, in order to address the shortfall of AI-relevant talent within Government.

(ii) Within 120 days of the date of this memorandum, the Department of State, DOD, DOJ, DOE, DHS, and IC elements shall each, in consultation with the Office of Management and Budget (OMB), identify education and training opportunities to increase the AI competencies of their respective workforces, via initiatives which may include training and skills-based hiring.

(d) To accelerate the use of AI in service of its national security mission, the United States Government needs coordinated and effective acquisition and procurement systems. This will require an enhanced capacity to assess, define, and articulate AI-related requirements for national security purposes, as well as improved accessibility for AI companies that lack significant prior experience working with the United States Government.

(e) Consistent with these goals:

(i)   Within 30 days of the date of this memorandum, DOD and ODNI, in coordination with OMB and other agencies as appropriate, shall establish a working group to address issues involving procurement of AI by DOD and IC elements and for use on NSS. As appropriate, the working group shall consult the Director of the NSA, as the National Manager for NSS, in developing recommendations for acquiring and procuring AI for use on NSS.

(ii)   Within 210 days of the date of this memorandum, the working group described in subsection 4.1(e)(i) of this section shall provide written recommendations to the Federal Acquisition Regulatory Council (FARC) regarding changes to existing regulations and guidance, as appropriate and consistent with applicable law, to promote the following objectives for AI procured by DOD and IC elements and for use on NSS:

(A) Ensuring objective metrics to measure and promote the safety, security, and trustworthiness of AI systems;

(B) Accelerating the acquisition and procurement process for AI, consistent with the Federal Acquisition Regulation, while maintaining appropriate checks to mitigate safety risks;

(C) Simplifying processes such that companies without experienced contracting teams may meaningfully compete for relevant contracts, to ensure that the United States Government has access to a wide range of AI systems and that the AI marketplace is competitive;

(D) Structuring competitions to encourage robust participation and achieve best value to the Government, such as by including requirements that promote interoperability and prioritizing the technical capability of vendors when evaluating offers;

(E) Accommodating shared use of AI to the greatest degree possible and as appropriate across relevant agencies; and

(F) Ensuring that agencies with specific authorities and missions may implement other policies, where appropriate and necessary.

(iii) The FARC shall, as appropriate and consistent with applicable law, consider proposing amendments to the Federal Acquisition Regulation to codify recommendations provided by the working group pursuant to subsection 4.1(e)(ii) of this section that may have Government-wide application.

(iv) DOD and ODNI shall seek to engage on an ongoing basis with diverse United States private sector stakeholders — including AI technology and defense companies and members of the United States investor community — to identify and better understand emerging capabilities that would benefit or otherwise affect the United States national security mission.

(f) The United States Government needs clear, modernized, and robust policies and procedures that enable the rapid development and national security use of AI, consistent with human rights, civil rights, civil liberties, privacy, safety, and other democratic values.

(g) Consistent with these goals:

(i) DOD and the IC shall, in consultation with DOJ as appropriate, review their respective legal, policy, civil liberties, privacy, and compliance frameworks, including international legal obligations, and, as appropriate and consistent with applicable law, seek to develop or revise policies and procedures to enable the effective and responsible use of AI, accounting for the following:

(A) Issues raised by the acquisition, use, retention, dissemination, and disposal of models trained on datasets that include personal information traceable to specific United States persons, publicly available information, commercially available information, and intellectual property, consistent with section 9 of Executive Order 14110;

(B) Guidance that shall be developed by DOJ, in consultation with DOD and ODNI, regarding constitutional considerations raised by the IC's acquisition and use of AI;

(C) Challenges associated with classification and compartmentalization;

(D) Algorithmic bias, inconsistent performance, inaccurate outputs, and other known AI failure modes;

(E) Threats to analytic integrity when employing AI tools;

(F) Risks posed by a lack of safeguards that protect human rights, civil rights, civil liberties, privacy, and other democratic values, as addressed in further detail in subsection 4.2 of this section;

(G) Barriers to sharing AI models and related insights with allies and partners; and

(H) Potential inconsistencies between AI use and the implementation of international legal obligations and commitments.

(ii) As appropriate, the policies described in subsection 4.1(g) of this section shall be consistent with direction issued by the Committee on NSS and DOD governing the security of AI used on NSS, policies issued by the Director of National Intelligence governing adoption of AI by the IC, and direction issued by OMB governing the security of AI used on non-NSS.

(iii) On an ongoing basis, each agency that uses AI on NSS shall, in consultation with ODNI and DOD, take all steps appropriate and consistent with applicable law to accelerate responsible approval of AI systems for use on NSS and accreditation of NSS that use AI systems.

(h) The United States' network of allies and partners confers significant advantages over competitors. Consistent with the 2022 National Security Strategy or any successor strategies, the United States Government must invest in and proactively enable the co-development and co-deployment of AI capabilities with select allies and partners.

(i) Consistent with these goals:

(i) Within 150 days of the date of this memorandum, DOD, in coordination with the Department of State and ODNI, shall evaluate the feasibility of advancing, increasing, and promoting co-development and shared use of AI and AI-enabled assets with select allies and partners. This evaluation shall include:

(A) A potential list of foreign states with which such co-development or co-deployment may be feasible;

(B) A list of bilateral and multilateral fora for potential outreach;

(C) Potential co-development and co-deployment concepts;

(D) Proposed classification-appropriate testing vehicles for co-developed AI capabilities; and

(E) Considerations for existing programs, agreements, or arrangements to use as foundations for future co-development and co-deployment of AI capabilities.

(j) The United States Government needs improved internal coordination with respect to its use of and approach to AI on NSS in order to ensure interoperability and resource sharing consistent with applicable law, and to reap the generality and economies of scale offered by frontier AI models.

(k) Consistent with these goals:

(i) On an ongoing basis, DOD and ODNI shall issue or revise relevant guidance to improve consolidation and interoperability across AI functions on NSS. This guidance shall seek to ensure that the United States Government can coordinate and share AI-related resources effectively, as appropriate and consistent with applicable law. Such work shall include:

(A) Recommending agency organizational practices to improve AI research and deployment activities that span multiple national security institutions. In order to encourage AI adoption for the purpose of national security, these measures shall aim to create consistency to the greatest extent possible across the revised practices.

(B) Steps that enable consolidated research, development, and procurement for general-purpose AI systems and supporting infrastructure, such that multiple agencies can share access to these tools to the extent consistent with applicable law, while still allowing for appropriate controls on sensitive data.

(C) Aligning AI-related national security policies and procedures across agencies, as practicable and appropriate, and consistent with applicable law.

(D) Developing policies and procedures, as appropriate and consistent with applicable law, to share information across DOD and the IC when an AI system developed, deployed, or used by a contractor demonstrates risks related to safety, security, and trustworthiness, including to human rights, civil rights, civil liberties, or privacy.

4.2. *Strengthening AI Governance and Risk Management*. (a) As the United States Government moves swiftly to adopt AI in support of its national security mission, it must continue taking active steps to uphold human rights, civil rights, civil liberties, privacy, and safety; ensure that AI is used in a manner consistent with the President's authority as Commander in Chief to decide when to order military operations in the Nation's defense; and ensure that military use of AI capabilities is accountable, including through such use during military operations within a responsible human chain of command and control. Accordingly, the United States

Government must develop and implement robust AI governance and risk management practices to ensure that its AI innovation aligns with democratic values, updating policy guidance where necessary. In light of the diverse authorities and missions across covered agencies with a national security mission and the rapid rate of ongoing technological change, such AI governance and risk management frameworks shall be:

(i) Structured, to the extent permitted by law, such that they can adapt to future opportunities and risks posed by new technical developments;

(ii) As consistent across agencies as is practicable and appropriate in order to enable interoperability, while respecting unique authorities and missions;

(iii) Designed to enable innovation that advances United States national security objectives;

(iv) As transparent to the public as practicable and appropriate, while protecting classified or controlled information;

(v) Developed and applied in a manner and with means to integrate protections, controls, and safeguards for human rights, civil rights, civil liberties, privacy, and safety where relevant; and

(vi) Designed to reflect United States leadership in establishing broad international support for rules and norms that reinforce the United States' approach to AI governance and risk management.

(b) Covered agencies shall develop and use AI responsibly, consistent with United States law and policies, democratic values, and international law and treaty obligations, including international humanitarian and human rights law. All agency officials retain their existing authorities and responsibilities established in other laws and policies.

(c) Consistent with these goals:

(i) Heads of covered agencies shall, consistent with their authorities, monitor, assess, and mitigate risks directly tied to their agency's development and use of AI. Such risks may result from reliance on AI outputs to inform, influence, decide, or execute agency decisions or actions, when used in a defense, intelligence, or law enforcement context, and may impact human rights, civil rights, civil liberties, privacy, safety, national security, and democratic values. These risks from the use of AI include the following:

(A) Risks to physical safety: AI use may pose unintended risks to human life or property.

(B) Privacy harms: AI design, development, and operation may result in harm, embarrassment, unfairness, and prejudice to individuals.

(C) Discrimination and bias: AI use may lead to unlawful discrimination and harmful bias, resulting in, for instance, inappropriate surveillance and profiling, among other harms.

(D) Inappropriate use: operators using AI systems may not fully understand the capabilities and limitations of these technologies, including systems used in conflicts. Such unfamiliarity could impact operators' ability to exercise appropriate levels of human judgment.

(E) Lack of transparency: agencies may have gaps in documentation of AI development and use, and the public may lack access to information about how AI is used in national security contexts because of the necessity to protect classified or controlled information.

(F) Lack of accountability: training programs and guidance for agency personnel on the proper use of AI systems may not be sufficient, including to mitigate the risk of overreliance on AI systems (such as "automation bias"), and accountability mechanisms may not adequately address possible intentional or negligent misuse of AI-enabled technologies.

(G) Data spillage: AI systems may reveal aspects of their training data — either inadvertently or through deliberate manipulation by malicious actors — and data spillage may result from AI systems trained on classified or controlled information when used on networks where such information is not permitted.

(H) Poor performance: AI systems that are inappropriately or insufficiently trained, used for purposes outside the scope of their training set, or improperly integrated into human workflows may exhibit poor performance, including in ways that result in inconsistent outcomes or unlawful discrimination and harmful bias, or that undermine the integrity of decision-making processes.

(I) Deliberate manipulation and misuse: foreign state competitors and malicious actors may deliberately undermine the accuracy and efficacy of AI systems, or seek to extract sensitive information from such systems.

(d) The United States Government's AI governance and risk management policies must keep pace with evolving technology.

(e) Consistent with these goals:

(i)   An AI framework, entitled "Framework to Advance AI Governance and Risk Management in National Security" (AI Framework), shall further implement this subsection. The AI Framework shall be approved by the NSC Deputies Committee through the process described in National Security Memorandum 2 of February 4, 2021 (Renewing the National Security Council System), or any successor process, and shall be reviewed periodically through that process. This process shall determine whether adjustments are needed to address risks identified in subsection 4.2(c) of this section and other topics covered in the AI Framework. The AI Framework shall serve as a national security-focused counterpart to OMB's Memorandum M-24-10 of March 28, 2024 (Advancing Governance, Innovation, and Risk Management for Agency Use of Artificial Intelligence), and any successor OMB policies. To the extent feasible, appropriate, and consistent with applicable law, the AI Framework shall be as consistent as possible with these OMB policies and shall be made public.

(ii)   The AI Framework described in subsection 4.2(e)(i) of this section and any successor document shall, at a minimum, and to the extent consistent with applicable law, specify the following:

(A)   Each covered agency shall have a Chief AI Officer who holds primary responsibility within that agency, in coordination with other responsible officials, for managing the agency's use of AI, promoting AI innovation within the agency, and managing risks from the agency's use of AI consistent with subsection 3(b) of OMB Memorandum M-24-10, as practicable.

(B) Covered agencies shall have AI Governance Boards to coordinate and govern AI issues through relevant senior leaders from the agency.

(C) Guidance on AI activities that pose unacceptable levels of risk and that shall be prohibited.

(D) Guidance on AI activities that are "high impact" and require minimum risk management practices, including for high-impact AI use that affects United States Government personnel. Such high-impact activities shall include AI whose output serves as a principal basis for a decision or action that could exacerbate or create significant risks to national security, international norms, human rights, civil rights, civil liberties, privacy, safety, or other democratic values. The minimum risk management practices for high-impact AI shall include a mechanism for agencies to assess AI's expected benefits and potential risks; a mechanism for assessing data quality; sufficient test and evaluation practices; mitigation of unlawful discrimination and harmful bias; human training, assessment, and oversight requirements; ongoing monitoring; and additional safeguards for military service members, the Federal civilian workforce, and individuals who receive an offer of employment from a covered agency.

(E) Covered agencies shall ensure privacy, civil liberties, and safety officials are integrated into AI governance and oversight structures. Such officials shall report findings to the heads of agencies and oversight officials, as appropriate, using existing reporting channels when feasible.

(F) Covered agencies shall ensure that there are sufficient training programs, guidance, and accountability processes to enable proper use of AI systems.

(G) Covered agencies shall maintain an annual inventory of their high-impact AI use and AI systems and provide updates on this inventory to agency heads and the APNSA.

(H) Covered agencies shall ensure that whistleblower protections are sufficient to account for issues that may arise in the development and use of AI and AI systems.

(I) Covered agencies shall develop and implement waiver processes for high-impact AI use that balance robust implementation of risk mitigation measures in this memorandum and the AI Framework with the need to utilize AI to preserve and advance critical agency missions and operations.

(J) Covered agencies shall implement cybersecurity guidance or direction associated with AI systems issued by the National Manager for NSS to mitigate the risks posed by malicious actors exploiting new technologies, and to enable interoperability of AI across agencies. Within 150 days of the date of this memorandum, and periodically thereafter, the National Manager for NSS shall issue minimum cybersecurity guidance and/or direction for AI used as a component of NSS, which shall be incorporated into AI governance guidance detailed in subsection 4.2(g)(i) of this section.

(f) The United States Government needs guidance specifically regarding the use of AI on NSS.

(g) Consistent with these goals:

(i) Within 180 days of the date of this memorandum, the heads of the Department of State, the Department of the Treasury, DOD, DOJ, Commerce, DOE, DHS, ODNI (acting on behalf of the 18 IC elements), and any other covered agency that uses AI as part of a NSS (Department Heads) shall issue or update guidance to their components/sub-agencies on AI governance and risk management for NSS, aligning with the policies in this subsection, the AI Framework, and other applicable policies. Department Heads shall review their respective guidance on an annual basis, and update such guidance as needed. This guidance, and any updates thereto, shall be provided to the APNSA prior to issuance. This guidance shall be unclassified and made available to the public to the extent feasible and appropriate, though it may have a classified annex. Department Heads shall seek to harmonize their guidance, and the APNSA shall convene an interagency meeting at least annually for the purpose of harmonizing Department Heads' guidance on AI governance and risk management to the extent practicable and appropriate while respecting the agencies' diverse authorities and missions. Harmonization shall be pursued in the following areas:

(A) Implementation of the risk management practices for high-impact AI;

(B) AI and AI system standards and activities, including as they relate to training, testing, accreditation, and security and cybersecurity; and

(C) Any other issues that affect interoperability for AI and AI systems.

Sec. 5. *Fostering a Stable, Responsible, and Globally Beneficial International AI Governance Landscape.* (a) Throughout its history, the United States has played an essential role in shaping the international order to enable the safe, secure, and trustworthy global adoption of new technologies while also protecting democratic values. These contributions have ranged from establishing nonproliferation regimes for biological, chemical, and nuclear weapons to setting the foundations for multi-stakeholder governance of the Internet. Like these precedents, AI will require new global norms and coordination mechanisms, which the United States Government must maintain an active role in crafting.

(b) It is the policy of the United States Government that United States international engagement on AI shall support and facilitate improvements to the safety, security, and trustworthiness of AI systems worldwide; promote democratic values, including respect for human rights, civil rights, civil liberties, privacy, and safety; prevent the misuse of AI in national security contexts; and promote equitable access to AI's benefits. The United States Government shall advance international agreements, collaborations, and other substantive and norm-setting initiatives in alignment with this policy.

(c) Consistent with these goals:

(i) Within 120 days of the date of this memorandum, the Department of State, in coordination with DOD, Commerce, DHS, the United States Mission to the United Nations (USUN), and the United States Agency for International Development

(USAID), shall produce a strategy for the advancement of international AI governance norms in line with safe, secure, and trustworthy AI, and democratic values, including human rights, civil rights, civil liberties, and privacy. This strategy shall cover bilateral and multilateral engagement and relations with allies and partners. It shall also include guidance on engaging with competitors, and it shall outline an approach to working in international institutions such as the United Nations and the Group of 7 (G7), as well as technical organizations. The strategy shall:

(A) Develop and promote internationally shared definitions, norms, expectations, and standards, consistent with United States policy and existing efforts, which will promote safe, secure, and trustworthy AI development and use around the world. These norms shall be as consistent as possible with United States domestic AI governance (including Executive Order 14110 and OMB Memorandum M-24-10), the International Code of Conduct for Organizations Developing Advanced AI Systems released by the G7 in October 2023, the Organization for Economic Cooperation and Development Principles on AI, United Nations General Assembly Resolution A/78/L.49, and other United States-supported relevant international frameworks (such as the Political Declaration on Responsible Military Use of AI and Autonomy) and instruments. By discouraging misuse and encouraging appropriate safeguards, these norms and standards shall aim to reduce the likelihood of AI causing harm or having adverse impacts on human rights, democracy, or the rule of law.

(B) Promote the responsible and ethical use of AI in national security contexts in accordance with democratic values and in compliance with applicable international law. The strategy shall advance the norms and practices established by this memorandum and measures endorsed in the Political Declaration on Responsible Military Use of AI and Autonomy.

Sec. 6. Ensuring Effective Coordination, Execution, and Reporting of AI Policy. (a) The United States Government must work in a closely coordinated manner to make progress on effective and responsible AI adoption. Given the speed with which AI technology evolves, the United States Government must learn quickly, adapt to emerging strategic developments, adopt new capabilities, and confront novel risks.

(b) Consistent with these goals:

(i) Within 270 days of the date of this memorandum, and annually thereafter for at least the next 5 years, the heads of the Department of State, DOD, Commerce, DOE, ODNI (acting on behalf of the IC), USUN, and USAID shall each submit a report to the President, through the APNSA, that offers a detailed accounting of their activities in response to their taskings in all sections of this memorandum, including this memorandum's classified annex, and that provides a plan for further action. The Central Intelligence Agency (CIA), NSA, the Defense Intelligence Agency (DIA), and NGA shall submit reports on their activities to ODNI for inclusion in full as an appendix to ODNI's report regarding IC activities. NGA, NSA, and DIA shall submit their reports as well to DOD for inclusion in full as an appendix to DOD's report.

(ii)   Within 45 days of the date of this memorandum, the Chief AI Officers of the Department of State, DOD, DOJ, DOE, DHS, OMB, ODNI, CIA, DIA, NSA, and NGA, as well as appropriate technical staff, shall form an AI National Security Coordination Group (Coordination Group). Any Chief AI Officer of an agency that is a member of the Committee on National Security Systems may also join the Coordination Group as a full member. The Coordination Group shall be co-chaired by the Chief AI Officers of ODNI and DOD. The Coordination Group shall consider ways to harmonize policies relating to the development, accreditation, acquisition, use, and evaluation of AI on NSS. This work could include development of:

(A) Enhanced training and awareness to ensure that agencies prioritize the most effective AI systems, responsibly develop and use AI, and effectively evaluate AI systems;

(B) Best practices to identify and mitigate foreign intelligence risks and human rights considerations associated with AI procurement;

(C) Best practices to ensure interoperability between agency deployments of AI, to include data interoperability and data sharing agreements, as appropriate and consistent with applicable law;

(D) A process to maintain, update, and disseminate such trainings and best practices on an ongoing basis;

(E) AI-related policy initiatives to address regulatory gaps implicated by executive branch-wide policy development processes; and

(F) An agile process to increase the speed of acquisitions, validation, and delivery of AI capabilities, consistent with applicable law.

(iii) Within 90 days of the date of this memorandum, the Coordination Group described in subsection (b)(ii) of this section shall establish a National Security AI Executive Talent Committee (Talent Committee) composed of senior AI officials (or designees) from all agencies in the Coordination Group that wish to participate. The Talent Committee shall work to standardize, prioritize, and address AI talent needs and develop an updated set of Government-wide procedures for attracting, hiring, developing, and retaining AI and AI-enabling talent for national security purposes. The Talent Committee shall designate a representative to serve as a member of the AI and Technology Talent Task Force set forth in Executive Order 14110, helping to identify overlapping needs and address shared challenges in hiring.

(iv)   Within 365 days of the date of this memorandum, and annually thereafter for at least the next 5 years, the Coordination Group described in subsection (b)(ii) of this section shall issue a joint report to the APNSA on consolidation and interoperability of AI efforts and systems for the purposes of national security.

Sec. 7. *Definitions.* (a) This memorandum uses definitions set forth in section 3 of Executive Order 14110. In addition, for the purposes of this memorandum:

(i)   The term "AI safety" means the mechanisms through which individuals and organizations minimize and mitigate the potential for harm to individuals and society that can result from the malicious use, misapplication, failures, accidents, and unintended behavior of AI models; the systems that integrate them; and the ways in which they are used.

(ii)   The term "AI security" means a set of practices to protect AI systems — including training data, models, abilities, and lifecycles — from cyber and physical attacks, thefts, and damage.

(iii)   The term "covered agencies" means agencies in the Intelligence Community, as well as all agencies as defined in 44 U.S.C. 3502(1) when they use AI as a component of a National Security System, other than the Executive Office of the President.

(iv)   The term "Critical Technical Artifacts" (CTAs) means information, usually specific to a single model or group of related models that, if possessed by someone other than the model developer, would substantially lower the costs of recreating, attaining, or using the model's capabilities. Under the technical paradigm dominant in the AI industry today, the model weights of a trained AI system constitute CTAs, as do, in some cases, associated training data and code. Future paradigms may rely on different CTAs.

(v)   The term "frontier AI model" means a general-purpose AI system near the cutting-edge of performance, as measured by widely accepted publicly available benchmarks, or similar assessments of reasoning, science, and overall capabilities.

(vi)　The term "Intelligence Community" (IC) has the meaning provided in 50 U.S.C. 3003.

(vii)　The term "open-weight model" means a model that has weights that are widely available, typically through public release.

(viii)　The term "United States Government" means all agencies as defined in 44 U.S.C. 3502(1).

Sec. 8. *General Provisions.* (a) Nothing in this memorandum shall be construed to impair or otherwise affect:

(i)　the authority granted by law to an executive department or agency, or the head thereof; or

(ii)　the functions of the Director of the Office of Management and Budget relating to budgetary, administrative, or legislative proposals.

(b)　This memorandum shall be implemented consistent with applicable law and subject to the availability of appropriations.

(c)　This memorandum is not intended to, and does not, create any right or benefit, substantive or procedural, enforceable at law or in equity by any party against the United States, its departments, agencies, or entities, its officers, employees, or agents, or any other person.

White House, Executive Order 14141, Advancing United States Leadership in Artificial Intelligence Infrastructure (2025)

By the authority vested in me as President by the Constitution and the laws of the United States of America, it is hereby ordered as follows:

Section 1. *Purpose*. Artificial intelligence (AI) is a defining technology of our era. Recent advancements in AI demonstrate its rapidly growing relevance to national security, including with respect to logistics, military capabilities, intelligence analysis, and cybersecurity. Building AI in the United States will help prevent adversaries from gaining access to, and using, powerful future systems to the detriment of our military and national security. It will also enable the United States Government to continue harnessing AI in service of national-security missions while preventing the United States from becoming dependent on other countries' infrastructure to develop and operate powerful AI tools.

Advances at the frontier of AI will also have significant implications for United States economic competitiveness. These imperatives require building AI infrastructure in the United States on the time frame needed to ensure United States leadership over competitors who, already, are racing to take the lead in AI development and adoption. Building AI in the United States requires enormous private-sector investments in infrastructure, especially for the advanced computing clusters needed to train AI models and the energy infrastructure needed to power this work. Already, AI's electricity and computational needs are vast, and they are set to surge in the years ahead. This work also requires secure, reliable supply chains for critical components needed to build AI infrastructure, from construction materials to advanced electronics.

This order sets our Nation on the path to ensure that future frontier AI can, and will, continue to be built here in the United States. In building domestic AI infrastructure, our Nation will also advance its leadership in the clean energy technologies needed to power the future economy, including geothermal, solar, wind, and nuclear energy; foster a vibrant, competitive, and open technology ecosystem in the United States, in which small companies can compete alongside large ones; maintain low consumer electricity prices; and help ensure that the development of AI infrastructure benefits the workers building it and communities near it.

With this order, I provide a plan for protecting national security, preserving our economic competitiveness, revitalizing our energy infrastructure, and ensuring United States leadership in AI.

Sec. 2. *Policy.* It is the policy of the United States to enable the development and operation of AI infrastructure, including data centers, in the United States in accordance with five guiding principles. When undertaking the actions set forth in this order, executive departments and agencies (agencies) shall adhere to these principles, as appropriate and consistent with applicable law:

(a) The development of AI infrastructure should advance United States national security and leadership in AI. Meeting this goal will require steps by the Federal Government, in collaboration with the private sector, to advance AI development and use AI for future national-security missions, including through the work described in National Security Memorandum 25 of October 24, 2024 (Advancing the United States' Leadership in Artificial Intelligence; Harnessing Artificial Intelligence to Fulfill National Security Objectives; and Fostering the Safety, Security, and Trustworthiness of Artificial Intelligence) (NSM-25). It will also require the use of safeguards to improve the cyber, supply-chain, and physical security of the laboratories at which powerful AI is developed, stored, and used. Additionally, protecting United States national security will require further work to evaluate and manage risks related to the powerful capabilities that future frontier AI may possess.

(b) The development of AI infrastructure should advance United States economic competitiveness, including by fostering a vibrant technology ecosystem. Already, AI is creating new jobs and industries, and its effects are being felt in sectors across the economy. The Federal Government must ensure that the United States remains competitive in the global economy, including through harnessing the benefits of this technology for all Americans. It must also promote a fair, open, and competitive AI ecosystem so that small developers and entrepreneurs can continue to drive innovation—a priority highlighted in both Executive Order 14110 of October 30, 2023 (Safe, Secure, and Trustworthy Development and Use of Artificial Intelligence), and NSM-25—as well as to support secure, reliable supply-chain infrastructure for AI activities.

(c) The United States can and should lead the world in operating the next generation of AI data centers with clean power. Meeting this goal will require building on recent successes to modernize our Nation's energy infrastructure; improve permitting processes; and support investments in, and expeditious development of, both currently available and emerging clean energy technologies, such as geothermal energy, nuclear energy, and long-duration energy storage used to store clean energy, as well as relevant supply chains. The United States must not be surpassed in its support for the development, commercialization, and operation of clean energy technologies at home and abroad, and the rapid buildout of AI infrastructure offers another vital opportunity to accelerate and deploy these energy technologies. To help ensure that new data center electricity demand does not take clean power away from other end users, result in resource adequacy issues, or increase grid emissions, the construction of AI infrastructure must be matched with new, clean electricity generation resources.

(d) The development of AI infrastructure should proceed without raising energy costs for American consumers and businesses, and it should have strong community support. The companies developing, commercializing, and deploying AI must finance the cost of building the infrastructure needed for AI operations, including the development of next-generation power infrastructure built for these operations.

(e) The development of AI infrastructure should benefit those working to build it. Meeting this goal will require high labor standards and safeguards for the buildout of AI infrastructure, consultation and close collaboration with communities affected by this infrastructure's development and operation, and continuous work to mitigate risks and potential harms. The American people more broadly must safely enjoy the gains and opportunities from technological innovation in the AI ecosystem.

Sec. 3. *Definitions*. For purposes of this order:

(a) The term "agency" means each agency described in 44 U.S.C. 3502(1), except for the independent regulatory agencies described in 44 U.S.C. 3502(5).

(b) The term "AI data center" means a data center used primarily with respect to developing or operating AI.

(c) The term "AI infrastructure" refers collectively to AI data centers, generation and storage resources procured to deliver electrical energy to data centers, and transmission facilities developed or upgraded for the same purpose.

(d) The term "AI model" means a component of an information system that implements AI technology and uses computational, statistical, or machine-learning techniques to produce outputs from a given set of inputs.

(e) The term "clean energy" or "clean energy generation resources" means generation resources that produce few or no emissions of carbon dioxide during operation, including when paired with clean storage technologies. This term includes geothermal, nuclear fission, nuclear fusion, solar, wind, hydroelectric, hydrokinetic (including tidal, wave, and current), and marine energy; and carbon capture, utilization, and storage technologies (for which the carbon capture equipment meets the definition set forth in 26 C.F.R. 1.45Q-2(c)) that operate with fossil fuel generation resources, that achieve carbon dioxide capture rates of 90 percent or higher on an annual basis, and that permanently sequester the captured carbon dioxide.

(f) The term "clean power" means electricity generated by the generation resources described in subsection (e) of this section.

(g) The term "clean repowering" means the practice of siting new clean generation sources at a site with an existing point of interconnection and generation sources operating with fossil fuels, such that some output or capacity from existing generation sources is replaced by the new clean generation sources.

(h) The term "critical electric infrastructure information" has the same meaning as set forth in 18 C.F.R. 388.113(c).

(i) The term "data center" means a facility used to store, manage, process, and disseminate electronic information for a computer network, and it includes any facility that is composed of one or more permanent or semi-permanent structures, or that is a dedicated space within such structure, and operates persistently in a fixed location; that is used for the housing of information technology equipment, including servers, mainframe computers, high-performance computing devices, or data-storage devices; and that is actively used for the hosting of information and information systems that are accessed by other systems or by users on other devices.

(j) The term "distributed energy resource" has the same meaning as set forth in 18 C.F.R. 35.28(b)(10).

(k) The term "Federal Permitting Agencies" refers to the agency members of the Federal Permitting Improvement Steering Council (Permitting Council) established under section 41002 of the Fixing America's Surface Transportation (FAST) Act, 42 U.S.C. 4370m-1, as well as any other agency with authority to issue a Federal permit or approval required for the development or operation of AI infrastructure.

(l) The term "Federal Risk and Authorization Management Program" refers to the program established to provide an approach for the adoption and use of cloud services by the Federal Government, as codified in 44 U.S.C. 3607-3616 (as enacted by the FedRAMP Authorization Act, section 5921 of Public Law 117-263).

(m) The term "frontier AI data center" means an AI data center capable of being used to develop, within a reasonable time frame, an AI model with characteristics related either to performance or to the computational resources used in its development that approximately match or surpass the state of the art at the time of the AI model's development.

(n) The term "frontier AI infrastructure" means AI infrastructure for which the relevant data center is a frontier AI data center.

(o) The term "frontier AI training" refers to the act of developing an AI model with characteristics related either to performance or to the computational resources used in its development that approximately match or surpass the state of the art at the time of the AI model's development.

(p) The term "generation resource" means a facility that produces electricity.

(q) The terms "interconnection," "interconnection facilities," and "point of interconnection" refer to facilities and equipment that physically and electrically connect generation resources or electrical load to the electric grid for the purpose of the delivery of electricity, for which grid operators have granted all appropriate approvals required for those facilities and equipment to operate.

(r) The term "lab-security measures" refers to steps to detect, prevent, or mitigate physical, cyber, or other threats to the operation of a data center, to the integrity of information or other assets stored within it, or of unauthorized access to such information or assets.

(s) The term "leading-edge logic semiconductors" refers to semiconductors produced at high volumes using extreme ultraviolet lithography tools as defined by the CHIPS Incentives Program Notice of Funding Opportunity, 2023-NIST-CHIPS-CFF-01.

(t) The term "model weight" means a numerical parameter within an AI model that helps determine the model's outputs in response to inputs.

(u) The term "new source review" refers to the permitting program with this name in   parts 51 or 52.

(v) The term "non-Federal parties" refers to private-sector entities that enter into a contract with the Department of Defense or the Department of Energy pursuant to section 4(g) of this order.

(w) The term "priority geothermal zone" refers to lands with high potential for the development of geothermal power generation resources, as designated by the Secretary of the Interior, including pursuant to section 4(c) of this order.

(x) The term "project labor agreement" means a pre-hire collective bargaining agreement that establishes the terms and conditions of a construction project.

(y) The term "surplus interconnection service" has the same meaning as set forth in Federal Energy Regulatory Commission Order No. 845.

(z) The terms "transmission facilities" and "transmission infrastructure" mean equipment or structures, including transmission lines and related facilities, used for the purpose of delivering electricity.

(aa) The term "transmission organization" refers to a Regional Transmission Organization or an Independent System Operator.

(bb) The term "transmission provider" means an entity that manages or operates transmission facilities for the delivery of electric energy used primarily by the public and that is not a transmission organization.

(cc) The term "waters of the United States" has the same meaning as set forth in 33 C.F.R. 328.3(a).

Sec. 4. *Establishing Federal Sites for AI Infrastructure*. (a) By February 28, 2025, the Secretary of Defense and the Secretary of Energy shall, if possible, each identify a minimum of 3 sites on Federal land managed by their respective agencies that may be suitable for the agencies to lease to non-Federal entities for the construction and operation of a frontier AI data center, as well as for the construction and operation of clean energy facilities to serve the data center, by the end of 2027. In identifying these sites, each Secretary shall, as feasible and appropriate, seek to prioritize sites that possess the following characteristics, as consistent with the objective of fully permitting and approving work to construct a frontier AI data center at each site by the end of 2025:

(i) inclusion of sufficient terrain with appropriate land gradients, soil durability, and other topographical characteristics for frontier AI data centers;

(ii) minimized adverse effects from AI infrastructure development or operation on local communities' health, wellbeing, and resource access; natural or cultural resources; threatened or endangered species; and harbors or river improvements not associated with hydropower generation resources;

(iii) proximity to any communities seeking to host AI infrastructure, including for reasons related to local workers' access to jobs involved in designing, building, maintaining, and operating data centers;

(iv) ready access and proximity to high-voltage transmission infrastructure that minimizes the scale of, cost of, and timeline to develop any transmission upgrades or development needed to interconnect AI infrastructure, in consideration of access and proximity to:

(A) high-capacity transmission infrastructure with unused capacity, as identified by collection activities described in section 6 of this order;

(B) any planned generation facilities that can enable delivery of electricity to an AI data center on the site managed by each Secretary's respective agency, that possess an executed interconnection agreement with a transmission provider, that do not possess an executed power purchase agreement, and for which construction has not yet begun;

(C) any lands that the Secretary of the Interior identifies pursuant to subsection (c) of this section; and

(D) any power generation facilities with high clean repowering potential;

(v) location within geographic areas that are not at risk of persistently failing to attain National Ambient Air Quality Standards, and where the total cancer risk from air pollution is at or below the national average according to the Environmental Protection Agency's (EPA's) 2020 AirToxScreen;

(vi) lack of proximity to waters of the United States for purposes of permitting requirements;

(vii) lack of extensive restrictions on land uses associated with constructing and operating AI infrastructure or on access to necessary rights-of-way for such activities;

(viii) ready access to high-capacity telecommunications networks;

(ix) suitability for the development of access roads or other temporary infrastructure necessary for the construction of AI infrastructure; and

(x) absence of other characteristics that would, if the site was used or repurposed for AI infrastructure, compromise a competing national security concern as determined by the relevant Secretary in consultation with the Assistant to the President for National Security Affairs.

(b) By March 15, 2025, the Secretary of the Interior, acting through the Director of the Bureau of Land Management (BLM), in consultation with the Secretary of

Defense, the Secretary of Energy, and the Chair of the Federal Energy Regulatory Commission, shall identify sites managed by BLM that the Secretary of the Interior, acting through the Director of BLM, deems may be suitable for granting or issuing rights of way to private-sector entities to construct and operate additional clean energy facilities that are being or may be built as components of frontier AI infrastructure developed pursuant to this section. In performing this work, the Secretary of the Interior, in consultation with the Secretary of Defense and the Secretary of Energy, shall take steps to ensure where feasible and appropriate that any such sites identified under this subsection include sufficient acreage for developing clean generation resources that can deliver sufficient electricity to each site identified under subsection (a) of this section for matching the capacity needs of frontier AI data centers on the latter sites. The sites identified under this subsection shall include any land managed by the Department of the Interior that is within a region designated by the Secretary of the Interior under subsection (c) of this section, or a region preliminarily identified as a candidate for such designation. In determining the suitability of sites, the Secretary of the Interior, acting through the Director of BLM, shall prioritize identification of sites that:

(i) contain completed, permitted, or planned clean generation projects that can enable delivery of electricity as described in this subsection and possess an executed interconnection agreement with a transmission provider;

(ii) have been allocated as available for solar applications in the *Final Programmatic Environmental Impact Statement and Proposed Resource Management Plan Amendments for Utility-Scale Solar Energy Development*, published by BLM, or that have otherwise been allocated as available for clean-energy applications in a BLM resource management plan;

(iii) have reasonable access to and are located nearby existing high-voltage transmission lines that have at least one gigawatt of additional capacity available, or for which such capacity can be reasonably developed through reconductoring, grid-enhancing technologies, or transmission upgrades;

(iv) possess the characteristics described in subsections (a)(i)-(x) of this section, in a manner that is consistent with the objective of fully permitting and approving work to construct utility-scale power facilities on a timeline that allows for the operation of those facilities by the end of 2027 or as soon as feasible thereafter; and

(v) possess other characteristics conducive to enabling new clean power development at such sites to contribute to lower regional electricity prices or to bring other community benefits.

(c) By March 15, 2025, the Secretary of the Interior, acting through the Director of BLM and in consultation with the Secretary of Energy, shall, if possible, designate at least five regions composed of lands or subsurface areas managed by the Department of the Interior as Priority Geothermal Zones (PGZs). The Secretary of the Interior shall designate those regions based on their potential for geothermal power generation resources, including hydrothermal and next-generation geothermal power and thermal storage; diversity of geological characteristics; and possession of the characteristics described in subsections (a)(i)-(x) and (b)(i)-(v) of this section.

(d) The Secretary of Defense, the Secretary of Energy, and the Secretary of the Interior shall each make a legal determination as to whether each site identified pursuant to subsections (a) and (b) of this section is available for lease or for the issuance of a right of way, as appropriate, pursuant to the authority of the Secretary that made the identification, and as to whether the Secretary has the legal authority to lease or grant a right of way over or upon each site identified for the construction of frontier AI infrastructure. For purposes of this order, a site shall be considered "cleared" under this subsection if the relevant Secretary has determined that the site is available for lease and the Secretary concerned has the authority to lease it.

(e) By March 31, 2025, the Secretary of Defense and the Secretary of Energy, in coordination with the heads of any other agencies that either Secretary deems appropriate, shall coordinate to design, launch, and administer competitive public solicitations of proposals from non-Federal entities to lease Federal land to construct frontier AI infrastructure, including frontier AI data centers, on sites identified under subsection (a) of this section and cleared under subsection (d) of this section, if any. When issuing the solicitations, the Secretaries shall announce the sites identified under subsection (a) of this section and cleared under subsection (d) of this section, if any, and additional relevant information including the sites' geographic coordinates, technical characteristics, proximity to sites identified consistent with subsection (b) of this section and cleared under subsection (d) of this section, if any, and other relevant information. The solicitations shall, to the extent consistent with applicable law and to the extent the Secretaries agree that such requirements promote national defense, national security, or the public interest, as appropriate, require applicants to identify particular sites on which they propose to construct and operate frontier AI infrastructure; submit a detailed plan specifying proposed timelines, financing methods, and technical construction plans associated with such construction work, including a contingency plan for decommissioning infrastructure on Federal sites; submit a plan that describes proposed frontier AI training work to occur at the site once operational; submit a plan for detailing the extent of the use of high labor and construction standards as described in subsection (g)(viii) of this section; and submit a plan with proposed lab-security measures, including personnel and material access requirements, that could be associated with the operation of frontier AI infrastructure. These requirements should be designed to ensure adequate collection of information from applicants regarding the criteria in subsections (g)(i)-(xvi) of this section. The solicitations shall close within 30 days of their issuance.

(f) By March 31, 2025, the Secretary of the Interior, in consultation with the Secretary of Defense and the Secretary of Energy, shall publicize the sites identified under subsection (b) of this section and cleared under subsection (d) of this section, if any, and additional relevant information including the sites' geographic coordinates, technical characteristics, proximity to sites identified consistent with subsection (a) of this section and cleared under subsection (d) of this section, if any, and other relevant information.

(g) By June 30, 2025, the Secretary of Defense and the Secretary of Energy shall announce any winning proposals identified through solicitations described in subsection (e) of this section. In selecting any winning proposals, the Secretary of Defense and the Secretary of Energy shall, in consultation with each other, assign winners the opportunity to apply for any Federal permits needed to build and operate frontier AI infrastructure pursuant to the frameworks described in subsection (h) of this section on any sites included in the solicitations issued under subsection (e) of this section, as the Secretaries deem appropriate. The Secretaries shall consult with the Attorney General on the implications of selections on the competition and market-structure characteristics of the broader AI ecosystem. The Chair of the Federal Trade Commission is encouraged to participate in these consultations. The Secretaries shall, to the extent consistent with applicable law and to the extent that the Secretaries assess that the requirement promotes national defense, national security, or the public interest, as appropriate, select at least one proposal developed and submitted jointly by a consortium of two or more small- or medium-sized organizations—as determined by those organizations' market capitalization, revenues, or similar characteristics—provided that the Secretaries receive at least one such proposal that meets the appropriate qualifications. The Secretaries shall provide technical assistance, as appropriate, to small- or medium-sized organizations seeking to submit proposals. The criteria for selecting winning proposals shall include, at a minimum, consideration of the following characteristics of the applicants and any identified partner organizations, to the extent consistent with applicable law and to the extent that the Secretaries agree that the listed characteristics promote national defense, national security, or the public interest, as appropriate:

(i) proposed financing mechanisms and sources of funds secured or likely to be secured for work to be performed at the site;

(ii) plans for ensuring high-quality AI training operations to be executed at the site by the applicant or third-party partners;

(iii) plans for maximizing energy, water, and other resource efficiency, including waste-heat utilization in constructing and operating the AI data center at the site, the strength of the proposed energy master plan for the site, and the quality of analysis of potential strains on local communities;

(iv) safety and security measures, including cybersecurity measures, proposed to be implemented at the site, and capabilities for such implementation;

(v) capabilities and acumen of applicable AI scientists, engineers, and other workforce essential to the operation of AI infrastructure;

(vi) plans for commercializing or otherwise deploying or advancing deployment of appropriate intellectual property, including AI model weights, developed at the site, as well as plans for commercializing or otherwise deploying or advancing deployment of innovations related to power generation and transmission infrastructure developed in the course of building or operating AI infrastructure;

(vii) plans to help ensure that the construction and operation of AI infrastructure does not increase electricity costs to other ratepayers or water costs to consumers, including, as appropriate, through appropriate proposed or recommended future engagement with any applicable regulatory authorities and State, Tribal, or local governments;

(viii) plans to use high labor standards that help ensure continuous and high-quality work performed on the site, such as paying prevailing wages; hiring registered apprentices; promoting positive labor-management relations through a project labor agreement; and otherwise adopting high job quality and labor standards for the construction and operations workforce as set forth in Executive Order 14126 of September 6, 2024 (Investing in America and Investing in American Workers), and a plan to address labor-related risks associated with the development and use of AI;

(ix) design features and operational controls and plans that mitigate potential environmental effects and implement strong community health, public safety, and environmental protection measures;

(x) other benefits to the community and electric grid infrastructure surrounding the site;

(xi) experience completing comparable construction projects;

(xii) experience in compliance with Federal, State, and local permits and environmental reviews relevant to construction and operation of AI infrastructure or, in the alternative, other evidence of an ability to obtain and comply with such permits or reviews in an efficient manner;

(xiii) the presence of organizational and management structures to help ensure sound governance of work performed at the site;

(xiv) the effect of the selection of an applicant on the emergence of an interoperable, competitive AI ecosystem;

(xv) whether an applicant has already been assigned an opportunity, or is being assigned another opportunity, to build a frontier AI data center on a Federal site through the solicitation process described in this section; and

(xvi) other considerations of national defense, national security, or the public interest, including economic security, as the Secretary of Defense and the Secretary of Energy deem appropriate.

(h) By June 30, 2025, the Secretary of Defense and the Secretary of Energy, in consultation with the Secretary of the Interior, shall each develop a framework through which any winning applicants selected under subsection (g) of this section may apply to lease sites respectively identified under subsection (a) of this section, and cleared under subsection (d) of this section, to construct and operate AI infrastructure, and by which the applicants may own the AI infrastructure facilities on those sites, subject to the conditions described in subsections (i)-(x) of this subsection. To the extent that the Secretaries assess that it is consistent with national defense, national security, or the public interest, as appropriate, these frameworks shall allow for winning applicants to cooperate with other appropriate private-sector entities on construction and operation activities, including through contracting and subcontracting relationships, and the frameworks shall not require that parties proposing to own AI infrastructure be identical to those proposing to operate the infrastructure or perform work at the sites on which the infrastructure is located. Actions taken by Federal entities pursuant to the frameworks shall conform to any applicable requirements of Appendix B of Office of Management and Budget (OMB) Circular A- 11 and any other appropriate budget-scoring practices; applicable in-kind consideration shall be taken into account in calculating the cost to lessees of any such leases. As part of the foregoing work, the Secretary of Defense and the Secretary of Energy shall, to the extent consistent with their respective authorities and with national defense, national security, or the public interest, as appropriate, require lease or contract terms that accomplish the following:

(i) establish a target of the applicant's beginning construction of a frontier AI data center by January 1, 2026, and commencing full-capacity operation of the AI infrastructure by December 31, 2027, subject to fulfillment of relevant statutory and regulatory requirements, and in a manner consistent with opportunities to operate the infrastructure at or below full capacity at an earlier date;

(ii) require that, concurrent with operating a frontier AI data center on a Federal site, non-Federal parties constructing, owning, or operating AI infrastructure have procured sufficient new clean power generation resources with capacity value to meet the frontier AI data center's planned electricity needs, including by providing power that matches the data center's timing of electricity use on an hourly basis and is deliverable to the data center;

(iii) clarify that non-Federal parties bear all responsibility for paying any costs that parties to the frameworks described in subsection (h) of this section, as well as transmission providers or transmission organizations or other entities not party to the contract, incur from work pursuant to it, including costs of work performed by agencies to complete necessary environmental reviews, any costs related to the procurement of clean power generation resources and capacity in accordance with subsection (g)(ii) of this section, any costs of decommissioning AI infrastructure on Federal sites, any costs of developing transmission infrastructure needed to serve a frontier AI data center on a Federal site, and the fair market value of leasing and using applicable Federal lands;

(iv) require adherence to technical standards and guidelines for cyber, supply-chain, and physical security for protecting and controlling any facilities, equipment, devices, systems, data, and other property, including AI model weights, that are developed, acquired, modified, used, or stored at the site or in the course of work performed on the site. The Secretary of Commerce, acting through the Director of the National Institute of Standards and Technology (NIST) and the Director of the AI Safety Institute (AISI) at NIST, in consultation with the Secretary of Defense, the Secretary of Energy, and the Director of National Intelligence, shall identify available standards and guidelines to which adherence shall be required under this subsection. The identified standards should reflect and incorporate guidelines and best practices developed by the Secretary of Commerce, acting through the Director of NIST, pursuant to Executive Order 14028 of May 12, 2021 (Enhancing United States Cybersecurity), and Executive Order 14110 of November 1, 2023 (Safe, Secure, and Trustworthy Development and Use of Artificial Intelligence). The Secretary of Commerce, acting through the Director of AISI at NIST, shall support the ongoing improvement of the framework described in this subsection by developing security guidelines for frontier AI training and operation and, as part of this work, shall comprehensively evaluate the security implications of publicly available AI models that the Secretary of Commerce, acting through the Director of AISI at NIST, deems globally significant;

(v) require that non-Federal parties owning or operating frontier AI data centers sign a memorandum of understanding with the Secretary of Commerce, acting through the Director of AISI at NIST, to facilitate collaborative research and evaluations on AI models developed, acquired, modified, run, or stored at the site or in the course of work performed on the site, for the purpose of assessing the national-security or other significant risks of those models;

(vi) require non-Federal parties to report information about investments or financial capital from any person used or involved in the development (including construction), ownership, or operation of AI infrastructure on the site and in the development, operation, or use of AI models operating in such AI infrastructure, as appropriate to evaluate risks to national security; and require non-Federal parties to limit the involvement in any such activities of, or the use or involvement in any such activities of investments or financial capital from, any person whom the Secretaries of Defense or Energy deem appropriate on national security grounds;

(vii) require non-Federal parties owning or operating AI data centers on Federal sites to take appropriate steps to advance the objective of harnessing AI, with appropriate safeguards, for purposes of national security, military preparedness, and intelligence operations, including with respect to the objectives and work outlined in NSM-25. Such steps shall, as consistent with applicable legal authorities, include collaborating with the Federal Government on regularly recurring assessments of the national-security implications of AI models developed on Federal sites, as appropriate. In addition, as appropriate and consistent with any relevant Federal procurement laws and regulations, the non-Federal parties shall be required to commit to providing access to such models, and critical resources derivative of such models, to the Federal Government for national-security applications at terms at least no less favorable than current market rates, consistent with NSM-25 and the associated Framework to Advance AI Governance and Risk Management in National Security. To the extent feasible, AI models and resources derived from them shall be developed and provided to the Federal Government in a manner that prevents vendor lock-in and supports interoperability, including as consistent with the measures in section 5 of OMB Memorandum M-24-18;

(viii) require that non-Federal parties owning or operating frontier AI data centers on Federal sites develop plans to make available computational resources that are not dedicated to supporting frontier AI training, or otherwise allocated under another provision, for commercial use by startups and small firms on nondiscriminatory terms and in a manner that minimizes barriers to interoperability, entry, or exit for users;

(ix) require non-Federal parties owning or operating AI infrastructure on Federal sites to explore the availability of clean energy resources—such as geothermal power generation resources and thermal storage, long-duration storage paired with clean energy, and carbon capture and sequestration as described in section 3(e) of this order, as well as beneficial uses of waste heat—at any appropriate sites that those parties lease for purposes of constructing frontier AI data centers on Federal sites or procuring power generation capacity to serve these data centers; and

(x) require AI developers owning and operating frontier AI data centers on Federal sites either to procure, for use in the development of their data centers, an appropriate share (as measured by monetary value) of leading-edge logic semiconductors fabricated in the United States to the maximum extent practicable; or to develop and implement a plan, subject to the respective approval of the Secretary of Defense or the Secretary of Energy, to qualify leading-edge logic semiconductors fabricated in the United States for use in the developer's data centers as soon as practicable. The Secretary of Defense and the Secretary of Energy shall develop any such requirements—including any determinations about amounts of leading-edge logic semiconductors that may be considered "appropriate"—in consultation with the Secretary of Commerce.

(i) Within 1 year of the date of this order and consistent with applicable law, the Secretary of Defense, in consultation with the Secretary of Commerce, the Secretary of Energy, the Secretary of Homeland Security, the Director of National Intelligence, and the Assistant to the President for National Security Affairs, shall issue regulations that prescribe heightened safeguards to protect computing hardware acquired, developed, stored, or used on any sites on which frontier AI infrastructure is located and that are managed by the Department of Defense, as needed to implement or build upon the objectives of, or the requirements established pursuant to, subsection 4(g)(iv). The regulations shall include requirements to conform with appropriate high-impact level standards identified through the Federal Risk and Authorization Management Program, and they shall further provide for appropriate penalties consistent with applicable authorities. No less than annually the Secretary of Defense, in consultation with the aforementioned individuals, shall review the need for updates to the regulations, and promulgate any necessary revisions. The Secretary of Energy shall impose substantively the same requirements with respect to frontier AI infrastructure on sites managed by the Department of Energy, to the extent authorized by law.

(j) To enable the use—for advancing geothermal power development, including the development of thermal storage—of Federal lands already subject to leases:

(i) Within 180 days of the date of this order, the Secretary of the Interior shall establish a program with personnel dedicated to providing technical assistance for, streamlining, and otherwise advancing direct-use leasing of geothermal projects on BLM lands, including as consistent with the policies set forth in 43 C.F.R. subpart 3205, and leases of geothermal projects on lands subject to mining claims or under an oil and gas lease.

(ii) When issuing leases and related authorizations for geothermal projects, the Secretary of the Interior shall consider the extent to which the requirements of the National Environmental Policy Act (NEPA), 42 U.S.C. 4321 *et seq.*, the Endangered Species Act, 16 U.S.C. 1531 *et seq.*, and other appropriate statutes have been satisfied by prior analyses of the lease area.

(k) In performing the work described in section 4 of this order, including as related to the selection and management of sites, the head of each respective Federal agency shall:

(i) consult, as appropriate and consistent with applicable law, Executive Order 13175 of November 6, 2000 (Consultation and Coordination with Indian Tribal Governments), and the Presidential Memorandum of November 30, 2022 (Uniform Standards for Tribal Consultation), with Tribal Nations for which such work may have implications or who otherwise request such consultation;

(ii) seek input from, as appropriate and consistent with applicable law and Administration policies, with State and local governments and other stakeholders and communities for which such work may have implications; and

(iii) consider taking actions that present the greatest opportunities to support the goals described in *Safely and Responsibly Expanding U.S. Nuclear Energy: Deployment Targets and A Framework for Action* (November 2024).

Sec. 5. *Protecting American Consumers and Communities.* (a) Within 180 days of the date of this order, the Secretary of Energy, in consultation with the Chair of the Council of Economic Advisors and the heads of other agencies that the Secretary deems appropriate, shall submit a report to the President on the potential effects of AI data centers on electricity prices for consumers and businesses. This report shall include electricity-rate-structure best practices for appropriate Federal agencies, State regulators, and transmission providers and transmission organizations to promote procurement of clean energy generation resources as components of AI infrastructure without increasing costs for other customers through cost-allocation processes or other mechanisms—particularly in regions that have or are expected to have high concentrations of AI infrastructure—as well as regional analyses of key data center hubs. The report shall further account for any existing approaches developed by Federal agencies to engage transmission providers and State regulators regarding electricity prices. After submitting the report, the Secretary of Energy shall engage appropriate private-sector entities, to include the winning applicants selected under subsection 4(g) of this order, on the report's findings and recommendations.

(b) The Secretary of Energy shall provide technical assistance to State public utility commissions to consider rate structures, including clean transition tariffs and any other appropriate structures identified under subsection (a) of this section, to enable new AI infrastructure to use clean energy without causing unnecessary increases in electricity or water prices.

(c) The Secretary of Energy and the heads of other appropriate agencies as the Secretary of Energy deems appropriate, shall coordinate to expand research-and-development efforts related to AI data center efficiency. Supported research and development shall cover, as appropriate, efficiency considerations associated with data center buildings, including the data center shell; electrical systems; heating, ventilation, and cooling infrastructure; software; and beneficial use cases for wastewater heat from data center operations. As part of this work, the Secretary of Commerce and the Secretary of Energy shall submit a report to the President identifying appropriate ways that agencies can advance industry-wide data center energy efficiency through research and development, including server consolidation; hardware efficiency; virtualization; optimized cooling and airflow management; and power management, monitoring, and capacity planning.

(d) In implementing this order with respect to AI infrastructure on Federal sites, the heads of relevant agencies shall prioritize taking appropriate measures to keep electricity costs low for households, consumers, and businesses.

(e) Within 180 days of the date of this order, the Director of OMB, in consultation with the Chair of the Council on Environmental Quality (CEQ), shall evaluate best practices for public participation and governmental engagement in the development of potential siting and energy-related infrastructure for data centers, to include practices for seeking input on potential health, safety, and environmental impacts and mitigation measures for nearby communities. The Director shall present recommendations to the Secretary of Defense and the Secretary of Energy, who shall—as feasible and appropriate, and to advance the goals of assuring effective governmental engagement and meaningful public participation—implement and incorporate these recommendations into their siting and related decision-making processes regarding AI infrastructure.

Sec. 6. *Facilitating Electric Grid Interconnections for Federal Sites.* (a) Within 60 days of the date of this order, for the purpose of supporting any winning applicants of the solicitations described in subsection 4(e) of this order, the Secretary of Energy shall establish requirements for transmission providers and transmission organizations to report to the Secretary information regarding surplus interconnection service; available transmission capacity for interconnecting generators; opportunities for clean repowering; and proposed, planned, or initiated projects to build clean power generation capacity for which construction is not complete, but which have executed generation interconnection agreements. Information requested regarding these proposed, planned, or initiated projects shall include the size, location, and generation technology for each such clean power generation project, as well as the status and estimated cost of any transmission upgrades necessary to enable that project's interconnection consistent with the interconnection agreement. The Secretary shall facilitate communication, as appropriate, among the owners of such surplus interconnection service, facilities with opportunities for clean repowering, or clean power generator projects and winning applicants to the solicitations described in subsection 4(e) of this order. The Secretary shall further establish appropriate requirements for transmission providers and transmission organizations to continue reporting information described in this subsection on an ongoing basis, and in any event no less than annually.

(b) Within 120 days of the date of this order, the Secretary of Energy shall identify and communicate, as appropriate, a prioritized list of underutilized points of interconnection that are relevant to AI infrastructure on Federal sites and that demonstrate the highest potential for uses associated with AI infrastructure. In developing this list, the Secretary shall direct transmission providers and transmission organizations to identify areas of the transmission network best suited to serve as points of interconnection for either data centers or other AI infrastructure that will use electricity from the transmission system—and locations best suited for interconnection of clean generators to serve such data centers—considering criteria such as minimizing the need for transmission upgrades necessary to accommodate such interconnection and access to clean energy generation resources.

(c) By June 30, 2025, the Secretary of Energy, in coordination with the Secretary of Defense and in consultation, as appropriate, with the Secretary of the Interior and the Secretary of Agriculture, shall engage with transmission providers and transmission organizations owning, operating, or maintaining transmission infrastructure located near Federal sites selected for AI infrastructure to identify any grid upgrades, deployment of advanced transmission technologies such as high-performance conductors or grid-enhancing technologies, operational changes, or other steps expected to be required for extending interconnection services to AI infrastructure by the end of 2027. Such engagements shall continue as the parties deem appropriate, and they shall prioritize, as appropriate, efforts to enable use of surplus interconnection services, clean repowering, and other methods of accelerated shifts toward clean power and beneficial use of waste heat. The engagements shall also include consideration of ways that the performance of such work as described in this subsection can most contribute to lower regional electricity prices.

(d) The Secretary of Energy shall conduct an analysis of currently available transmission infrastructure serving potential sites, and the likely cost and feasibility of, and timeline for, developing additional such infrastructure needed for constructing and operating a frontier AI data center on sites identified under subsection 4(a) of this order, and cleared under subsection 4(d) of this order, including by providing the frontier AI data center with clean energy and capacity. The Secretary shall identify and collect from transmission providers and transmission organizations information that the Secretary deems necessary for the analysis required under this subsection. The Secretary shall, as appropriate, treat such information as critical electric infrastructure information.

Sec. 7. *Expeditiously Processing Permits for Federal Sites*. (a) The heads of Federal Permitting Agencies shall prioritize work and exercise all applicable authorities, as appropriate, to expedite the processing of permits and approvals required for the construction and operation of AI infrastructure on Federal sites, with the goal of issuing all permits and approvals required for construction by the end of 2025 or as soon as they can be completed consistent with applicable law. As part of this work, the Permitting Council may provide coordination of permitting for AI infrastructure on Federal sites, as appropriate and to the extent that the relevant developers of AI infrastructure submit a notice of the initiation of a proposed covered project under 42 U.S.C. 4370m-2 and the project is determined to be such a covered project by the Permitting Council.

(b) To facilitate expeditious implementation of the requirements under NEPA with respect to Federal sites:

(i) The Secretary of Defense, the Secretary of the Interior, and the Secretary of Energy shall identify, within their respective agencies, personnel dedicated to performing NEPA reviews of projects to construct and operate AI infrastructure on Federal sites.

(ii) The Secretary of Defense, in consultation with the Secretary of the Interior, the Secretary of Agriculture, the Secretary of Commerce, and the Secretary of Energy, shall undertake a programmatic environmental review, on a thematic basis, of the environmental effects—and opportunities to mitigate those effects—involved with the construction and operation of AI data centers, as well as of other components of AI infrastructure as the Secretary of Defense deems appropriate. The review shall conclude, with all appropriate documents published, on the date of the close of the solicitations described in subsection 4(e) of this order, or as soon thereafter as possible. The review shall, as applicable, incorporate by reference previously developed environmental studies, surveys, and impact analyses, including the analysis described in subsection 4(b)(ii) of this order.

(iii) After the conclusion of the programmatic review described in subsection (b)(ii) of this section, the Secretary of Defense, the Secretary of the Interior, the Secretary of Energy, and the heads of other relevant agencies, as appropriate, shall commence any further environmental reviews that are required under NEPA for the construction and operation of AI infrastructure on Federal sites, including by applying any available categorical exclusions. Such reviews shall, as appropriate, build on or incorporate by reference the programmatic environmental review conducted under subsection (b)(ii) of this section, as well as any other studies, surveys, and impact analyses that the Secretaries deem appropriate.

(c) To advance expeditious preconstruction permitting and ensure full compliance with air-quality permit requirements for AI infrastructure, the Administrator of the EPA, in consultation with the Secretary of Defense and the Secretary of Energy, shall:

(i) within 30 days of the selection of winning applications under subsection 4(g) of this order, engage State and local permitting authorities with jurisdiction over sites selected for AI infrastructure, as appropriate, to enhance relevant authorities' understanding of the technical characteristics of AI infrastructure projects as relevant to new source reviews under the Clean Air Act, 42 U.S.C. 7401 *et seq.*, and to enhance the public's understanding of the same, as well as to facilitate the acquisition of information by AI developers operating on Federal sites regarding best practices for expeditiously obtaining air-quality permits;

(ii) continue engagements with State and local permitting authorities, and provide technical assistance to AI developers operating on Federal sites, on an ongoing basis and as appropriate, to help advance expeditious conclusion of, and compliance with, new source reviews; and

(iii) following the acquisition of all preconstruction air-quality permits by developers, take steps to ensure, on an ongoing basis and as appropriate, that AI developers operating on Federal sites adhere to all requirements of operational air-quality permits applicable to their respective projects; that information needed to demonstrate compliance, possibly including air-monitoring data, is made publicly available and regularly updated; and that best practices are identified for air-emissions reduction and air-quality monitoring regarding AI infrastructure on Federal sites.

(d) To help ensure expeditious permitting or permission processes related to waters of the United States and harbor and river improvements, the Secretary of Defense shall prioritize work, as appropriate, to process applications for permits administered by the United States Army Corps of Engineers (USACE) under the Clean Water Act, 33 U.S.C. 1251 *et seq.*, and to process applications for permission for appropriate projects under section 14 of the Act of March 3, 1899 (33 U.S.C. 408), as consistent with the statutes' requirements, in order to render determinations on any such permits or permissions associated with AI infrastructure on Federal sites by the end of 2025, or as soon as feasible consistent with statutory requirements. The Secretary shall, consistent with applicable law, prioritize allocation of resources toward USACE district offices, and direct the allocation of resources within such offices, as needed to comply with this directive. The Secretary shall further apply all general permits applicable to AI infrastructure where appropriate to promote expeditious permitting on such Federal sites.

(e) Within 30 days of the selection of any winning applications under subsection 4(g) of this order, the Secretary of Defense and the Secretary of Energy shall initiate Tribal consultations as applicable and appropriate based on the sites selected. Upon receipt of sufficient project information, the Secretary of Defense and the Secretary of Energy shall further initiate consultations with the Secretary of the Interior, acting through the Director of the United States Fish and Wildlife Service (USFWS), to ensure that the construction and operation of AI infrastructure on each site that is identified under subsection 4(a) of this order, cleared under subsection 4(d) of this order, and subsequently chosen as the location for the construction and operation of AI infrastructure pursuant to a winning application under subsection 4(g) of this order are not likely to jeopardize the continued existence of any endangered species or threatened species or result in the destruction or adverse modification of a critical habitat of such species. The Secretary of Defense and the Secretary of Energy shall conclude such consultations with USFWS, to the maximum extent practicable, within 90 days of the initiation of such consultations when feasible and consistent with statutory requirements.

(f) To advance the development of geothermal energy production and thermal storage, including in support of AI infrastructure on Federal sites:

(i) Within 60 days of the date of this order, the Secretary of the Interior shall undertake a programmatic environmental review, on a thematic basis, of the environmental impacts and associated mitigations involved with the construction and operation of a geothermal power plant.

(ii) By the date on which the review described in subsection (f)(i) of this section is completed, the Secretary of the Interior shall establish a target cumulative capacity of permitted or operational geothermal projects by a year that the Secretary shall designate.

(iii) Within 60 days of the date of this order, the Secretary of the Interior shall assess existing categorical exclusions that are listed in the NEPA procedures of other agencies and could apply to actions taken in connection with geothermal energy development. The Secretary shall propose adopting such categorical exclusions as the Secretary, after consultation with the heads of agencies whose NEPA procedures list the categorical exclusions, deems appropriate, and, after considering all comments received through applicable public comment processes, take any actions to adopt categorical exclusions that are appropriate given the received comments, as consistent with the requirements of NEPA and 40 C.F.R. parts 1500-1508. The Secretary shall prioritize the expeditious permitting of geothermal projects, including the application of any appropriate categorical exclusions adopted under this subsection, on PGZs. The Secretary shall prioritize work to expeditiously permit geothermal projects on PGZs above the work described in subsection (f)(i) of this section.

(iv) When issuing leases and related authorizations for geothermal projects on PGZs, the Secretary of the Interior shall fulfill the requirements of NEPA and the Endangered Species Act in a manner that allows for the earliest possible operation of geothermal power plants consistent with applicable law.

(v) The Secretary of Defense, the Secretary of the Interior, and the Secretary of Energy shall, as appropriate, coordinate to determine and clarify appropriate procedures for the execution of leases or subleases for developing or expanding clean energy generation resources, including geothermal energy generation resources, on withdrawn lands subject to the jurisdiction of the Department of Defense or the Department of Energy.

Sec. 8. *Ensuring Adequate Transmission Infrastructure for Federal Sites.* (a) The Secretary of Energy, in consultation with the Secretary of Defense and the Secretary of the Interior, shall take steps to enable AI infrastructure on Federal sites to have reliable access to transmission facilities adequate for the operation of frontier AI data centers by the end of 2027.

(b) To promote any needed upgrades and development of transmission infrastructure that is located on or that is necessary to support Federal sites with AI infrastructure, the Secretary of Energy, in consultation with the Secretary of the Interior, acting through the Director of BLM and the Director of USFWS, shall:

(i) by September 30, 2025, identify and initiate use of all appropriate authorities to construct, finance, facilitate, and plan such upgrades and development, including through the Transmission Infrastructure Program administered by the Western Area Power Administration; and

(ii) prioritize the allocation of staff and resources for developing transmission infrastructure needed to support AI infrastructure on Federal sites—and in doing so, as appropriate, allocate relevant staff and resources from any component within the Department of Energy for this purpose—consistent with the requirements and objectives of this order and applicable law.

(c) Because of the importance of frontier AI infrastructure, including transmission capacity, to the defense industrial base, critical infrastructure, and military preparedness:

(i) The Secretary of Energy shall consider expected use of frontier AI data centers on Federal sites as part of the Secretary's triennial study of electric transmission capacity constraints and congestion under section 216(a)(1) of the Federal Power Act (16 U.S.C. 824p(a)(1)).

(ii) Consistent with the requirements of section 216(a)(2) of the Federal Power Act (16 U.S.C. 824p(a)(2)), and based on any findings made in future studies of electric transmission capacity constraints and congestion as described in subsection (c)(i) of this section, the Secretary shall consider whether to designate geographic areas around frontier AI infrastructure on Federal sites as national interest electric transmission corridors.

(d) The Secretary of Energy shall, as appropriate, help ensure that transmission facilities upgraded or developed to support AI data centers on Federal sites:

(i) are designed to support all reasonably foreseeable electric loads, including through the deployment of grid-enhancing technologies, high-performance conductors, and other advanced transmission technologies, including those described in the Department of Energy's *Innovative Grid Deployment* Liftoff report, that will increase the capabilities of the transmission facilities on a timely and cost-effective basis; and

(ii) conform to conductor efficiency standards or other technical standards or criteria that the Secretary determines will optimize facilities' performance and cost-effectiveness.

(e) To improve the timely availability of critical grid equipment for frontier AI infrastructure, such as electrical transformers, circuit breakers, switchgears, and cables, and to protect electricity consumers from exposure to rising equipment prices:

(i) Within 90 days of the date of this order, the Secretary of Defense, the Secretary of Commerce, and the Secretary of Energy shall jointly consult with domestic suppliers of such technologies on the expected needs of AI infrastructure on Federal sites, suppliers' current production plans, and opportunities for Government support in helping suppliers meet market demands.

(ii) Within 180 days of the date of this order, the Secretary of Energy shall facilitate industry-led convenings on transformers and other critical grid components, which shall include appropriate representatives from agencies, transmission providers and transmission organizations, domestic suppliers of transformers, data center developers, and other private-sector organizations. On an ongoing basis, the Secretary, after consulting with participants in the industry-led convenings, shall:

(A) on at least an annual basis, develop and publish supply and demand forecasts for transformers, including forecasts for different transformer variants and analyses of supply and demand trends under different future scenarios, which shall include scenarios for growth in electricity demand from AI infrastructure and other sources of demand; and

(B) consider and, as appropriate, execute purchases of transformers and other critical grid components in order to provide demand certainty for domestic manufacturers to invest in capacity for meeting the needs of AI infrastructure. Any decision to execute such purchases shall be based on economic or other industry data, including the capacity utilization of domestic suppliers of transformers or other components, that the Secretary deems relevant to evaluating the status of the domestic industry. The Secretary shall subsequently execute sales of any purchased transformers or other critical grid components at times that the Secretary deems appropriate based on such data.

(f) Within 180 days of the date of this order, the Secretary of Energy shall establish requirements for transmission providers and transmission organizations to report to the Secretary transmission-related information to assist in siting and accelerating the interconnection of generation resources to serve frontier AI data centers on sites identified under section 4(a) of this order and cleared under subsection 4(d) of this order. Such information may include data on transmission congestion to help identify where additional transmission investments could enable the development of additional transmission capacity to serve such AI data centers.

(g) Within 180 days of the date of this order, the heads of agencies that possess loan or loan-guarantee authorities shall evaluate whether any such authorities could be used to support the development of AI infrastructure on Federal sites—including the production of critical grid equipment as described in subsection (e) of this section, or other actions to strengthen the AI infrastructure supply chain. In cases in which any authorities are available and appropriate for this purpose, the heads of relevant agencies shall provide that information to developers of AI infrastructure on Federal sites or other appropriate private-sector entities.

Sec. 9. *Additional Efforts to Improve Permitting and Power Procurement Nationwide.* (a) The heads of Federal Permitting Agencies shall designate, with respect to each of their component agencies, dedicated staff to handle all matters related to permits and approvals for AI infrastructure. Such designations shall include personnel dedicated to coordinating with and addressing the needs of applicants for permits under the respective agency's purview. In designating such personnel, the heads of Federal Permitting Agencies shall, as appropriate, implement staffing arrangements and other mechanisms that accelerate permitting for AI infrastructure to the maximum extent possible.

(b) To improve review practices pursuant to NEPA:

(i) Within 60 days of the date of this order, the heads of Federal Permitting Agencies, in coordination with the Chair of CEQ, shall assess existing categorical exclusions and identify opportunities to establish new categorical exclusions to support AI infrastructure on Federal sites, consistent with the requirements of NEPA and 40 C.F.R. parts 1500-1508. The heads of agencies whose NEPA regulations include categorical exclusions related to fiber-optic cables are encouraged, in undertaking these assessments, to evaluate whether such categorical exclusions may be applied to the development of fiber-optic cables as used for AI infrastructure.

(ii) Within 120 days of the date of this order, the heads of Federal Permitting Agencies shall, as appropriate and consistent with applicable law, propose any new categorical exclusions and, after considering all comments received through applicable public comment processes, take any actions to establish categorical exclusions that are appropriate given the received comments.

(iii) Within 120 days of the date of this order, and consistent with the directives described in section 7 of this order, the Secretary of Defense, the Secretary of the Interior, the Secretary of Agriculture, and the Secretary of Energy shall identify any existing categorical exclusions that are listed in the NEPA procedures of other agencies and that are relevant to the development of clean energy, electric transmission, or AI data centers and take any appropriate steps to adopt such categorical exclusions where appropriate and consistent with the requirements of NEPA and 40 C.F.R. parts 1500-1508. The Secretary of Defense, the Secretary of the Interior, the Secretary of Agriculture, and the Secretary of Energy shall take any appropriate steps to adopt and apply such categorical exclusions to AI infrastructure on Federal sites where consistent with the requirements of NEPA and 40 C.F.R. parts 1500-1508.

(c) Within 180 days of the date of this order, the Secretary of Energy shall issue a request for information on opportunities for accelerated interconnection at existing power plants, including as related to surplus interconnection service and clean repowering. The request shall seek details on the ownership of such plants with surplus interconnection service and the plants' suitability for colocation of new clean power generation resources with shared grid access.

(d) Within 90 days of the date of this order, the Secretary of Energy shall issue a request for information from private-sector entities including transmission providers, transmission organizations, and clean energy developers regarding load interconnection processes. The Secretary shall subsequently engage with transmission providers and transmission organizations regarding best practices to improve the transparency and efficiency of such processes, including through adopting new technologies, software, and procedures. The Secretary shall provide technical assistance and financial assistance to facilitate such adoption, as appropriate. The Secretary shall publish a report describing the results of this work within 1 year of the date of this order.

(e) To promote the expeditious, responsible development of nuclear power generation resources, the Secretary of Defense and the Secretary of Energy shall:

(i) seek to facilitate the deployment of additional nuclear power and, as relevant, supply-chain services on lands owned by, respectively, the Department of Defense and the Department of Energy—including Department of Defense installations and sites owned or managed by the Department of Energy National Laboratories—by, as appropriate and consistent with applicable law, identifying opportunities for such deployment on specific lands to the extent such opportunities exist and, in the case of the Secretary of Energy only, by evaluating whether financial support for such deployment is appropriate;

(ii) within 180 days of the date of this order, coordinate to publish a joint list of ten high-priority sites—or, if fewer than ten appropriate sites exist, as many sites as possible—which may overlap with sites identified and cleared under section 4 of this order, that are most conducive to expeditious, safe, and responsible deployment of additional nuclear power capacity readily available to serve AI data center electricity demand by December 31, 2035, taking into account factors including Federal, State, Tribal, and local ordinances; permitting and other regulatory requirements; water access; climate resilience and natural-hazard risks; and transmission and interconnection dynamics; and

(iii) within 1 year of the date of this order, publish either a joint plan or their own respective plans describing how each Secretary will facilitate deployment of additional nuclear power capacity as described in this subsection on any such sites. Any such plan shall address selection of appropriate nuclear reactor technologies; the licensing and permitting of relevant technologies or facilities; the approach that each Secretary would take to ensure the safe and responsible transportation of uranium and any other radioactive material to the site; the approach that each Secretary would take to ensure the safe and responsible storage or disposal of any spent nuclear fuel; remediation of the site after the plant ceases operation as needed; and any other steps necessary to ensure the deployment will protect public health, safety, and the environment, consistent with all applicable legal requirements and the principles of the document entitled *Safely and Responsibly Expanding U.S. Nuclear Energy: Deployment Targets and a Framework for Action* (November 2024); and

(iv) when carrying out actions under this subsection, comply with the directives of section 4(k) of this order.

(f) Within 180 days of the date of this order, the Secretary of Commerce, in consultation with the Secretary of Defense, the Secretary of Energy, and the White House Council on Supply Chain Resilience, shall submit a report to the President on supply chain risks applicable to the United States data center industry. The report shall include analysis of supply chain risks associated with the materials used to construct and maintain data centers, the electronics necessary to operate a data center, and emerging data center technologies, as well as recommended steps for the Federal Government to take to address identified risks. The report shall also include analysis on supply chain risks applicable to the generation and transmission infrastructure needed to power AI data centers. On an ongoing basis, as appropriate, the Secretary of Commerce shall engage with the private sector to identify emerging supply chain risks that have the potential to undermine the success of the United States AI infrastructure industry—with such success defined to include the industry's commercialization of emerging technologies—and to recommend policy solutions to address identified risks.

(g) Within 180 days of the date of this order, to promote the expeditious, responsible development and deployment of distributed energy solutions that support the development and operation of AI infrastructure, the Secretary of Energy shall develop model contracts for using distributed energy resources (DERs) to increase the local grid's capacity to support AI infrastructure. In developing such contracts, the Secretary shall consider options for cost-effective uses of DERs, including distribution-sited generation resources, energy storage assets, and opportunities for flexible management of electricity demand. The model contracts shall, as appropriate, include clauses providing for the owners of data centers to finance costs incurred by other entities in developing, installing, and operating DERs, consistent with the objective of utilities accounting for these financing activities when processing data center owners' interconnection applications.

(h) By July 31, 2025, the Permitting Council shall engage with developers of AI infrastructure to advance their understanding of resources available under title 41 of the Fixing America's Surface Transportation Act (Public Law 114-94) to accelerate permitting processes and reviews for clean energy projects that are part of AI infrastructure on Federal sites. As part of this work, the Permitting Council, in consultation with the White House Task Force on AI Datacenter Infrastructure announced on October 29, 2024, shall endeavor to engage small developers of AI infrastructure.

(i) Within 180 days of the date of this order, the Secretary of the Army, acting through the Chief of Engineers and Commanding General of the USACE, shall, consistent with applicable law, assess existing nationwide permits (NWPs) to determine how they may be applied to facilitate the construction of AI data centers and develop and publish a list of NWPs that could facilitate such construction. The Secretary of the Army, acting through the Chief of Engineers and Commanding General of the USACE, shall, as appropriate and consistent with applicable law, subsequently establish such new NWPs as expediently as possible.

(j) Within 60 days of the date of this order, the Secretary of Energy shall release for public comment draft reporting requirements for AI data centers covering all phases of AI data centers' development and operation—including material extraction, component fabrication, transportation, construction, operation, recycling, and retirement —regarding embodied greenhouse gas emissions, water usage, and excess heat or energy expenditures, as distinct from operational intensity of greenhouse gas emissions.

(k) Within 60 days of the date of this order, the Secretary of Energy, in coordination with the Administrator of the EPA and the Chair of CEQ, shall establish a grand challenge, serving as a call to voluntary action for appropriate private-sector and other stakeholders, for the purpose of:

(i) setting targets for minimizing the power usage effectiveness ratio and water usage effectiveness ratio of AI data centers, with a goal of bringing the power usage effectiveness ratio of AI data centers on Federal sites below 1.1;

(ii) promoting best practices for the beneficial use of waste heat and other efforts to maximize efficiency;

(iii) promoting best practices for data center energy management and sustainable design and operational practices for data centers that avoid or reduce adverse effects on natural and cultural resources and communities, and that protect public health and the environment;

(iv) raising AI developer and user awareness regarding the comparative energy intensities of different computational tasks; and

(v) developing best practices and standards for software and algorithmic efficiency.

Sec. 10. *Engagement Abroad.* (a) Within 90 days of the date of this order, the Secretary of State, in consultation with the Secretary of Defense, the Secretary of Commerce, the Secretary of Energy, the Administrator of the United States Agency for International Development, the Assistant to the President for National Security Affairs, and the heads of other relevant agencies as the Secretary of State may deem appropriate, shall develop a plan for engaging allies and partners on accelerating the buildout of trusted AI infrastructure around the world. Such a plan shall include measures to advance collaboration on the global buildout of trusted AI infrastructure; mitigate and prevent harms to local and affected communities; engage the private sector and investor community to identify and mitigate barriers to AI infrastructure investments; support the deployment of commercially available reliable clean power sources and the development and commercialization of emerging clean energy technologies, such as small modular nuclear reactors; exchange best practices for permitting, power procurement, and cultivating talent to build, operate, and maintain trusted AI infrastructure; and strengthen cyber, physical, and supply chain security safeguards related to AI infrastructure. Within 1 year of the date of this order, the Secretary of State shall submit to the Assistant to the President for National Security Affairs a report on actions taken pursuant to this plan.

(b) Within 120 days of the date of this order, the Assistant to the President for National Security Affairs shall convene heads of appropriate agencies, to include the Secretary of State, the Secretary of the Treasury, the Secretary of Commerce, the Secretary of Energy, the Chief Executive Officer of the United States International Development Finance Corporation, and the President of the Export-Import Bank of the United States, to identify and implement actions to facilitate United States exports and engagements abroad related to advanced nuclear technologies and relevant supply-chain services.

Sec. 11. *General Provisions*. (a) Nothing in this order shall be construed to impair or otherwise affect:

(i) the authority granted by law to an executive department or agency, or the head thereof; or

(ii) the functions of the Director of the Office of Management and Budget relating to budgetary, administrative, or legislative proposals.

(b) This order shall be implemented consistent with applicable law and subject to the availability of appropriations.

(c) This order is not intended to, and does not, create any right or benefit, substantive or procedural, enforceable at law or in equity by any party against the United States, its departments, agencies, or entities, its officers, employees, or agents, or any other person.

JOSEPH R. BIDEN, JR.

THE WHITE HOUSE,
January 14, 2025

White House, Executive Order 14179, Removing Barriers to American Leadership in Artificial Intelligence (2025)

By the authority vested in me as President by the Constitution and the laws of the United States, it is hereby ordered as follows:

Section 1. *Purpose.* The United States has long been at the forefront of artificial intelligence (AI) innovation, driven by the strength of our free markets, world-class research institutions, and entrepreneurial spirit. To maintain this leadership, we must develop AI systems that are free from ideological bias or engineered social agendas. With the right Government policies, we can solidify our position as the global leader in AI and secure a brighter future for all Americans. This order revokes certain existing AI policies and directives that act as barriers to American AI innovation, clearing a path for the United States to act decisively to retain global leadership in artificial intelligence.

Sec. 2. *Policy.* It is the policy of the United States to sustain and enhance America's global AI dominance in order to promote human flourishing, economic competitiveness, and national security.

Sec. 3. *Definition.* For the purposes of this order, "artificial intelligence" or "AI" has the meaning set forth in 15 U.S.C. 9401(3).

Sec. 4. *Developing an Artificial Intelligence Action Plan.* (a) Within 180 days of this order, the Assistant to the President for Science and Technology (APST), the Special Advisor for AI and Crypto, and the Assistant to the President for National Security Affairs (APNSA), in coordination with the Assistant to the President for Economic Policy, the Assistant to the President for Domestic Policy, the Director of the Office of Management and Budget (OMB Director), and the heads of such executive departments and agencies (agencies) as the APST and APNSA deem relevant, shall develop and submit to the President an action plan to achieve the policy set forth in section 2 of this order.

Sec. 5. *Implementation of Order Revocation.* (a) The APST, the Special Advisor for AI and Crypto, and the APNSA shall immediately review, in coordination with the heads of all agencies as they deem relevant, all policies, directives, regulations, orders, and other actions taken pursuant to the revoked Executive Order 14110 of October 30, 2023 (Safe, Secure, and Trustworthy Development and Use of Artificial

Intelligence). The APST, the Special Advisor for AI and Crypto, and the APNSA shall, in coordination with the heads of relevant agencies, identify any actions taken pursuant to Executive Order 14110 that are or may be inconsistent with, or present obstacles to, the policy set forth in section 2 of this order. For any such agency actions identified, the heads of agencies shall, as appropriate and consistent with applicable law, suspend, revise, or rescind such actions, or propose suspending, revising, or rescinding such actions. If in any case such suspension, revision, or rescission cannot be finalized immediately, the APST and the heads of agencies shall promptly take steps to provide all available exemptions authorized by any such orders, rules, regulations, guidelines, or policies, as appropriate and consistent with applicable law, until such action can be finalized.

(b) Within 60 days of this order, the OMB Director, in coordination with the APST, shall revise OMB Memoranda M-24-10 and M-24-18 as necessary to make them consistent with the policy set forth in section 2 of this order.

Sec. 6. *General Provisions*. (a) Nothing in this order shall be construed to impair or otherwise affect:

(i) the authority granted by law to an executive department or agency, or the head thereof; or

(ii) the functions of the Director of the Office of Management and Budget relating to budgetary, administrative, or legislative proposals.

(b) This order shall be implemented consistent with applicable law and subject to the availability of appropriations.

(c) This order is not intended to, and does not, create any right or benefit, substantive or procedural, enforceable at law or in equity by any party against the United States, its departments, agencies, or entities, its officers, employees, or agents, or any other person.

DONALD J. TRUMP

THE WHITE HOUSE,
January 23, 2025

Comparison: Definitions of "AI" and "AI System"

China Electronics Standardization Institute, *Artificial Intelligence Standardization White Paper* (2018)

> *AI is the theories, technologies, methods, and application systems for using digital computers or digital computer-controlled machines to simulate, extend, and expand human intelligence, perceive environments and acquire knowledge, and use knowledge to obtain the best results.* (p. 5)

OECD AI Principles / G20 AI Guidelines (2019, updated 2024)

> *AI system: An AI system is a machine-based system that, for explicit or implicit objectives, infers, from the input it receives, how to generate outputs such as predictions, content, recommendations, or decisions that can influence physical or virtual environments. Different AI systems vary in their levels of autonomy and adaptiveness after deployment.* (Preamble)

UNESCO, Recommendation on the Ethics of AI (2022)

> *AI systems are information-processing technologies that integrate models and algorithms that produce a capacity to learn and to perform cognitive tasks leading to outcomes such as prediction and decision-making in material and virtual environments. AI systems are designed to operate with varying degrees of autonomy by means of knowledge modelling and representation and by exploiting data and calculating correlations. AI systems may include several methods, such as but not limited to:*

> *(i) machine learning, including deep learning and reinforcement learning;*

> *(ii) machine reasoning, including planning, scheduling, knowledge representation and reasoning, search, and optimization.*

> *AI systems can be used in cyber-physical systems, including the Internet of things, robotic systems, social robotics, and human-computer interfaces, which involve control, perception, the processing of data collected by sensors, and the operation of actuators in the environment in which AI systems work.*

(Section I(2))

EU AI Act, Regulation 2024/1689 (2024)

'AI system' means a machine-based system that is designed to operate with varying levels of autonomy and that may exhibit adaptiveness after deployment, and that, for explicit or implicit objectives, infers, from the input it receives, how to generate outputs such as predictions, content, recommendations, or decisions that can influence physical or virtual environments; (Article 3(1))

Council of Europe, Framework Convention on AI (2024)

For the purposes of this Convention, "artificial intelligence system" means a machine-based system that, for explicit or implicit objectives, infers, from the input it receives, how to generate outputs such as predictions, content, recommendations or decisions that may influence physical or virtual environments. Different artificial intelligence systems vary in their levels of autonomy and adaptiveness after deployment. (Article 2)

AI Basic Act (Republic of Korea 2024)

"AI system" refers to an artificial intelligence-based system that deduces outputs such as predictions, recommendations, and decisions that affect real and virtual environments for a given goal with various levels of autonomy and adaptability. (Article 2)

An Act Concerning Consumer Protections in Interactions with Artificial Intelligence Systems (Colorado, U.S. 2024)

"artificial intelligence system" means any machine-based system that, for any explicit or implicit objective, infers from the inputs the system receives how to generate outputs, including content, decisions, predictions, or recommendations, that can influence physical or virtual environments. (Section 6-1-1701(2))

White House, Executive Order 14110, The Safe, Secure, and Trustworthy Development and Use of Artificial Intelligence (2023)

The term "AI system" means any data system, software, hardware, application, tool, or utility that operates in whole or in part using AI. (Section 3(e))

White House, Executive Order 14179, Removing Barriers to American Leadership in Artificial Intelligence (2025)

For the purposes of this order, "artificial intelligence" or "AI" has the meaning set forth in 15 U.S.C. 9401(3). (Section 3)

The term "artificial intelligence" means a machine-based system that can, for a given set of human-defined objectives, make predictions, recommendations or decisions influencing real or virtual environments. Artificial intelligence systems use machine and human-based inputs to- (A) perceive real and virtual environments; (B) abstract such perceptions into models through analysis in an automated manner; and (C) use model inference to formulate options for information or action. (15 U.S.C. 9401(3))

European Law Institute, *The concept of 'AI system' under the new AI Act: Arguing for a Three-Factor Approach (2025)*

(I) The amount of data or domain-specific knowledge that went into development: the data or domain-specific expert knowledge that went into the system's development, ranging from data-driven models to those leveraging expert rule-based programming.

(II) The extent to which know-how is created during operation: the presence of goal-oriented optimisation or search algorithms in the operation of the system, differentiating simple forward computations from systems capable of generating new knowledge on 'how to solve a problem;

(III) The degree of formal indeterminacy of outputs: the formal indeterminacy in the system's outputs, where the system handles tasks that would, if performed by humans, involve discretion and require subjective judgement or creative interpretation.

(Executive summary)

AI Resources

National Strategies & Guidelines

Argentina

National Plan of Artificial Intelligence (2019), https://oecd-opsi.org/wp-content/uploads/2021/02/Argentina-National-AI-Strategy.pdf

Argentina Digital Agenda 2030 (2018), https://www.boletinoficial.gob.ar/detalleAviso/primera/195154/20181105

National Plan 2030 for Science, Technology, and Innovation (2022), https://www.argentina.gob.ar/sites/default/files/plan_nacional_de_cti_2030.pdf

Australia

Artificial Intelligence Roadmap (2019), https://www.csiro.au/en/research/technology-space/ai/artificial-intelligence-roadmap

AI Ethic Principles (2019), https://www.industry.gov.au/publications/australias-artificial-intelligence-ethics-principles/australias-ai-ethics-principles

Australia's Artificial Intelligence Action Plan (2021), https://webarchive.nla.gov.au/awa/20220816053410/https:/www.industry.gov.au/data-and-publications/australias-artificial-intelligence-action-plan

Voluntary AI Safety Standard (2024), https://www.industry.gov.au/publications/voluntary-ai-safety-standard

Austria

Artificial Intelligence Mission Austria 2030 (2021) ("AIM AT 2030"), https://www.digitalaustria.gv.at/eng/strategy/strategy-AI-AIM-AT-2030.html

Harnessing the Potential of AI: AI Implementation Plan (2024), https://www.digitalaustria.gv.at/eng/strategy/strategy-AI-AIM-AT-2030.html#ai-implementation-plan

Azerbaijan

AZS ISO/IEC TR 24028:2024 "Information Technology – Artificial Intelligence – Overview of Trustworthiness in Artificial Intelligence" (2024), ttps://e-standart.gov.az/Standard/Details/9a0521b7-fa40-4e82-ad29-0503abad1921

AZS ISO/IEC TR 24372:2024 "Information Technology – Artificial Intelligence (AI) – Overview of Computational Approaches for AI Systems" (2024), https://e-standart.gov.az/Standard/Details/b40a3eae-8b66-478f-bfc0-cfb649861af8

Bahrain

Artificial Intelligence Regulation Law (2024)

Bangladesh

National Strategy for Artificial Intelligence (2020), https://ictd.portal.gov.bd/sites/default/files/files/ictd.portal.gov.bd/policies/e57f1366_a62c_4d1a_8369_a9d3bc156cd5/National%20Strategy%20for%20Artificial%20Intellgence%20-%20Bangladesh%20.pdf

National Strategy for Artificial Intelligence Bangladesh 2019-2024, https://ictd.portal.gov.bd/sites/default/files/files/ictd.portal.gov.bd/page/6c9773a2_7556_4395_bbec_f132b9d819f0/Draft%20-%20Mastering%20National%20Strategy%20for%20Artificial%20Intellgence%20-%20Bangladesh.pdf

National AI Policy (2024), https://ictd.portal.gov.bd/sites/default/files/files/ictd.portal.gov.bd/page/6c9773a2_7556_4395_bbec_f132b9d819f0/National_AI_Policy_2024_DRAFT.pdf

Belgium

National Convergence Plan for the Development of Artificial Intelligence (2022), https://bosa.belgium.be/sites/default/files/content/documents/DTdocs/AI/Nationaal_convergentieplan_voor_de_ontwikkeling_van_artificiele_intelligentie.pdf

Artificial Intelligence Systems and the GDPR A Data Protection Perspective (2024), https://www.autoriteprotectiondonnees.be/publications/artificial-intelligence-systems-and-the-gdpr---a-data-protection-perspective.pdf

Brazil

Brazil Artificial Intelligence Strategy (2021), https://www.gov.br/mcti/pt-br/acompanhe-o-mcti/transformacaodigital/arquivosinteligenciaartificial/ebia-portaria_mcti_4-979_2021_anexo1.pdf

Brazilian AI Plan 2024-2028 (2024), https://www.gov.br/mcti/pt-br/acompanhe-o-mcti/noticias/2024/07/plano-brasileiro-de-ia-tera-supercomputador-e-investimento-de-r-23-bilhoes-em-quatro-anos/ia_para_o_bem_de_todos.pdf

Canada

Artificial Intelligence and Data Act (AIDA) (draft 2022), https://ised-isde.canada.ca/site/innovation-better-canada/en/artificial-intelligence-and-data-act-aida-companion-document

Pan-Canadian Artificial Intelligence Strategy (2017, renewed 2022), https://ised-isde.canada.ca/site/ai-strategy/en

Guide on the Use of Generate Artificial Intelligence (2024),

https://www.canada.ca/en/government/system/digital-government/digital-government-innovations/responsible-use-ai/guide-use-generative-ai.html *Directive on Automated Decision-Making* (2023), https://www.tbs-sct.canada.ca/pol/doc-eng.aspx?id=32592

Office of the Privacy commissioner of Canada, *Recommended Legal Framework for Police Agencies' Use of Facial Recognition Joint Statement by Federal, Provincial and Territorial Privacy Commissioners* (2022), https://www.priv.gc.ca/en/opc-actions-and-decisions/advice-to-parliament/2022/s-d_prov_20220502/

Voluntary Code of Conduct on the Responsible Development and Management of Advanced Generative AI Systems (2023), https://ised-isde.canada.ca/site/ised/en/voluntary-code-conduct-responsible-development-and-management-advanced-generative-ai-systems

Chile

National Artificial Intelligence Policy (2025),

https://www.diariooficial.interior
.gob.cl/publicaciones/2025/01/2
8/44060/01/2600578.pdf
AI Action Plan (2024),
https://drive.google.com/file/d/1
YhKNO8zFyHnqN8FuiPinB22
KKVXmplQp/

China

*New Generation Artificial
Intelligence Development Plan*
(2017),
https://digichina.stanford.edu/wo
rk/full-translation-chinas-new-
generation-artificial-intelligence-
development-plan-2017/

Cyberspace Administration of
China, *Measures for Labelling
Artificial Intelligence Generated
Synthetic Content* (2024),
https://www.cac.gov.cn/2023-
04/11/c_1682854275475410.ht
m

Cyberspace Administration of
China, *Cybersecurity
technology—Labelling method
for content generated by
artificial intelligence* (2024),
https://www.cac.gov.cn/2024-
09/14/c_1728000676244628.ht
m

*Ethical Norms for New
Generation AI* (2021),
https://cset.georgetown.edu/wp-
content/uploads/t0400_AI_ethic
al_norms_EN.pdf

*Guidelines for the Construction
of a National New Generation
Artificial Intelligence Standards
System* (2021),
https://cset.georgetown.edu/wp-
content/uploads/t0401_AI_stand
ards_guidelines_EN.pdf

*Provisions on the Management
of Algorithmic
Recommendations in Internet
Information Services* (2022),
https://www.cac.gov.cn/2022-
01/04/c_1642894606364259.ht
m

China Electronics
Standardization
Institute, *Artificial Intelligence
Standardization White Paper*
(2021),
https://cset.georgetown.edu/wp-
content/uploads/t0393_AI_white
_paper_EN.pdf

China Electronics
Standardization
Institute, *Artificial Intelligence
Standardization White Paper*
(2018),
https://www.cesi.cn/images/edit
or/20180124/201801241355287
42.pdf

Colombia

*National Intelligence Policy
COPNES 4144* (2025),

https://colaboracion.dnp.gov.co/
CDT/Conpes/Econ%C3%B3mic
os/4144.pdf

*Roadmap for the Development
and Application of Artificial
Intelligence in Colombia* (2024),
https://minciencias.gov.co/sites/
default/files/upload/noticias/hoja
_de_ruta_adopcion_etica_y_sost
enible_de_inteligencia_artificial
_colombia_0.pdf

*Ethical Framework for Artificial
Intelligence in Colombia* (2021),
https://minciencias.gov.co/sites/
default/files/marco-etico-ia-
colombia-2021.pdf

Costa Rica

*National Strategy for Artificial
Intelligence of Costa Rica*
(2024),
https://www.micitt.go.cr/sites/de
fault/files/2024-
10/Estrategia%20Nacional%20d
e%20Inteligencia%20Artificial
%20de%20Costa%20Rica%20E
SP.pdf

Czech Republic

*National Artificial Intelligence
Strategy of the Czech Republic*
(2019),
https://www.mpo.cz/assets/en/gu
idepost/for-the-media/press-
releases/2019/5/NAIS_eng_web.
pdf

*Analysis of the Development
Potential of Artificial
Intelligence in the Czech
Republic* (2018),
https://vlada.gov.cz/assets/evrop
ske-zalezitosti/aktualne/AI-
Summary-Report.pdf

Denmark

*National Strategy for Artificial
Intelligence* (2019),
https://en.digst.dk/media/lz0fxbt
4/305755_gb_version_final-
a.pdf

*Responsible Use of AI Assistants
in Public and Private Sectors*
(2024), https://table.media/wp-
content/uploads/2024/11/131420
36/AI-
assistants_Whitepaper_EN.pdf

Dominican Republic

National Innovation Policy 2030
(2021),
https://innovacionrd.gob.do/#pol
itica

Egypt

Egypt National AI Strategy
(2025),
https://mcit.gov.eg/Upcont/Docu
ments/Publications_672021000_
Egypt-National-AI-Strategy-
English.pdf

Egyptian Charter for Responsible AI (2023), https://aicm.ai.gov.eg/en/Resou ces/EgyptianCharterForResponsi bleAIEnglish-v1.0.pdf

Estonia

Estonia's National Artificial Intelligence Strategy (2019), https://www.kratid.ee/_files/ugd/ 980182_8d0df96fd41145739dff 2595e0ab3e8d.pdf

Estonia's National Artificial Intelligence Strategy (2022), https://www.kratid.ee/_files/ugd/ 980182_4434a890f1e64c66b119 0b0bd2665dc2.pdf

AI Use Cases in the Public Sector (2022), https://www.kratid.ee/kasutusloo d-kratid

Ethiopia

A National AI Policy – Ethiopia (2024), https://www.lawethiopia.com/im ages/Policy_documents/Ethiopia n%20ai%

Finland

National AI Strategy (2017), https://julkaisut.valtioneuvosto.fi /bitstream/handle/10024/80849/ TEMrap_41_2017_Suomen_tek o%C3%A4lyaika.pdf

National Regulation on Automated Decision-making in Administration (2020), https://julkaisut.valtioneuvosto.fi /bitstream/handle/10024/162355/ OM_2020_14_S0.pdf?

France

France 2030: National Strategy for Artificial Intelligence (2024) ("Ai for Humanity"), https://www.entreprises.gouv.fr/ priorites-et-actions/autonomie-strategique/soutenir-linnovation-dans-les-secteurs-strategiques-de-6

AI for Humanity (2021), https://www.inria.fr/sites/default /files/2021-06/PNRIA-Flyer_National_EN.pdf

Cedric Villani, F*or a Meaningful Artificial Intelligence* (2018), https://www.jaist.ac.jp/~bao/AI/ OtherAIstrategies/MissionVillan i_Report_ENG-VF.pdf

Priority Research Programme and Equipment on AI (PEPR AI) (2024), https://www.enseignementsup-recherche.gouv.fr/fr/pepr-intelligence-artificielle-la-recherche-francaise-la-pointe-95298

Germany

Artificial Intelligence Strategy for the German Federation (2020), https://www.ki-strategie-deutschland.de/files/downloads/Fortschreibung_KI-Strategie_engl.pdf

Opinion of the Data Ethics Commission (2019), https://www.bfdi.bund.de/SharedDocs/Downloads/EN/Datenschutz/Data-Ethics-Commission_Opinion.pdf?

German Bundestag The Study Commission on Artificial Intelligence – Social Responsibility and Economic, Social and Ecological Potential (2020), https://www.btg-bestellservice.de/pdf/81021700.pdf

Ghana

National Artificial Intelligence Strategy (2023-2033), https://www.slideshare.net/slideshow/ghana-s-national-artificial-intelligence-strategy-2023-2033-pdf/270928634#3

Hong Kong

Ethical Artificial Intelligence Framework (2024),

https://www.digitalpolicy.gov.hk/en/our_work/data_governance/policies_standards/ethical_ai_framework/doc/Ethical_AI_Framework.pdf

Policy Statement on Responsible Application of Artificial Intelligence in the Financial Market (2024), https://gia.info.gov.hk/general/202410/28/P2024102800154_475819_1_1730087238713.pdf

Artificial Intelligence: Model Personal Data Protection Framework (2024), https://www.pcpd.org.hk/english/resources_centre/publications/files/ai_protection_framework.pdf

Guidance on Ethical Development and Use of AI (2021), https://www.pcpd.org.hk/english/resources_centre/publications/files/guidance_ethical_e.pdf
Hungary

Hungary's Artificial Intelligence Strategy 2020-2030 (2020), https://ai-hungary.com/files/e8/dd/e8dd79bd380a40c9890dd2fb01dd771b.pdf

India

National Strategy for Artificial Intelligence – AI for All (2023), https://www.niti.gov.in/sites/def ault/files/2023-03/National-Strategy-for-Artificial-Intelligence.pdf

Responsible AI #AIFORALL – Approach Document for India: Part 1 Principles of Responsible AI (2021), https://www.niti.gov.in/sites/def ault/files/2021-02/Responsible-AI-22022021.pdf

Responsible AI # AIFORALL – Approach Document for India: Part 2 Operationalizing Principles Responsible AI (2022), https://www.niti.gov.in/sites/def ault/files/2021-08/Part2-Responsible-AI-12082021.pdf

Indonesia

National Strategy for Artificial Intelligence Indonesia 2020-2045 (2020), https://ai-innovation.id/images/gallery/ebo ok/stranas-ka.pdf

Iran

National Artificial Intelligence Strategy (2024),

https://en.itrc.ac.ir/page/innovati on-and-development-artificial-intelligence-center

Ireland

AI- Here for Good – A National Artificial Intelligence Strategy (2024), https://enterprise.gov.ie/en/publi cations/publication-files/national-ai-strategy-refresh-2024.pdf

AI Standards & Assurance Roadmap (2023), https://www.nsai.ie/images/uplo ads/general/NSAI_AI_report_di gital_links.pdf

Israel

Israel's Policy on Artificial Intelligence Regulation and Ethics (2023), https://www.gov.il/BlobFolder/p olicy/ai_23/he/2023%20Artificia l%20Intelligence%20Regulation %20and%20Ethics%20Policy% 20Principles%20Document.pdf

White Paper AI Principles of Policy, Regulation, and Ethics (2020), https://www.gov.il/he/pages/mos t-news20223110

Ministry of Health Guiding Principles for the Development of Machine Learning-Based

Technologies (2023), https://www.gov.il/en/pages/digital-medical-technology-gmlp-1

Italy

Italian Strategy for Artificial Intelligence 2024-2026 (2024), https://www.agid.gov.it/en/news/the-italian-strategy-for-artificial-intelligence

Strategic Programme on Artificial Intelligence 2022 – 2024 (2021), https://docs.italia.it/italia/mid/programma-strategico-nazionale-per-intelligenza-artificiale-en-docs/en/bozza/index.html

Agency for Digital Italy Artificial Intelligence at the Service of the Citizen (2018), https://libro-bianco-ia.readthedocs.io/en/latest/

Japan

Artificial Intelligence Law (20205)

AI Strategy (2022), https://www8.cao.go.jp/cstp/ai/aistratagy2022en.pdf

Governance Guidelines for Implementing AI Principles (Ver 1.1) (2022),

https://www.meti.go.jp/shingikai/mono_info_service/ai_shakai_jisso/pdf/20220128_2.pdf

AI guidelines for Business (Ver 1.0) (2024), https://www.meti.go.jp/shingikai/mono_info_service/ai_shakai_jisso/pdf/20240419_9.pdf

Japan Cabinet Office Social Principles of Human Centric AI (2019), https://www8.cao.go.jp/cstp/english/humancentricai.pdf

Jamaica

National Artificial Intelligence Taskforce's - Policy Recommendations (2025), https://opm.gov.jm/wp-content/uploads/2025/02/National-Artificial-Intelligence-Task-Force-Policy-Recommendations-Final-1.pdf

Kazakhstan

Decree of the President of the Republic of Kazakhstan No. 674 – "On approval of the Concept of legal policy of the Republic of Kazakhstan until 2030" (2021), https://adilet.zan.kz/rus/docs/U2100000674#z133

Kuwait

Kuwait National AI Strategy 2025-2028 (2025), https://cait.gov.kw/media/filer_p ublic/3f/b4/3fb49a45-4a78-4489-8898-b68e2bd260ca/kuwait_national_strategy.pdf

Lithuania

Lithuania Artificial Intelligence Strategy - A Vision for the Future (2019), https://eimin.lrv.lt/uploads/eimin /documents/files/DI_strategija_E NG(1).pdf

Govtech Sandbox, https://govtechlab.lt/govtech-sandbox/

Luxembourg

Artificial Intelligence: A Strategic Vision for Luxembourg (2020), https://innovative-initiatives.public.lu/sites/default/ files/2020-09/AI_EN_0.pdf

The Data-Driven Innovation Strategy for the Development of a Trusted and Sustainable Economy in Luxembourg (2019), https://gouvernement.lu/dam-assets/fr/publications/rapport-etude-analyse/minist-economie/The-Data-driven-Innovation-Strategy.pdf

Malaysia

Artificial Intelligence Roadmap 2021 – 2025 (2021), https://mastic.mosti.gov.my/publ ication/artificial-intelligence-roadmap-2021-2025/

The National Guidelines on AI Governance and Ethics (2024), https://mastic.mosti.gov.my/publ ication/the-national-guidelines-on-ai-governance-ethics/

Malta

Malta Towards Trustworthy AI – Malta's Ethical AI Framework (2019), https://malta.ai/wp-content/uploads/2019/10/Malta_ Towards_Ethical_and_Trustwort hy_AI_vFINAL.pdf

Malta – The Ultimate AI Launchpad – A Strategy and a Vision for Artificial Intelligence in Malta 2030 (2019), https://malta.ai/wp-content/uploads/2019/11/Malta_ The_Ultimate_AI_Launchpad_v Final.pdf

Mauritius

Mauritius Artificial Intelligence Strategy (2018), https://ncb.govmu.org/ncb/strate gicplans/MauritiusAIStrategy20 18.pdf

Mexico

Agenda Nacional Mexicana de Inteligencia Artificial (2020), https://36dc704c-0d61-4da0-87fa-917581cbce16.filesusr.com/ugd/7be025_6f45f669e2fa4910b32671a001074987.pdf

Netherlands

Strategic Action Plan for AI (2019), https://oecd.ai/en/wonk/documents/netherlands-strategic-action-plan-for-ai-2019

Government-Wide Vision on Generative AI (2024), https://www.government.nl/documents/parliamentary-documents/2024/01/17/government-wide-vision-on-generative-ai-of-the-netherlands

AI and Algorithmic Risks Report 2024), https://autoriteitpersoonsgegevens.nl/en/system/files?file=2024-09/AI%20%26%20Algorithmic%20Risks%20Report%20Netherlands%20-%20Summer%202024.pdf

Nigeria

National Artificial Intelligence Strategy (2024),

https://ncair.nitda.gov.ng/wp-content/uploads/2024/08/National-AI-Strategy_01082024-copy.pdf

National Digital Economy Policy and Strategy 2020-2030 (2019), https://www.ncc.gov.ng/docman-main/industry-statistics/policies-reports/883-national-digital-economy-policy-and-strategy/file

Norway

National Strategy for Artificial Intelligence (draft 2023), https://moitt.gov.pk/SiteImage/Misc/files/National%20AI%20Policy%20Consultation%20Draft%20V1.pdf

Pakistan

National AI Policy (draft 2023), https://moitt.gov.pk/SiteImage/Misc/files/National%20AI%20Policy%20Consultation%20Draft%20V1.pdf?utm_source=chatgpt.com

Peru

National Artificial Intelligence Strategy 2021-2026 (2021), https://wp.oecd.ai/app/uploads/2021/12/Peru_National_Artificial_Intelligence_Strategy_2021-2026.pdf

Draft Regulations of Law No. 31814 (2024), https://insightplus.bakermckenzie.com/bm/technology-media-telecommunications_1/peru-the-draft-regulations-of-the-law-that-promotes-the-use-of-ai-have-been-published_1

Philippines

National AI Strategy Roadmap 2.0 (NAISR 2.0), https://www.dti.gov.ph/archives/news-archives/dti-drives-ai-innovation-national-roadmap-2-0-cair-launch/

NPC Guidelines on the Application of the Republic Act No. 10173 or the Data Privact Act of 2012 (DPA, and the Issuances of the Commission toArtificial Intelligence Systems processing Personal Data (2024), https://privacy.gov.ph/wp-content/uploads/2025/02/Advisory-2024.12.19-Guidelines-on-Artificial-Intelligence-w-SGD.pdf

Poland

Policy for the Development of AI in Poland from 2020 (2020), https://wp.oecd.ai/app/uploads/2021/12/Poland_Policy_for_Artificial_Intelligence_Development_in_Poland_from_2020_2020.pdf

Portugal

AI Portugal 2030 (2019), https://www.incode2030.gov.pt/wp-content/uploads/2023/07/Estrategia-de-Inteligencia-artificial.aspx_.pdf?

Qatar

National AI Strategy (2019), https://www.mcit.gov.qa/wp-content/uploads/sites/4/2025/01/national_artificial_intelligence_strategy_for_qatar_2019_0.pdf?

Guidelines for Secure Adoption and Usage of AI (2024), https://assurance.ncsa.gov.qa/sites/default/files/publications/policy/2024/CSSP_Guidelines_for_Secure_Usage_and_Adoption_of_Artificial_intelligence-Eng-v1.0_2.pdf?

Republic of Korea

Mid- to Long-Term Master Plan in Preparation for the Intelligent Information Society (2016), https://k-erc.eu/wp-content/uploads/2017/12/Master-Plan-for-the-intelligent-information-society.pdf

National Strategy for Artificial Intelligence (2019), https://www.msit.go.kr/bbs/view.do?sCode=eng&nttSeqNo=9&bbsSeqNo=46&mId=10&mPid=9

National Guidelines for AI Ethics (2020), https://ai.kisdi.re.kr/eng/main/contents.do?menuNo=500011

Russia

National Strategy for Artificial Intelligence Development in the Russian Federation up to 2030 (2019), https://cset.georgetown.edu/wp-content/uploads/Decree-of-the-President-of-the-Russian-Federation-on-the-Development-of-Artificial-Intelligence-in-the-Russian-Federation-.pdf

Rwanda

National AI Policy (2023), https://www.minict.gov.rw/index.php?eID=dumpFile&t=f&f=67550&token=6195a53203e197efa47592f40ff4aaf24579640e

Saudi Arabia

SDAIA National Strategy for Data and AI (NSDAI) (2020), https://sdaia.gov.sa/en/SDAIA/SdaiaStrategies/Pages/NationalStrategyForDataAndAI.aspx

SDAIA Deepfake Guidelines Version 1 (2024), https://istitlaa.ncc.gov.sa/en/transportation/ndmo/deepfakesguidelines/Documents/SDAIA_Deepfakes%20Guidelines.pdf

SDAIA AI Ethics Principles (2023), https://sdaia.gov.sa/en/SDAIA/about/Documents/ai-principles.pdf

SDAIA Generative Artificial Intelligence Guidelines (2024), https://sdaia.gov.sa/en/SDAIA/about/Files/GenerativeAIPublicEN.pdf

Singapore

Singapore National AI Strategy (2023), https://file.go.gov.sg/nais2023.pdf

Singapore draft Model AI Governance Framework for Generative AI (2024), https://aiverifyfoundation.sg/resources/mgf-gen-ai/#proposed-model-governance-framework-for-generative-ai

Singapore AI Safety Red Teaming Challenge (2025), https://www.imda.gov.sg/-/media/imda/files/about/emerging-tech-and-research/artificial-intelligence/singapore-ai-safety-red-teaming-challenge-evaluation-report.pdf

Slovenia

National Programme to Promote the Development and Use of AI by 2024 (2021), https://www.gov.si/assets/ministrstva/MDP/National_Programme_for_AI_2025.pdf

South Africa

National AI Policy Framework (2024), https://www.policyvault.africa/policy/south-africa-national-artificial-intelligence-ai-policy-framework-2024/

Spain

National AI Strategy 2021-2025 (2020), https://portal.mineco.gob.es/RecursosArticulo/mineco/ministerio/ficheros/National-Strategy-on-AI.pdf

Sweden

An AI Strategy for Sweden, https://strategy.ai.se/

Switzerland

National AI Strategy (2021), https://www.sbfi.admin.ch/dam/sbfi/en/dokumente/2021/05/leitlinien-ki.pdf.download.pdf/leitlinien-ki_e.pdf

Guidelines on Artificial Intelligence for the Federal Administration (2020), https://www.sbfi.admin.ch/dam/sbfi/en/dokumente/2021/05/leitlinien-ki.pdf.download.pdf/leitlinien-ki_e.pdf

Thailand

National AI Strategy and Action Plan 2022-2027 (2022), https://ai.in.th/en/about-ai-thailand/

Trinidad and Tobago

National Digital Transformation Strategy 2023-2026, https://mdt.gov.tt/digital-transformation-strategy/

Draft Voluntary National Standard related to Information Technology – Artificial Intelligence (2024), https://gottbs.com/2024/07/31/for-public-comment-draft-voluntary-national-standard-related-to-information-technology-artificial-intelligence/

Tunisia

National AI Strategy Unlocking Tunisia's Capabilities & Potential (2018) https://www.anpr.tn/national-ai-strategy-unlocking-tunisias-capabilities-potential/

Türkiye

National AI Strategy 2021-2025 (2021), https://wp.oecd.ai/app/uploads/2021/12/Turkey_National_Artificial_Intelligence_Strategy_2021-2025.pdf

KVKK Recommendations on the Protection of Personal Data in the Field of Artificial Intelligence, https://www.kvkk.gov.tr/SharedFolderServer/CMSFiles/58678459-eba4-451a-a2f3-c1baf17b90f5.pdf

Ukraine

National Strategy for the Development of AI 2021-2030 (2021),

https://wp.oecd.ai/app/uploads/2021/12/Ukraine_National_Strategy_for_Development_of_Artificial_Intelligence_in_Ukraine_2021-2030.pdf

White Paper on AI Regulation in Ukraine (2024), https://thedigital.gov.ua/storage/uploads/files/page/community/docs

UAE

UAE National Strategy for AI 2031 (2021), https://ai.gov.ae/wp-content/uploads/2021/07/UAE-National-Strategy-for-Artificial-Intelligence-2031.pdf

UAE Charter for the Development and Use of AI (2024), https://ai.gov.ae/wp-content/uploads/2024/07/UAEAI-Methaq-EN2-3.pdf

Smart Dubai AI Ethics Principles, & Guidelines (2019), https://www.digitaldubai.ae/docs/default-source/ai-ethics-resources/ai-ethics.pdf?

Deepfake Guide (2021), https://ai.gov.ae/wp-content/uploads/2021/07/AI-DeepFake-Guide-EN-2021.pdf

Adoption Guideline in Government Services (2023),

https://ai.gov.ae/wp-content/uploads/2023/03/AI-Report-EN-V7.pdf

United Kingdom

National AI Strategy (2022), https://www.gov.uk/government/publications/national-ai-strategy/national-ai-strategy-html-version

Policy Paper – A Pro-Innovation Approach to AI Regulation (2023), https://www.gov.uk/government/publications/ai-regulation-a-pro-innovation-approach/white-paper

ICO, *Big Data, Artificial Intelligence, Machine Learning and Data Protection* (2017), ttps://ico.org.uk/media2/migrated/2013559/big-data-ai-ml-and-data-protection.pdf

AI Opportunities Action Plan (2025), https://www.gov.uk/government/publications/ai-opportunities-action-plan/ai-opportunities-action-plan
AI Security Institute Inspect AI, https://inspect.ai-safety-institute.org.uk/

United States

A Blueprint for an AI Bill of Rights (2022), ttps://bidenwhitehouse.archives.gov/ostp/ai-bill-of-rights/

Executive Order 1407, Advancing Effective, Accountable Policing and Criminal Justice Practices to Enhance Public Trust and Public Safety (2022), https://www.govinfo.gov/content/pkg/FR-2022-05-31/pdf/2022-11810.pdf

Framework to Advance AI Governance and Risk Management in National Security (2024), https://ai.gov/wp-content/uploads/2024/10/NSM-Framework-to-Advance-AI-Governance-and-Risk-Management-in-National-Security.pdf

NIST, *Artificial Intelligence Risk Management Framework (AI RMF 1.0)* (2023), https://nvlpubs.nist.gov/nistpubs/ai/NIST.AI.100-1.pdf

OMB, *Advancing Governance, Innovation, and Risk Management for Agency Use of Artificial Intelligence* (2024)

OMB, *Advancing the Responsible Acquisition of Artificial Intelligence in Government* (2024), https://bidenwhitehouse.archives.gov/wp-content/uploads/2024/10/M-24-18-AI-Acquisition-Memorandum.pdf

CFPB, DOJ, EEOC, FTC, *Joint Statement on Enforcement Efforts Against Discrimination and Bias in Automated Systems* (2023), https://www.ftc.gov/system/files/ftc_gov/pdf/EEOC-CRT-FTC-CFPB-AI-Joint-Statement%28final%29.pdf

Bipartisan House Task Force Report on Artificial Intelligence (2024), *https://republicans-science.house.gov/_cache/files/a/a/aa2ee12f-8f0c-46a3-8ff8-8e4215d6a72b/6676530F7A30F243A24E254F6858233A.ai-task-force-report-final.pdf*

Uruguay

*National AI Strategy for the Digital Government (*2019*),* https://www.gub.uy/agencia-gobierno-electronico-sociedad-informacion-conocimiento/comunicacion/publicaciones/ia-strategy-english-version/ia-strategy-english-version/ai-strategy-for

National AI Strategy for the Artificial Intelligence 2024-2030 (2024*),* https://www.gub.uy/agencia-gobierno-electronico-sociedad-informacion-conocimiento/comunicacion/publicaciones/estrategia-nacional-inteligencia-artificial-2024-2030

Vietnam

National Strategy for AI Research, Development and Application through 2030 (2021), https://english.luatvietnam.vn/decision-no-127-qd-ttg-dated-january-26-2021-of-the-prime-minister-on-the-promulgation-of-the-national-strategy-on-research-development-and-applica-197755-doc1.html

Decision No. 1290/QD-BKHCN (2024), https://www.most.gov.vn/vn/tin-tuc/24448/bo-khcn-huong-dan-nguyen-tac-nghien-cuu--phat-trien-tri-tue-nhan-tao-co-trach-nhiem.aspx

Zambia

Artificial Intelligence Strategy (2024), https://www.mots.gov.zm/?p=4492

International Organizations' Statements and Reports

African Union, *Continental Artificial Intelligence Strategy* (Africa 2024), https://au.int/en/documents/2024 0809/continental-artificial-intelligence-strategy

African Union, *Statement of H.E Prof. Sarah Anyang Agbor, AUC Commissioner for HRST on Artificial Intelligence (AI)* (Africa 2019), https://au.int/en/videos/2019082 8/he-prof-sarah-anyang-agbor-auc-commissioner-hrst-artificial-intelligence-ai

ASEAN, *ASEAN Guide on AI Governance and Ethics* (Asia 2024), https://asean.org/wp-content/uploads/2024/02/ASEA N-Guide-on-AI-Governance-and-Ethics_beautified_201223_v2.pd f

ASEAN, *Expanded ASEAN Guide on AI Governance and Ethics – Generative AI* (Asia 2024), https://asean.org/wp-content/uploads/2025/01/Expand ed-ASEAN-Guide-on-AI-Governance-and-Ethics-Generative-AI.pdf

Centro Latinoamericano de Administración para el Desarrollo, *Ibero American Charter on Artificial Intelligence in Civil Service* (Central America2024), https://clad.org/wp-content/uploads/2024/03/CIIA-EN-03-2024.pdf

COE, *Hudeira Methodology* (Global 2024), https://rm.coe.int/cai-2024-16rev2-methodology-for-the-risk-and-impact-assessment-of-arti/1680b2a09f

COE, *The Council of Europe and Artificial Intelligence* (Global 2023), https://rm.coe.int/brochure-artificial-intelligence-en-march-2023-print/1680aab8e6

COE, *Human Rights by Design: Future Proofing Human Rights Protection in the Era of AI* (Global 2023), https://rm.coe.int/follow-up-recommendation-on-the-2019-report-human-rights-by-design-fut/1680ab2279

COE, Recommendation *CM/Rec (2024)5*
of the Committee of Ministers to member States
regarding the ethical and organisational aspects of the use of artificial intelligence and related digital technologies by prison and probation services (Global 2024),
https://search.coe.int/cm?i=0900 001680b1d0e4

COE, *The Application of Artificial Intelligence in Healthcare and its Impact on the 'Patient-Doctor' Relationship* (Global 2025),
https://rm.coe.int/cdbio-2023-7-rev3-ai-report-new-banner-e/1680b1bfe4

EDPB, *Opinion 11/2024 on the Use of Facial Recognition to Streamline Airport Passengers' Flow* (EU 2024),
https://www.edpb.europa.eu/syst em/files/2024-05/edpb_opinion_202411_facial recognitionairports_en.pdf

EDPB, *Opinion 28/2024 on Certain Data Protection Aspects Related to the Processing of Personal Data in the Context of AI Models* (EU 2024),
https://www.edpb.europa.eu/syst em/files/2024-12/edpb_opinion_202428_ai-models_en.pdf

EDPB Support Pool of Experts Programme, Kris Shrishak, *AI-Complex Algorithms and effective Data Protection Supervision: Bias Evaluation* (EU 2025),
https://www.edpb.europa.eu/syst em/files/2025-01/d1-ai-bias-evaluation_en.pdf

EDPB Support Pool of Experts Programme, *Effective Implementation of Data Subjects' Rights* (EU 2025),
https://www.edpb.europa.eu/syst em/files/2025-01/d2-ai-effective-implementation-of-data-subjects-rights_en.pdf

EDPB, *Opinion 28/2024 on certain data protection aspects related to the processing of personal data in the context of AI models* (EU 2024),
https://www.edpb.europa.eu/our-work-tools/our-documents/opinion-board-art-64/opinion-282024-certain-data-protection-aspects_en

EDPB, *Statement 3/2024 on Data Protection Authorities'' Role in the Artificial Intelligence Act Framework* (EU 2024),
https://www.edpb.europa.eu/our-work-tools/our-documents/statements/statement-32024-data-protection-authorities-role-artificial_en

EDPS, *First EDPS Orientations for Ensuring Data Protection Compliance When Using Generative AI Systems* (EU 2024), https://www.edps.europa.eu/system/files/2024-06/24-06-03_genai_orientations_en.pdf

European Commission, *Living Guidelines on the Responsible Use of Generative AI in Research* (EU 2024), https://research-and-innovation.ec.europa.eu/document/download/2b6cf7e5-36ac-41cb-aab5-0d32050143dc_en?filename=ec_rtd_ai-guidelines.pdf

European Commission, *General-Purpose AI Code of Practice* (draft) (EU 2025), https://digital-strategy.ec.europa.eu/en/policies/ai-code-practice

European Commission, *Commission Guidelines on prohibited artificial intelligence practices established by Regulation* (draft) (EU 2025), https://digital-strategy.ec.europa.eu/en/library/commission-publishes-guidelines-prohibited-artificial-intelligence-ai-practices-defined-ai-act?

European Commission, *Harmonised Standards for European AI Act* (EU 2024), https://publications.jrc.ec.europa.eu/repository/handle/JRC139430

European Parliament, *Addressing AI Risks on the Workplace: Workers and Algorithms* (EU 2024), https://www.europarl.europa.eu/RegData/etudes/BRIE/2024/762323/EPRS_BRI(2024)762323_EN.pdf

European Union Agency for Cybersecurity, *Cybersecurity of AI and Standardisation* (EU 2023), https://www.enisa.europa.eu/sites/default/files/publications/Cybersecurity%20of%20AI%20and%20Standardisation.pdf

G7, *Ministerial Declaration* (Global 2024), https://assets.innovazione.gov.it/1710505409-final-version_declaration.pdf

G7, *Ministerial Declaration* (Global 2024), https://www.gov.uk/government/publications/g20-ministerial-declaration-maceio-13-september-2024/g20-ministerial-declaration-13-september-2024

G7, *Digital Competition Communiqué* (Global 2024), https://en.agcm.it/dotcmsdoc/pressrelease/G7%202024%20-%20Digital%20Competition%20Communiqu%C3%A9.pdf

G7, *Privacy in the Age of Data* (Global 2024), https://www.priv.gc.ca/en/opc-news/news-and-announcements/2024/communique-g7_241011/

G7, *Statement on AI and Children* (Global 2024), https://www.priv.gc.ca/en/opc-news/speeches-and-statements/2024/s-d_g7_20241011_child-ai/

G20, *Rio de Janeiro Leadership Declaration* (Global 2024), https://www.g20.utoronto.ca/2024/241118-declaration.html

G20, *Ministerial Declaration* (Global 2024), https://www.gov.uk/government/publications/g20-ministerial-declaration-maceio-13-september-2024/g20-ministerial-declaration-13-september-2024

GPA, *Declaration on Ethics and Data Protection in Artificial Intelligence* (Global 2018), https://globalprivacyassembly.org/wp-content/uploads/2018/10/20180922_ICDPPC-40th_AI-Declaration_ADOPTED.pdf

GPA, *Resolution on Principles and Expectations for the Appropriate Use of Personal Information in Facial Recognition Technology* (Global 2022), https://globalprivacyassembly.org/wp-content/uploads/2022/11/15.1.c.Resolution-on-Principles-and-Expectations-for-the-Appropriate-Use-of-Personal-Information-in-Facial-Recognition-Technolog.pdf

GPA, *Resolution on Generative Artificial Intelligence Systems* (Global 2023), https://globalprivacyassembly.org/wp-content/uploads/2023/10/5.-Resolution-on-Generative-AI-Systems-101023.pdf

GPAI, *Algorithmic Transparency in the Public Sector* (Global 2024), https://gpai.ai/projects/responsible-ai/algorithmic-transparency-in-the-public-sector/algorithmic-transparency-in-the-public-sector.pdf

NATO, *An Artificial Intelligence Strategy for NATO* (NorAm & Europe, 2021), https://www.nato.int/docu/review/articles/2021/10/25/an-artificial-intelligence-strategy-for-nato/index.html

NATO, *Revised Artificial Intelligence Strategy* (NorAm & Europe 2024), https://www.nato.int/cps/en/natohq/official_texts_227237.htm

OAS, *Declaration and Plan of Action: Towards the Safe, Secure and Trustworthy Development and Deployment of Artificial Intelligence in the Americas, the Importance of Governance, Regulatory, and Institutional Frameworks* (Western hemisphere 2024) HTTPS://WWW.OAS.ORG/ext/en/main/calendar/event/moduleid/7596/id/731/lang/1/controller/item/action/eventdownload

OECD Hiroshima AI Process, *International Guiding Principles for All AI Actors and for Organizations Developing Advanced AI Systems* (Global 2023), https://www.soumu.go.jp/hiroshimaaiprocess/pdf/document04_en.pdf

OECD Hiroshima AI Process, *International Code of Conduct for Organizations Developing Advanced AI Systems* (Global 2023), https://www.soumu.go.jp/hiroshimaaiprocess/pdf/document05_en.pdf

OECD Hiroshima AI Process, *Towards a G7 Common Understanding of Generative AI* (Global 2023), https://www.oecd.org/en/publications/2023/09/g7-hiroshima-process-on-generative-artificial-intelligence-ai_8d19e746.html

OECD, *A Blueprint for Building National Compute Capacity for Artificial Intelligence* (Global 2023), https://www.oecd.org/en/publications/a-blueprint-for-building-national-compute-capacity-for-artificial-intelligence_876367e3-en.html

OECD, *AI Language Models* (Global 2023), https://www.oecd.org/en/publications/ai-language-models_13d38f92-en.html

OECD, *Artificial Intelligence in Science* (Global 2023), https://www.oecd.org/en/publications/artificial-intelligence-in-science_a8d820bd-en.html

OECD, *The Supply, Demand, and Characteristics of the AI Workforce Across OECD Countries* (Global 2023), https://www.oecd.org/en/publications/the-supply-demand-and-characteristics-of-the-ai-workforce-across-oecd-countries_bb17314a-en.html

OECD, *Governing with Artificial Intelligence* (Global 2024), https://www.oecd.org/en/publications/governing-with-artificial-intelligence_26324bc2-en.html

OECD, *AI, Data Governance, and Privacy* (Global 2024), https://www.oecd.org/content/dam/oecd/en/publications/reports/2024/06/ai-data-governance-and-privacy_2ac13a42/2476b1a4-en.pdf

OECD, *Assessing Potential Future Artificial Intelligence Risks, Benefits, and Policy Imperatives* (Global 2024), https://www.oecd.org/en/publications/assessing-potential-future-artificial-intelligence-risks-benefits-and-policy-imperatives_3f4e3dfb-en.html

REAIM, *Responsible AI in Military Domain*, Blueprint for Action (Global 2024), https://www.reaim2024.kr/

UN Institute for Disarmament Research, *Draft Guidelines for the Development of a National Strategy on AI in Security and Defence* (Global 2024), https://unidir.org/publication/draft-guidelines-for-the-development-of-a-national-strategy-on-ai-in-security-and-defence/

UNESCO, OECD, G7, *G7 Toolkit for Artificial Intelligence in the Public Sector* (Global 2024), https://www.oecd.org/en/publications/g7-toolkit-for-artificial-intelligence-in-the-public-sector_421c1244-en.html

UNESCO, *Foundation Models Such as ChatGPT Through the prism of the UNESCO Recommendation on the Ethics of Artificial Intelligence* (Global 2023), https://unesdoc.unesco.org/ark:/48223/pf0000385629

UNESCO, *Open Data for AI: What Now?* (Global 2023), https://unesdoc.unesco.org/ark:/48223/pf0000385841

UNESCO, *Challenging Systematic Prejudices: An Investigation into Bias Against Women and Girls in Large Language Models* (Global 2024), https://unesdoc.unesco.org/ark:/48223/pf0000388971

UNESCO, *Guidance for Generative AI in Education and Research* (Global 2023), https://unesdoc.unesco.org/ark:/48223/pf0000386693

UNESCO, *Guidelines for the Governance of Digital Platforms* (Global 2023), https://unesdoc.unesco.org/ark:/48223/pf0000387339

UNESCO, *Readiness Assessment Methodology: A Tool of the Recommendation on the Ethics of Artificial Intelligence* (Global 2023), https://unesdoc.unesco.org/ark:/48223/pf0000385198

UN, *Governing AI for Humanity* (Global 2024), https://www.un.org/sites/un2.un.org/files/governing_ai_for_humanity_final_report_en.pdf

UN Office of the High Commissioner for Human Rights, *Taxonomy of Human Rights Risks Connected to Generative AI* (Global 2024), https://www.ohchr.org/sites/default/files/documents/issues/business/b-tech/taxonomy-GenAI-Human-Rights-Harms.pdf

WHO, *Regulatory Considerations on Artificial Intelligence for Health* (Global 2023), https://iris.who.int/handle/10665/373421

WHO, *Ethics and Governance of Artificial Intelligence for Health* (Global 2024), https://iris.who.int/bitstream/handle/10665/375579/9789240084759-eng.pdf?sequence=1

World Economic Forum,
Adopting AI Responsibility:
Guidelines for Procurement of
AI Solutions by the Private
Sector (Global 2023),
https://www3.weforum.org/docs/
WEF_Adopting_AI_Responsibl
y_Guidelines_for_Procurement_
of_AI_Solutions_by_the_Private
_Sector_2023.pdf

Other Reports

Access Now, *Regulatory Mapping on Artificial Intelligence in Latin America: Regional AI Public Policy Report* (2024), https://www.accessnow.org/wp-content/uploads/2024/07/TRF-LAC-Reporte-Regional-IA-JUN-2024-V3.pdf

Ada Lovelace Institute, *Inclusive AI Governance* (2023), https://www.adalovelaceinstitute.org/report/inclusive-ai-governance/

ADAPT Centre & Science Foundation Ireland, *Artificial intelligence Ethics: An Inclusive Global Discourse* (2021), https://www.mn.uio.no/ifi/english/research/groups/is/ifip-94/proceedings-virtual-conference-2021/all-papers/rocheetal.pdf

ACT-IAC, *White Paper: Ethical Application of AI Framework* (2020), https://www.actiac.org/documents/act-iac-white-paper-ethical-application-ai-framework

African Policy Research Institute, *AI in Africa* (2022), https://afripoli.org/uploads/publications/AI_in_Africa.pdf

AI Action Summit, *International AI Safety Report* (2025), https://assets.publishing.service.gov.uk/media/679a0c48a77d250007d313ee/International_AI_Safety_Report_2025_accessible_f.pdf

AI Ethics for Peace, *Rome Call for AI Ethics* 2024), https://www.romecall.org/wp-content/uploads/2022/03/RomeCall_Paper_web.pdf

Alan Turing Institute, *International AI Safety Report 2025* (2025), https://assets.publishing.service.gov.uk/media/679a0c48a77d250007d313ee/International_AI_Safety_Report_2025_accessible_f.pdf

Algorithmic Justice League, Change from the Outside: Towards Credible Third-Party Audits of AI Systems (2023), https://unesdoc.unesco.org/in/documentViewer.xhtml?v=2.1.196&id=p::usmarcdef_0000384787&file=/in/rest/annotationSVC/DownloadWatermarkedAttachment/

Algorithm Watch, *Algorithmic Transparency and Accountability in the World of Work* (2023), https://algorithmwatch.org/en/wp-content/uploads/2023/02/2023_AlgorithmWatch_ITUC_Report.pdf

Amnesty International, *Automated Apartheid* (2023), https://www.amnesty.org/en/documents/mde15/6701/2023/en/

Association for Computing Machinery, *Safer Algorithmic Systems* (2023), https://dl.acm.org/doi/pdf/10.1145/3582277

Asociación por los Derechos Civiles, IPANDETEC, et al., *Public Consultation Ethics and Data Protection in Artificial Intelligence: Continuing the Debate. A Contribution from Latin America and the Caribbean* (2019), https://www.ipandetec.org/centroamerica/public-consultation-ethics-data-protection-artificial-intelligence/

BEUC, *Proposal for an AI Liability Directive* (2023), https://www.beuc.eu/sites/default/files/publications/BEUC-X-2023-050_Proposal_for_an_AI_Liability_Directive.pdf

Berkman Klein Center, *Principled Artificial Intelligence: Mapping Consensus in Ethical and Rights-based Approaches to Principles for AI* (2020), https://papers.ssrn.com/sol3/papers.cfm?abstract_id=3518482

Berkman Klein Center, *Vectors of AI Governance - Juxtaposing the U.S. Algorithmic Accountability Act of 2022 with The EU Artificial Intelligence Act* (2023), https://papers.ssrn.com/sol3/papers.cfm?abstract_id=4476167

BEUC, *Artificial Intelligence: What Consumers Say* (2020), https://www.beuc.eu/sites/default/files/publications/beuc-x-2020-078_artificial_intelligence_what_consumers_say_report.pdf

CAIDP, *Artificial Index and Democratic Values Index* (2024), https://www.caidp.org/reports/

CAIDP, *Update – Volume 7* (2025), https://www.caidp.org/caidp-update/volume-7-2025/

CAIDP, *Making the AI Act work: How civil society can ensure Europe's new regulation serves people & society*, European AI and Society Fund (2023), https://europeanaifund.org/wp-content/uploads/2023/10/041023-FINAL-for-publication-EAISF-AIA-implmentation-report.pdf

Centre for Internet & Society India, *AI for Healthcare: Understanding Data Supply Chain and Auditability in India* (2024), https://cis-india.org/internet-governance/blog/ai-for-healthcare-understanding-data-supply-chain-and-auditability-in-india

Center for New American Security, *Future-Proofing Frontier AI Regulation* (2024), https://www.cnas.org/publications/reports/future-proofing-frontier-ai-regulation

CFR, *DeepSeek: Making Sense of the Reaction—and Overreaction* (2025), https://www.cfr.org/article/deepseek-making-sense-reaction-and-overreaction

CIFAR, *AICan: The Impact of the Pan-Canadian AI Strategy* (2023), https://cifar.ca/wp-content/uploads/2023/11/aican-impact-2023-eng.pdf

Corporate Europe Observatory, *The Lobbying Ghost in the Machine: Big Tech's Covert Defanging of the AI Act* (2023), https://corporateeurope.org/sites/default/files/2023-03/The%20Lobbying%20Ghost%20in%20the%20Machine.pdf

Cybersecurity Authorities of Australia, UK, France, Israel, et al., *Guidelines for Secure AI Development* (2023), https://www.ncsc.gov.uk/files/Guidelines-for-secure-AI-system-development.pdf

Cybersecurity Authorities of Canada, UK, US, & New Zealand, *Deploying AI Systems Securely* (2024), https://media.defense.gov/2024/Apr/15/2003439257/-1/-1/0/CSI-DEPLOYING-AI-SYSTEMS-SECURELY.PDF

Data & Society, *Governing with Algorithmic Impact Assessments: Six Observations* (2020), https://papers.ssrn.com/sol3/papers.cfm?abstract_id=3584818

Derechos Digitales, *Latin America in a Glimpse 2024: Reflections for a Community-Based, Feminist AI* (2024), https://www.derechosdigitales.org/wp-content/uploads/Glimpse_2024_ENG.pdf

Encode Justice, *AI 2030* (2024), https://2030.encodeai.org/

Digital Asia Hub, *AI in Asia* (2022), https://www.digitalasiahub.org/2022/12/13/ai-in-asia/

France & China, *Joint Declaration Between the French Republic and the People's Republic of China on Artificial Intelligence and the Governance of Global Issues* (2024), https://www.elysee.fr/emmanuel-macron/2024/05/06/declaration-conjointe-entre-la-republique-francaise-et-la-republique-populaire-de-chine-sur-lintelligence-artificielle-et-la-gouvernance-des-enjeux-globaux

Freedom House, *The Repressive Power of Artificial Intelligence* (2023), https://freedomhouse.org/report/freedom-net/2023/repressive-power-artificial-intelligence

Encode Justice, *Bridging the International AI Governance Divide: Key Strategies for Including the Global South* (2025), https://encodeai.org/wp-content/uploads/2025/01/Encode-Justice-Report_Safety-Summit_Global-North.pdf

European Artificial Intelligence & Society Fund, *Report: Strategic actions for philanthropy and civil society on the enforcement of the EU AI Act* (2024), https://europeanaifund.org/news publications/report-strategic-actions-for-philanthropy-and-civil-society-on-the-enforcement-of-the-eu-ai-act/

European Law Institute, *ELI Guiding Principles and Model Rules on Algorithmic Contracts* (2025), https://www.europeanlawinstitute.eu/projects-publications/current-projects/current-projects/eli-guiding-principles-and-model-rules-on-algorithmic-contracts/

European Law Institute, *EU Consumer Law and Automated Decision-Making (ADM): Is EU Consumer Law Ready for ADM* (2023),

https://www.europeanlawinstitut
e.eu/fileadmin/user_upload/p_eli
/Publications/ELI_Interim_Repo
rt_on_EU_Consumer_Law_and_
Automated_Decision-
Making.pdf

European Law Institute, *ELI
Model Rules on Impact
Assessment of Algorithmic
Decision-Making Systems Used
by Public Administration*
(2022),
https://www.europeanlawinstitut
e.eu/projects-
publications/publications/eli-
model-rules-on-impact-
assessment-of-algorithmic-
decision-making-systems-usey-
by-public-adminstration/

European Law Institute, *Guiding
Principles for Automated
Decision-Making in the
EU* (2022), https://www.europea
nlawinstitute.eu/fileadmin/user_
upload/p_eli/Publications/ELI_I
nnovation_Paper_on_Guiding_P
rinciples_for_ADM_in_the_EU.
pdf

Future of Privacy Forum, *AI
Governance Behind the Scenes*
(2024), https://fpf.org/wp-
content/uploads/2024/12/FPF-
AI-Governance-Behind-the-
Scenes-2024.pdf

Future of Life Institute, *Safety
Standards Delivering*

*Controllable and Beneficial AI
Tools* (2025),
https://futureoflife.org/document
/safety-standards-delivering-
controllable-and-beneficial-ai-
tools/

Fundación Vía Libre,
*Generative AI errors are
different from human errors*
(2024),
https://www.vialibre.org.ar/los-
errores-de-la-ia-generativa-son-
distintos-a-los-errores-humanos/

Future Society, *What are your
ideas for shaping AI to serve the
public good?* (2024),
https://thefuturesociety.org/wp-
content/uploads/2024/12/AI-
Action-Summit-Consultation-
Report.pptx-1.pdf

Homo Digitales, *Artificial
Intelligence Act: Analysis of
provisions on the prohibited
practices of Article 5 of
Regulation 2024/1689* (2024),
https://homodigitalis.gr/posts/13
3791/

Humanising Machine
Intelligence, *HMI Policy Paper:
Legal Audit of AI in the Public
Sector* (2021),
https://static1.squarespace.com/s
tatic/5c105643ec4eb7d1a8c68c9
c/t/628ae3c3e1a7ce73c7d7cb01/
1653269450230/Legal+Audit+o
f+AI+in+the+Public+Sector.pdf

IEEE, *The Impact of Technology in 2025* (2024), https://transmitter.ieee.org/iot-2025/?_gl=1*1ry7xol*_gcl_au*MTQ4MTY4ODY4Ny4xNzM5MTk2ODI2

International Network of AI Safety Institutes, *Joint Statement on Risk Assessment of Advanced AI Systems* (2024), https://www.nist.gov/system/files/documents/2024/11/20/Joint%20Statement%20on%20Risk%20Assessment%20of%20Advanced%20AI%20Systems.pdf

International Association for Safe & Ethical AI, *IASEAI Issues Call to Action for Lawmakers, Academics, and the Public Ahead of AI Summit in Paris* (2025), https://www.iaseai.org/conference/statement

International Bar Association & CAIDP, *The Future is Now: Artificial Intelligence and the Legal Profession* (2024), https://www.ibanet.org/document?id=The-future-is-now-artificial-intelligence-legal-profession

KICTAnet, *Crafting Kenya's Artificial Intelligence Destiny: A Roadmap to Responsible AI Innovation* (2024),

https://www.kictanet.or.ke/crafting-kenyas-ai-destiny-a-roadmap-to-responsible-innovation/

Machine Intelligence Research Institute, Mechanisms *to Verify International Agreements About AI Development* (2024), https://techgov.intelligence.org/research/mechanisms-to-verify-international-agreements-about-ai-development

MIT Schwarzman College of Computing, *A Framework for U.S. AI Governance: Creating a Safe and Thriving AI Sector* (2023), https://computing.mit.edu/wp-content/uploads/2023/11/AIPolicyBrief.pdf

Nigerian Bar Association, *Guidelines for the Use of Artificial Intelligence in the Legal Profession in Nigeria* (2024), https://nbaslp.org/wp-content/uploads/2024/04/Guidelines-for-the-Use-of-Artificial-Intelligence-in-the-Nigerian-Legal-Profession.pdf

Paradigm Initiative, *A Human Centric Artificial Intelligence Policy in Nigeria* (2023), https://paradigmhq.org/report/a-human-centric-artificial-intelligence-policy-in-nigeria/

Partnership on AI, *Fairer Algorithmic Decision-making and its Consequences* (2021), https://partnershiponai.org/paper/fairer-algorithmic-decision-making-and-its-consequences/

Panoptykon, *YouTube Algorithm: What Does the Platform Declare? We Read the Report* (2025), https://panoptykon.org/youtube-algorytm-analiza-ryzyk-raport

Reporters without Borders, *Paris Charter on AI and Journalism* (2023), https://rsf.org/sites/default/files/medias/file/2023/11/Paris%20charter%20on%20AI%20in%20Journalism.pdf

Stanford Institute for Human Centered AI, *AI Index Report* (2024), https://aiindex.stanford.edu/report/

UK Law Society, *Algorithm Use in the Criminal Justice System Report* (UK 2019), https://www.lawsociety.org.uk/topics/research/algorithm-use-in-the-criminal-justice-system-report

UK Competition & Markets Authority, European Commission, U.S. DOJ, U.S. FTC, *Joint Statement on Competition in Generative AI Foundation Models and AI Products* (2024), https://competition-policy.ec.europa.eu/document/download/79948846-4605-4c3a-94a6-044e344acc33_en?filename=20240723_competition_in_generative_AI_joint_statement_COMP-CMA-DOJ-FTC.pdf

Organizations

Ada Lovelace Institute
Nuffield Foundation
100 St John Street
London, EC1M 4EH
hello@adalovelaceinstitute.org
https://www.adalovelaceinstitute
.org/

AkiraChix
616 Korongo Rd,
Nairobi, Kenya
info@akirachix.com
https://akirachix.com/

AI Transparency Institute
https://aitransparencyinstitute.co
m/

Alan Turing Institute
96 Euston Road
London NW1 2DB
United Kingdom
https://turing.ac.uk

Algorithmic Justice League
https://www.ajl.org/

Asociación por los Derechos Civiles
Tucumán 924 8 ° (C1049AAT)
Autonomous City of Buenos
Aires, Buenos Aires
Argentina
54-11-5236-0555
adc@adc.org.ar
https://adc.org.ar/

Berkman Klein Center for Internet & Society at Harvard University
1557 Massachusetts Avenue, 5th
Floor,
Cambridge, MA 02138
617- 495-7547
https://cyber.harvard.edu

CareMessage
San Francisco, CA
https://www.caremessage.org/

Carnegie Council for Ethics in International Affairs
Merrill House
170 East 64th Street
New York, NY 10065-7478
212-838-4122
info@cceia.org
https://www.carnegiecouncil.org

Center for AI and Digital Policy
1100 13th St NW
Ste 800
Washington, DC 20005
https://www.caidp.org/

Center for Internet & Society India
#32, 1st Floor, 2nd Block,
Austin Town, Viveka Nagar,
Bengaluru, Karnataka 560047.
https://cis-india.org/

Center for Research, Information, Technology and Advanced Computing (CRITAC)
Cape Coast
Ghana
233-20-269-8355
http://critacghana.com/critac/

Canadian Institute for Advanced Research (CIFAR)
MaRS Centre, West Tower
661 University Ave., Suite 505
Toronto, ON M5G 1M1
Canada
416-971-4251
info@cifar.ca

Coding Rights
Brazil
https://codingrights.org/en/

Colegio Profesional de Ingenieros Técnicos en Informática de Andalucía
PO Box No. 56 - 04080
Almería
Spain
950-70-00-45
https://www.cpitia.org/

Council of Europe
Avenue de l'Europe F-67075
Strasbourg Cedex, France
33-(0)3-88-41-20-00
https://www.coe.int/en/web/artificial-intelligence/

Council on Foreign Relations
58 East 68th Street
New York, NY 10065
212-434-9400
https://www.cfr.org/

CyberPeace Insitute
Avenue de Sécheron 15
1202 GENÈVE
Switzerland
https://cyberpeaceinstitute.org/contact/

Data Ethics 4 All
https://dataethics4all.org/

Data Ethics EU and InTouch AI EU
Denmark
info@dataethics.eu
https://dataethics.eu/t

Derechos Digitales
Diagonal Paraguay 458
Second floor
8330051 Santiago
Chile
56-2-2702-7108
https://www.derechosdigitales.org/

Digital Futures Lab
1275 Grande Carona, Aldona
Goa - 403508, India
hello@digitalfutureslab.in
https://digitalfutureslab.in/

Digital Asia Hub
Hong Kong
dlewis@digitalasiahub.org
https://www.digitalasiahub.org/

**El Centro de
Estudios Estratégicos de
Derecho de la Inteligencia
Artificial (CEEDIA)**
https://www.ceedia.org/en/about
-us/

**Electronic Privacy
Information Center**
1519 New Hampshire Avenue
NW
Washington, DC 20036
1-202-483-1140
info@epic.org
https://epic.org/algorithmic-
transparency/

Encode Justice
https://encodeai.org/

Enfold Proactive Health Trust
No 42, 3rd main,
1st cross, Domlur 2nd Stage,
Bengaluru- 560071, India
+91-99000-94251
info@enfoldindia.org
https://enfoldindia.org/

**European Artificial
Intelligence & Society Fund**
Network of European
Foundations
Philanthropy House
Rue Royale 94
1000 Brussels
Belgium
info@europeanaifund.org
https://europeanaifund.org/

**European Data Protection
Board**
Rue Montoyer 30
B-1000 Brussels
Belgium
https://www.edpb.europa.eu/edp
b_en

**European Data Protection
Supervisor**
Rue Montoyer 30
B-1000 Brussels
Belgium
32-2-283-19-00
edps@edps.europa.eu
https://www.edps.europa.eu/_en

European Law Institute
Schottenring 16, Top 175
1010 Vienna
Austria
43-1-4277-221-01
secretariat@europeanlawinstitut
e.eu
https://www.europeanlawinstitut
e.eu/

The Future Society
Paris, France
https://thefuturesociety.org/

G20
https://g20.org/

Gender Rights In Tech (GRIT)
Fish Hoek, WC, ZAF
South Africa
info@grit-gbv.org
https://www.grit-gbv.org/

Global AI Ethics Consortium
TUM School of Socieal Sciences
and Technology
Arcisstrasse 21
D-80333 Munich
Germany
ieai@sot.tum.de
https://www.ieai.sot.tum.de/glob
al-ai-ethics-consortium/

**Global Partnership on
Artificial Intelligence**
https://oecd.ai/en/

Global Privacy Assembly
secretariat@globalprivacyassem
bly.org
https://globalprivacyassembly.or
g/

Generation
Washington, D.C.
https://www.generation.org/cont
act-generation/

Hack Foundation
West Hollywood, CA
hcb@hackclub.com
1-800-625-HACK
https://the.hackfoundation.org/

Hiperderecho
Av. Santo Toribio 143, Piso 2
San Isidro, Lima 15073
Lima, Perú
hola@hiperderecho.org
https://hiperderecho.org/

iamtheCODE
Africa
https://www.iamthecode.org/

Instituto Educa Digital
Brazil
contato@educadigital.org.br
https://educadigital.org.br/

in SupplyHealth
Rhapta Road, Nairobi Kenya
254 739 393 844
info@insupplyhealth.com
https://insupplyhealth.com/conta
ct-us/

**International Association for
Safe & Ethical AI**
501 W Broadway
Suite 1540
San Diego, CA 92101
info@iaseai.org
https://www.iaseai.org/

International Bar Association
Chancery House
53-64 Chancery Lane

London
WC2A 1QS
United Kingdom
44-(0)20-7842-0090
iba@int-bar.org
https://www.ibanet.org/

Internet Lab
Av. Ipiranga, 344 - República
São Paulo - SP, 01046-001
Brazil
https://internetlab.org.br/en/

Ipandetec
Spaces Panama - Plaza 2000
Panama City
Panama
ipandetec@gmail.com
http://www.ipandetec.org/

Jacaranda Health
788 Jabavu Rd Nairobi Nairobi
County, Kenya
hello@jacarandahealth.org
+254 748 322 535
https://jacarandahealth.org/about

Jembi
Unit 3B, 5A-C,
Tokai on Main, 382 Main Road,
Tokai, Cape Town, South Africa
Tel: +27 21 701 0939
info@jembi.org
https://www.jembi.org/

Karisma Foundation
Diagonal 61B # 20-39
Bogota, Colombia
+57 (1) 257 6946
contacto@karisma.org.co
https://web.karisma.org.co/

KICTAnet
Kenya
https://www.kictanet.or.ke/

Karya
Dover, DE
operations@karya.in
https://karya.in/

Lawyer's Hub
Kenya
+254 784 840 228
info@lawyershubafrica.com
https://www.lawyershub.org/

**MIT Computer Science &
Artificial Intelligence
Laboratory**
32 Vassar St, Cambridge MA
02139
news@csail.mit.edu
https://www.csail.mit.edu/

OECD AI Policy Observatory
gpai@oecd.org
https://oecd.ai/en/

Ona Systems, Inc
One Padmore Place, 2nd floor
George Padmore Ln
Nairobi, Kenya
+254 710 632 598
https://ona.io/home/

OdiseIA
Paseo de Juan XXIII
Madrid
CP 28040
contacto@odiseia.org
https://www.odiseia.org/en/quie
nes-somos

Open Data Institute
4th Floor, Kings Place, 90 York
Way, London N1 9AG
https://theodi.org/about-the-odi/

Pacific Community
95 Promenade Roger Laroque
BP D5
98848 Noumea
New Caledonia
+687 26 20 00
spc@spc.int
https://www.spc.int/

**Red en Defensa de los
Derechos Digitales (R3D)**
Mexico
https://r3d.mx/

**Science For Africa Foundation
(SFA Foundation)**
Riverside Drive
Chiromo Lane, Nairobi,
Kenya
+254 705 199 199
info@scienceforafrica.foundatio
n
https://scienceforafrica.foundatio
n/

The Tech Interactive
201 S. Market St.
San Jose, CA 95113-2008
1-408-294-8324
info@thetech.org
https://www.thetech.org/

TEDIC
Paraguay
https://www.tedic.org/

Transform Health
Basel, CHE
https://transformhealthcoalition.
org/

Trusting News
Sarasota, FL
info@TrustingNews.org
https://trustingnews.org/about-
us/

UNESCO
7 Pl. de Fontenoy-Unesco
75007 Paris, France
https://www.unesco.org/en

**United Nations High-Level
Advisory Body on Artificial
Intelligence**
1 UN Plaza
New York, New York 10017
https://www.un.org/digital-
emerging-technologies/ai-
advisory-body

UN Human Rights Office
Palais Wilson
52 rue des Pâquis
CH-1201 Geneva
Switzerland
41-22-917-9220
ohchr-InfoDesk@un.org
https://www.ohchr.org/en/ohchr
_homepage

Voices of Venezuela
Calle 69 A # 9 – 66, Bogota
D.C, Colombia, 110111
+57 300 1496885
https://voicesofvenezuela.com/

World Data Lab
Vienna, 9, Austria
+43 676 619 4432
contact@worlddatalab.org
https://worlddatalab.org/

World Leadership Alliance
Club de Madrid
C. Mayor, 69,
Planta 1, 28013
Madrid, Spain
34-911-548 -30
clubmadrid@clubmadrid.org
https://clubmadrid.org/

CAIDP Books and Articles

Doaa Abu-Elyounes, *Contextual fairness: A legal and policy analysis of algorithmic fairness*, University of Illinois Journal of Tech and Policy (2020), https://papers.ssrn.com/sol3/Delivery.cfm?abstractid=3478296

Doaa Abu-Elyounes, *"Computer Says No!": The Impact of Automation on the Discretionary Power of Public Officers,* Vanderbilt Journal of Entertainment and Technology Law (2020), https://scholarship.law.vanderbilt.edu/cgi/viewcontent.cgi?article=1041&context=jetlaw

Doaa Abu-Elyounes, *Bail or Jail? Judicial versus Algorithmic Decision-Making in the Pretrial System*, Columbia Science and Technology Law Review (2020), https://journals.library.columbia.edu/index.php/stlr/article/download/6838/3595

Elena Abrusci, et al., *The Universal Declaration of Human Rights at 70: Putting Human Rights at the heart of the design, development and deployment of artificial intelligence*, The Human Rights, Big Data and Technology Project, University of Essex, (2018), https://www.hrbdt.ac.uk/the-universal-declaration-of-human-rights-at-70-putting-human-rights-at-the-heart-of-the-design-development-and-deployment-of-artificial-intelligence/

AI Now Institute, *Prohibiting Surveillance Prices and Wages* (2025). https://ainowinstitute.org/publication/ai-now-coauthors-report-on-surveillance-prices-and-wages

AI Now Institute, *What Can We Learn From the FDA Model for AI Regulation*? (2024). https://ainowinstitute.org/publication/what-can-we-learn-from-the-fda-model-for-ai-regulation

AI Now Institute, *Regulating Biometrics Global Approaches and Urgent Questions (*2020). https://ainowinstitute.org/publication/regulating-biometrics-global-approaches-and-open-questionsBiometric

Beena Ammanath, *Reimagining Education and Equity with AI*, Linkedin (2025), https://www.linkedin.com/pulse/reimagining-education-equity-ai-beena-ammanath-sr0jc/

Beena Ammanath, *Beena Ammanath on importance of trustworthiness and diversity in*

AI, Youtube (2023), https://www.youtube.com/watch?v=jFOsGHCUYe0

Beena Ammanath, *Trustworthy AI : A business guide for navigating Trust and Ethics in AI*, Wiley (2020), https://www.amazon.com/Trustworthy-AI-Business-Navigating-Ethics-ebook/dp/B09VSB88XQ/

Almudena Arpón de Mendívil, Marc Rotenberg, et al., *The Future is Now: Artificial Intelligence and the Legal Profession*, Journal of AI Law and Regulation (2024), https://aire.lexxion.eu/list/articles/author/Arp%C3%B3n%20de%20Mend%C3%ADvil%20Aldama,%20Almudena

Irena Barkane & Lolita Buka, "Prohibited AI surveillance practices in the Artificial Intelligence Act: promises and pitfalls in protecting fundamental rights" in *Critical Perspectives on Predictive Policing: Anticipating Proof?*, Edward Elgar Publishing (2025), https://doi.org/10.4337/9781035323036.00011

Francesca Bignami, *Artificial intelligence accountability of public administration*, The American Journal of Comparative Law (2022),

https://academic.oup.com/ajcl/article/70/Supplement_1/i312/6596541

Abeba Birhane, et al., *The Dark Side of Dataset Scaling: Evaluating Racial Classification in Multimodal Models*, Computers and Society (2024), https://arxiv.org/abs/2405.04623

Abeba Birhane, Jelle van Dijk, & Frank Pasquale *Debunking robot rights metaphysically, ethically, and legally*, Computers and Society (2024), https://arxiv.org/abs/2404.10072

Abeba Birhane, et al., *AI auditing: The broken bus on the road to AI accountability*, Computers and Society (2024), https://arxiv.org/abs/2401.14462

Abeba Birhane, et al., *Into the laion's den: Investigating hate in multimodal datasets*, Computers and Society (2024), https://arxiv.org/abs/2311.03449

Linda Bonyo, Lawyers Hub. *Artificial Intelligence and the Future of Judicial Systems in Africa Report* (2024). https://www.lawyershub.org/Digital%20Resources/Reports/AI%20-%20the%20Future%20of%20Judicial%20Systems%20in%20Africa%202024.pdf

Linda Bonyo, Lawyers Hub, *5 Years of AI Regulation in Africa* (2024). https://www.lawyershub.org/Digital%20Resources/Reports/5%20Years%20of%20AI%20Regulation%20in%20Africa.pdf

Linda Bonyo, Lawyers Hub. *Catalyzing Artificial Intelligence for Women's empowerment in Africa* (2024), https://www.lawyershub.org/Digital%20Resources/Articles/Catalysing%20Artificial%20Intelligence%20for%20Womens%20empowerment%20in%20Africa.pdf

Linda Bonyo, Lawyers Hub. *Africa Privacy Report 2023/2024* (2023). https://www.lawyershub.org/Digital%20Resources/Africa%20Privacy%20Report%202023-24/Africa%20Privacy%20Report%202023-2024.pdf

Anu Bradford, *The False Choice Between Digital Regulation and Innovation*, Northwestern University Law Review (2024), http://dx.doi.org/10.2139/ssrn.4753107

Anu Bradford, *Whose AI Revolution?*, Project Syndicate (2023),

https://www.project-syndicate.org/onpoint/ai-regulation-us-eu-china-challenges-opportunities-by-anu-bradford-2023-09

Anu Bradford, *The Race to Regulate Artificial Intelligence.* Foreign Affairs (2023). https://www.foreignaffairs.com/united-states/race-regulate-artificial-intelligence

Virginia Dignum, *Responsible Artificial Intelligence: How to Develop and Use AI in a Responsible Way*, Springer International Publishing (2019). https://doi.org/10.1007/978-3-030-30371-6

Virginia Dignum, et al., *AI4People—An Ethical Framework for a Good AI Society: Opportunities, Risks, Principles, and Recommendations*, Minds and Machines (2018), https://doi.org/10.1007/s11023-018-9482-5

Virginia Dignum & Andreas Theodorou, *Towards ethical and socio-legal governance in AI*, Nature Machine Intelligence (2020), https://doi.org/10.1038/s42256-019-0136-y

Ivan Fong, et al., *Artificial Intelligence: Legal Issues, Policy, and Practical Strategies*, ABA (2024), https://www.americanbar.org/products/inv/book/443758527/

Maura Grossman, Paul Grimm, & Gordon Cormack, Artificial Intelligence as Evidence, *Northwestern Journal of Technology and Intellectual Property* (2021), https://scholarlycommons.law.northwestern.edu/njtip/vol19/iss1/2

Hiroki Habuka, David de la Osa, *Shaping Global AI Governance: Enhancements and Next Steps for the G7 Hiroshima AI Process*, Center for Strategic and International Studies (2024), https://www.jstor.org/stable/resrep60369

David Evan Harris, *Generative AI, Democracy and Human Rights*, Centre for International Governance Innovation (2025), https://www.cigionline.org/publications/generative-ai-democracy-and-human-rights/

David Evan Harris, *Tech Companies Pledged to Protect Elections from AI — Here's How They Did*, Brennan Center for Justice (2025), https://www.brennancenter.org/our-work/research-reports/tech-companies-pledged-protect-elections-ai-heres-how-they-did

David Evan Harris, *Philanthropy's Urgent Opportunity to Create the Interim International AI Institution*, Centre for International Governance Innovation (2024) https://www.cigionline.org/publications/philanthropys-urgent-opportunity-to-create-the-interim-international-ai-institution/

David Evan Harris, *Not Open and Shut: How to Regulate Unsecured AI*, Centre for International Governance Innovation (2024), https://www.cigionline.org/articles/not-open-and-shut-how-to-regulate-unsecured-ai/

Gry Hasselbalch & Carolina Aguerre, *The EU Human-Centric Approach to AI: Foundational Elements for a Global Framework*, DataEthics.eu (2025), https://dataethics.eu/wp-content/uploads/2024/10/EU-HCA-Baseline-Study-for-publication.pdf

Gry Hasselbalch, *Data Ethics of Power: A Human Approach in the Big Data and AI Era*, Edward Elgar Publishing (2021), https://www.e-elgar.com/shop/usd/data-ethics-of-power-9781802203103.html?srsltid=AfmBOorhsSil6idkFBKV269UICJZpkBj4WMUWyibFcxjd64mJLOqjrgW

Merve Hickok, *From Trustworthy AI Principles to Public Procurement Practices*, De Gruyter (2024), https://www.degruyter.com/document/doi/10.1515/9783111250182/html?recommended=sidebar&lang=de

Merve Hickok, *Don't Let Governments Buy AI Systems That Ignore Human Rights,* Issues in Science and Technology (2024), https://www.proquest.com/openview/9ba4a45ef14a0cec0088eeec14667167/1?pq-origsite=gscholar&cbl=32581

Merve Hickok, *A policy primer and roadmap on AI worker surveillance and productivity scoring tools*, AI and Ethics (2023), https://link.springer.com/article/10.1007/s43681-023-00275-8

Merve Hickok, *Towards intellectual freedom in an AI Ethics Global Community*, AI and Ethics (2021), https://link.springer.com/article/10.1007/s43681-021-00052-5

Margaret Hu, "Biometric Surveillance and Big Data Governance" in *The Cambridge Handbook of Surveillance Law*, Cambridge University Press (2017), https://doi.org/10.1017/9781316481127.006

Margaret Hu, Eliott Behar, & Davi Ottenheimer, *National Security And Federalizing Data Privacy Infrastructure For AI Governance*, Fordham Law Review (2024), https://ir.lawnet.fordham.edu/flr/vol92/iss5/4/

Aziz Huq, *The Geopolitics of Digital Regulation*, University of Chicago (2024), https://doi.org/10.2139/ssrn.4946481

Bulelani Jill, *Africa: regulate surveillance technologies and personal data*, Nature (2022), https://pubmed.ncbi.nlm.nih.gov/35851875/

Len Kennedy, *Society's New Frontier— Cybersecurity, Privacy and Online Expression*, Cornell International Law Journal (2020), https://community.lawschool.cornell.edu/wp-content/uploads/2021/03/Kennedy-Introduction.pdf

Lorraine Kisselburgh & Marc Rotenberg, *Next Steps on the AI Bill Of Rights*, Washington Spectator (Nov. 2021), https://washingtonspectator.org/author/lorraine-marc/

Lorraine Kisselburgh, *VizScribe: A visual analytics approach to understand designer behavior*, International Journal of Human-Computer Studies (2017), https://lkisselburgh.net/pubs/471

Jan Kleijssen, "The Council of Europe's Convention on Artificial Intelligence and Human Rights, Democracy and the Rule of Law" in *AI Governance and Liability in Europe: A Primer* (2025), https://law-store.wolterskluwer.com/s/product/ai-governance-and-liability-in-europe-a-primer/

Jan Kleijssen, *l'Europe et intelligence artificielle, Les droits de l'homme, l'État de droit et la démocratie face aux défis du développement et de l'utilisation de l'intelligence artificielle*, Coe.int (2019), https://rm.coe.int/2019-ai-dbf-jk/168093c8e5

Eleni Kosta, *Algorithmic state surveillance: Challenging the notion of agency in human rights*, Regulation & Governance (2020), https://doi.org/10.1111/rego.12331

Katja Langenbucher, "Artificial Intelligence and Financial Services," in *The Cambridge Handbook of the Law, Ethics and Policy of Artificial Intelligence,* Cambridge University Press (2025), https://doi.org/10.1017/9781009367783.020

Seth Lazar, *Governing the Algorithmic City*, Philosophy & Public Affairs (2025). https://doi.org/10.1111/papa.12279

Gianclaudio Malgieri & Frank Pasquale, *Generative AI, explainability, and score-based natural language processing in benefits administration*, Journal of Cross-Disciplinary Research in Computational Law (2024), https://journalcrcl.org/crcl/article/view/59

Gianclaudio Malgieri & Margot Kaminski, *Impacted Stakeholder Participation in AI and Data Governance*, Social Science Research Network (2024), https://papers.ssrn.com/abstract=4836460

Gianclaudio Malgieri & Margot Kaminski, *Algorithmic impact assessments under the GDPR: Producing multi-layered explanations*, International Data Privacy Law (2021), https://doi.org/10.1093/idpl/ipaa020

Gianclaudio Malgieri, *Automated decision-making in the EU Member States: The right to explanation and other "suitable safeguards" in the national legislations*, Computer Law & Security Review (2019), https://doi.org/10.1016/j.clsr.2019.05.002

Alessandro Mantelero, *Beyond Data: Human Rights, Ethical and Social Impact Assessment in AI*, Springer Nature (2022), https://doi.org/10.1007/978-94-6265-531-7

Alessandro Mantelero & Maria Samantha Esposito, *An evidence-based methodology for human rights impact assessment (HRIA) in the development of AI data-intensive systems*, Computer Law & Security Review (2021), https://doi.org/10.1016/j.clsr.2021.105561

Alessandro Mantelero, *AI and Big Data: A blueprint for a human rights, social and ethical impact assessment*, Computer Law & Security Review (2018), https://doi.org/10.1016/j.clsr.2018.05.017

Valsamis Mitsilegas & Maria Bergström, *EU Law in the Digital Age*, Bloomsbury (2025), https://www.bloomsbury.com/us/eu-law-in-the-digital-age-9781509981182

Valsamis Mitsilegas, "The Digital Border and the Rule of Law: Lessons from the Establishment of the European Travel Information and Authorisation System (ETIAS)," in *EU Law in the Digital Age*. Bloomsbury (2025), https://www.bloomsbury.com/us/eu-law-in-the-digital-age-9781509981182/

Pablo Molina, "Protecting Data Privacy in Education" in *Privacy in the Modern Age: The Search for Solutions* (2015), https://www.amazon.com/Privac y-Modern-Age-Search-Solutions/dp/1620971070

Maria Murphy, *Will we have a say in Ireland's artificial intelligence strategy?*, RTE (2021), https://www.rte.ie/brainstorm/20 21/0218/1197882-ireland-artificial-intelligence-strategy-ai/

Maria Murphy, *Algorithmic Surveillance: The Collection Conundrum*, International Review of Law, Computers & Technology (2017), https://mural.maynoothuniversit y.ie/id/eprint/11681/

Maria Murphy, et al., *Algorithmic Governance: Developing a Research Agenda through the Power of Collective Intelligence,* Big Data & Society (2017), https://mural.maynoothuniversit y.ie/id/eprint/11679/

Angella Ndaka, Michael Zimba, & Maha Jouini, "Context-Aware Africa-Led Designing of Responsible Artificial Intelligence Technologies," *Trustworthy AI: African Perspectives*, Palgrave Macmillan (2025), https://doi.org/10.1007/ 978-3-031-75674-0_7

Chinasa T. Okolo, *Re-envisioning AI safety through global majority perspectives*, TechTank (2025), https://www.brookings.edu/articl es/a-new-writing-series-re-envisioning-ai-safety-through-global-majority-perspectives/

Chinasa T. Okolo, et al., *Options and Motivations for International AI Benefit Sharing,* Centre for the Governance of AI (2025), https://www.governance.ai/resea rch-paper/options-and-motivations-for-international-ai-benefit-sharing

Chinasa T. Okolo, et al., *Voice and Access in AI: Global AI Majority Participation in Artificial Intelligence Development and Governance*, Oxford Martin AI Governance Initiative (2024), https://www.oxfordmartin.ox.ac. uk/publications/voice-and-access-in-ai-global-ai-majority-participation-in-artificial-intelligence-development-and-governance

Chinasa T. Okolo, "Operationalizing Comprehensive Data Governance in Africa," in *African Journal of Sustainable Development (AJSD) Special Issue: "Redefining African*

Futures: The State, Resilience, and Pathways to Progress (2024), https://chinasatokolo.github.io/files/Okolo_AJSD_2024.pdf

Derya Ozkul, *Automating Immigration and Asylum: The Use of New Technologies in Migration and Asylum Governance in Europe*, Refugee Studies Centre, University of Oxford (2023), https://www.rsc.ox.ac.uk/files/files-1/automating-immigration-and-asylum_afar_9-1-23.pdf

Deya Ozkul & Francesca Palmiotto, *Climbing a Wall: Strategic Litigation Against Automated Systems in Migration and Asylum*, German Law Journal (2024), https://www.cambridge.org/core/journals/german-law-journal/article/climbing-a-wall-strategic-litigation-against-automated-systems-in-migration-and-asylum/8B6B45F793CC59C16B33E0BA58080355

Ursula Pachl, *How far will we — and the EU — let AI go?*, Eurobserver (2023), https://euobserver.com/digital/ar5440ab9e

Shobita Parthasarathy & Jared Katzman, "Bringing Power In: Rethinking Equity Solutions for AI," in *Realizing the Promise and Minimizing the Perils of Artificial Intelligence for the Scientific Community*, University of Pennsylvania Press (2024), https://www.scilit.com/publications/184e1da034a9f00a294c8c39d447ce68

Shobita Parthasarathy & Jared Katzman, *Bringing Communities In, Achieving AI for All*, Issues in Science and Technology (2024), https://doi.org/10.58875/SLRG2529

Shobita Parthasarathy, *Can Innovation Serve the Public Good?*, Boston Review (2023), https://www.bostonreview.net/articles/can-innovation-serve-the-public-good/

Shobita Parthasarathy, *A Tale of Two Perspectives on Innovation and Global Inequality*, Engaging Science, Technology, and Society (2023), https://estsjournal.org/index.php/ests/article/view/1369

Ceyhun Necati Pehlivan, Nikolaus Forgó, & Peggy Valcke, *AI Governance and Liability in Europe*, Wolters

Kluwer (2025), https://law-store.wolterskluwer.com/s/product/ai-governance-and-liability-in-europe-a-primer/01tPg000005OKCRIA4

Ceyhun Necati Pehlivan, Nikolaus Forgó, & Peggy Valcke, *The EU AI Act: A Commentary*, Wolters Kluwer (2024), https://law-store.wolterskluwer.com/s/product/the-eu-artificial-intelligence-ai-act-a-commentary/01tPg000007gkK9IAI

Tawana Petty, *Defending Black Lives Means Banning Facial Recognition*, WIRED (2020), https://www.wired.com/story/defending-black-lives-means-banning-facial-recognition/

Tawana Petty, et al., *Reclaiming our Data: Interim report, Detroit*, Our Data Our Bodies (2018), https://eprints.lse.ac.uk/89638

Oreste Pollicino & Giovanni De Gregorio, "Constitutional Law in the Algorithmic Society," in *Constitutional Challenges in the Algorithmic* Cambridge University Press (2021),

https://www.cambridge.org/core/books/constitutional-challenges-in-the-algorithmic-society/constitutional-law-in-the-algorithmic-society/969E0889109C8092AD0AB57019E507E3

Sushma Raman & William Schulz, *The Coming Good Society*, Harvard University Press (2020), https://www.hup.harvard.edu/books/9780674977082

Sushma Raman, John Shattuck, & Mathias Risse, Holding Together: *The Hijacking of Rights in America and How to Reclaim Them for Everyone,* New Press (2022), https://thenewpress.com/books/holding-together

Sushma Raman, *Defunding the Police Might Leave Americans More Surveilled and Less Secure*, Foreign Policy (2020), https://foreignpolicy.com/2020/08/25/defunding-the-police-might-leave-americans-more-surveilled-and-less-secure/

Gabriela Ramos, UNESCO, *Inclusive and Resilient Societies* (2022), https://en.unesco.org/inclusivepolicylab/e-teams/resilient-approaches-international-development

Sneha Revanur, *We Must Put an End to AI's Culture of Secrecy*, TIME (2024), https://time.com/6985855/openai-ai-companies-secrecy/

Sneha Revanur, *The Futurist Summit: The Chat GPT Generation with Sneha Revanur*, YouTube (2024), https://m.youtube.com/watch?v=ySmSihsZB1g

Sneha Revanur, *Artificial Intelligence in Policing Is the Focus of Encode Justice*, Teen Vogue (2021), https://www.teenvogue.com/story/artificial-intelligence-policing-encode-justice

Nagla Rizk, "Artificial Intelligence and Inequality in the Middle East: The Political Economy of Inclusion," in *The Oxford Handbook of Ethics of AI*, Oxford University Press (2020), https://doi.org/10.1093/oxfordhb/9780190067397.013.40

Carolina Rossini, Giovana Carneiro, & Thiago Gumaraes Moraes, "Agile governance for an agile future: Sandboxes for promoting responsible innovation," in *T20 2024 Task Force 05: Inclusive Digital Transformation* (2024),

https://www.t20brasil.org/media/documentos/arquivos/TF05_ST_03_Agile_Governance_fo66ccd375c4ae7.pdf

Marc Rotenberg, *Framework Convention on AI and Human Rights, Democracy, and the Rule of Law (Council Eur.)*, International Legal Materials, Cambridge University Press (2025)

Marc Rotenberg, *The Imperative for a UN Special Rapporteur on AI and Human Rights*, Journal of AI Law and Regulation (2024), https://doi.org/10.21552/aire/2024/1/13

Marc Rotenberg, "Human Rights Alignment: The Challenge Ahead for AI Lawmakers," in *Introduction to Digital Humanism*, Springer (2024), https://doi.org/10.1007/978-3-031-45304-5_38

Marc Rotenberg, *US Supreme Court: NetChoice Cases Explore AI and the First Amendment*, Journal of AI Law and Regulation (2024), https://aire.lexxion.eu/article/AIRE/2024/2/13

Marc Rotenberg & Christabel Randolph, *The AI Red Line Challenge*, Tech Policy Press (2024), https://www.techpolicy.press/the-ai-red-line-challenge/

Marc Rotenberg, "Artificial Intelligence and the Right to Algorithmic Transparency" in *Information and Communication Technologies and Human Rights*, Cambridge University Press (2022), https://www.cambridge.org/core/books/abs/cambridge-handbook-of-information-technology-life-sciences-and-human-rights/artificial-intelligence-and-the-right-to-algorithmic-transparency/A92EE127AF24D868066EC0AEAE3A370C

Marc Rotenberg, *The Law of Artificial Intelligence and the Protection of Fundamental Rights: The Role of the ELI Guiding Principles on Automated Decisionmaking in the EU*, European Law Institute Newsletter (2022), https://am.aals.org/wp-content/uploads/sites/4/2024/01/rotenberg_eli_newsletter_julyaug_2022.pdf

Marc Rotenberg & Merve Hickok, *Artificial Intelligence and Democratic Values: Next Steps for the United States*, Council on Foreign Relations (2022),

https://www.cfr.org/blog/artificial-intelligence-and-democratic-values-next-steps-united-states

Stuart Russell & Peter Norvig, *Artificial Intelligence: A Modern Approach* (2021), https://aima.cs.berkeley.edu/

Stuart Russell, et al., *"Trustworthy AI,"* in *Reflections on Artificial Intelligence for Humanity,* Springer Cham (2021), https://doi.org/10.1007/978-3-030-69128-8_2

Stuart Russell, *Human Compatible: Artificial Intelligence and the Problem of Control* (2020), https://www.amazon.com/Human-Compatible-Artificial-Intelligence-Problem/dp/0525558632/

Stuart Russell, et al., *Cooperative Inverse Reinforcement Learning*, NIPS (2016), https://proceedings.neurips.cc/paper_files/paper/2016/hash/c3395dd46c34fa7fd8d729d8cf88b7a8-Abstract.html

Stuart Russell, Daniel Dewey, Max Tegmark *Research priorities for robust and beneficial artificial intelligence*, AI Magazine (2015), https://onlinelibrary.wiley.com/doi/pdf/10.1609/aimag.v36i4.2577

Emma Ruttkamp-Bloem, "Epistemic Just and Dynamic AI Ethics in Africa," in *Responsible AI in Africa: Challenges and Opportunities*, Springer International Publishing (2023), https://doi.org/10.1007/978-3-031-08215-3

Emma Ruttkamp-Bloem, "The Quest for Actionable AI Ethics, in *Artificial Intelligence Research*, Springer International Publishing (2020), https://doi.org/10.1007/978-3-030-66151-9_3

Ingrid Schneider & Krishna Shrinivas, *A chance for India to shape a data governance regime. The crafting of the country's data governance must enable a secure, more egalitarian, and trustworthy digital future for all*, The Hindu (2024), https://www.thehindu.com/opinion/lead/a-chance-for-india-to-shape-a-data-governance-regime/article66615759.ece

Ingrid Schneider, "Data Stewardship by Data Trusts: A Promising Model for the Governance of the Data Economy?," in *Global*

Communication Governance at the Crossroads (2024), https://doi.org/10.1007/978-3-031-29616-1_19

Ingrid Schneider, "Digital Sovereignty and Governance in the Data Economy: Data Trusteeship Instead of Property Rights on Data," in *A Critical Mind* (2023), https://doi.org/10.1007/978-3-662-65974-8_15

Ingrid Schneider, Catharina Rudschies, & Judith Simon, "The Heterogeneity of AI Ethics Guidelines Examined: Varying Natures, Actors, and Perceptions," in *Artificial Intlligence – Ethics and Law* (2022), https://www.fis.uni-hamburg.de/en/publikationen/detail.html?id=0fc78169-a7b4-498d-b169-816538b7e15c

Nathalie Smuha, *Algorithmic Rule By Law: How Algorithmic Regulation in the Public Sector Erodes the Rule of Law*, Cambridge University Press (2024), https://doi.org/10.1017/9781009427500

Nathalie Smuha, *Beyond a Human Rights-Based Approach to AI Governance: Promise, Pitfalls, Plea. Philosophy & Technology* (2021), https://doi.org/10.1007/s13347-020-00403-w

Kutoma Wakunuma, et al., *Trustworthy AI: African Perspectives*, Palgrave Macmillan (2025), https://link.springer.com/book/10.1007/978-3-031-75674-0

Kutoma Wakunuma & Damian Eke, *Africa, ChatGPT, and Generative AI Systems: Ethical Benefits, Concerns, and the Need for Governance. Philosophies* (2024), https://www.mdpi.com/2409-9287/9/3/80

Kutoma Wakunuma, Damian Eke, & Schmidt Shilukobo Chintu, "Towards shaping the future of responsible AI in Africa," in *Responsible AI in Africa: Challenges and Opportunities*, Springer International Publishing Cham (2023), https://library.oapen.org/bitstream/handle/20.500.12657/60787/1/978-3-031-08215-3.pdf#page=185

Kutoma Wakunuma, Damian Eke, & Sunusika Akintoye, "Introducing responsible AI in Africa," in *Responsible AI in Africa: Challenges and*

Opportunities, Springer International Publishing Cham (2023), https://library.oapen.org/bitstream/handle/20.500.12657/60787/978-3-031-08215-3.pdf?sequence=1#page=22

Hannes Werthner, et al., *Introduction to Digital Humanism*, Springer Nature (2024), https://library.oapen.org/handle/20.500.12657/86926

Hannes Werthner, et al., *Perspectives on Digital Humanism*, Springer Nature (2022), https://link.springer.com/book/10.1007/978-3-030-86144-5

Larissa Zutter, Paula Soumaya Domit, & Rebecca Leeper, *US Tapping China: Trends of Surveillance and Digital Authoritarianism in the US and China* (2020), https://aiws.net/monitoring-and-judging/monitoring/us-tapping-china-trends-of-surveillance-and-digital-authoritarianism-in-the-us-and-china/

CAIDP Global Academic Network

Elena Abrusci, University of Brunel (United Kingdom)

Professor Alessandro Acquisti, Carnegie Mellon University (United States)

Professor, Marco Almada, University of Luxembourg (Luxembourg)

Urvashi Aneja, Digital Futures Lab (India)

Professor Ricardo Baeza-Yates,Northeastern University (Chile / Spain)

Professor Irena Barkane, University of Latvia (Latvia)

Professor Francesca Bignami, George Washington University (United States)

Professor Abeba Birhane, Trinity College Dublin (Ireland)

Professor Meredith Broussard, New York University (United States)

Professor Derrick L. Cogburn, American University (United States)

Professor Renee Cummings, University of Virginia (United States)

Professor Joseph David, Sapir College (Israel)

Professor Virgnia Dignum, Umeå University (Sweden)

Professor Laura K. Donohue, Georgetown Law (United States)

Hlengiwe Dube, Center for Human Rights (South Africa)

Zeynep Engin (United Kingdom)

Professor Giovanni De Gregorio, Católica Global School of Law (Italy)

Professor Maura Grossman, University of Waterloo (Canada)

Professor Hiroki Habuka, Kyoto University Graduate School of Law (Japan)

Professor David Evan Harris, Berkeley University (United States)

Gry Hasselbalch, Data Ethics EU and InTouchAI EU, (Denmark)

Professor Masao Horibe, Hitotsubashi University and Chuo University (Japan)

Professor Margaret Hu, Wiliam and Mary Law School (United States)

Professor Aziz Huq, University of Chicago (United States)

Professor Meltem Ineli-Ciger, Suleyman Demirel University (Turkey)

Malavika Jayaram, Berkman-Klein Center (Singapore)

Bulelani Jili, Harvard University (United States)

Professor Leyla Keser, Istanbul Bilgi University (Turkey)

Lorraine Kisselburgh, Purdue University (United States)

Professor Eleni Kosta, Tilburg University (the Netherlands)

Professor Katja Langenbucher, Goethe-University (Germany)

Professor Seth Lazar, Australian National University (Australia)

Professor Gianclaudio Malgieri, EDHEC Business School (France)

Professor Alessandro Mantelero, Polytechnic University of Turin (Italy)

Professor Wonki Min, President SUNY Korea (Korea)

Dean Valsamis Mitsilegas, Liverpool School of Law and Social Justice (Great Britain)

Professor Maria Murphy, Co-Director, CAIDP Global Academic Network, National University of Ireland, Maynooth (Ireland)

Francis Nyarai Ndende (South Africa)

Professor Hellen Nissenbaum, Cornell University (United States)

Chinasa T. Okolo, Brookings Institution (United States)

Shobita Parthasarathy, University of Michigan (United States)

Professor Bilyana Petkova, University of Graz (Austria)

Prof. Pascal Pichonnaz, University of Fribourg (Switzerland)

Professor Viviana Polisena, Universidad Católica de Córdoba (Argentina)

Professor Oreste Pollicino, Bocconi University (Italy)

Professor Derya Ozkul, University of Warwick (England)

Professor Nagla Rizk, The American University in Cairo (Egypt)

Lecturer Carolina A. Rossini, Boston University (United States)

Marc Rotenberg, President CAIDP (United States)

Professor Stuart Russell, University of California at Berkeley (United States)

Professor Emma Ruttkamp-Bloem, University of Pretoria (South Africa)

Professor Idoia Salazar, San Pablo CEU University (Spain)

Professor Edward Santow, University of Technology Sydney (Australia)

Professor Giovanni Sartor, University of Bologna / European University Institute (Italy)

Professor Ingrid Schneider, Universität Hamburg (Germany)

Professor John Shattuck, Tufts University (United States) Nathalie Smuha, KU Leuven (Belgium)

Professor Lee Tiedrich, Duke University (United States)

Professor Kutoma J. Wakunuma (United Kingdom)

Professor Hans Werther, TU Wien (Austria)

Professor Kyoko Yoshinaga, Keio University (Japan)

Dean Michael Zimba, Malawi Institute ofTechnology (Malawi)

CAIDP Fellows

Class of 2024

Amir Noy
April Yoder, PhD
Artem Kobrin
Brenda W. Maina.
Çiğdem Akın, LLM,MA
Dirk Brand
Julian Theseira
Monique Munarini
Nana Khechikashvili
Natalia Alarcón Rueda
Sharvari Dhote
Tatiana G. Zasheva

Class of 2023

Ayca Ariyoruk
Evelina Ayrapetyan
Daniela Constantin
Tamiko Eto
Dominique Greene-Sanders
Caroline Friedman Levy
Tina Lassiter
Ren Bin Lee Dixon
Claudio Mutua
Selin Ozbek Cittone
Anca Radu
Nayyara Rahman
Varsha Sewlal

Class of 2022

Rick Cai
Elisa Elhadj
Ivana Feldfeber
Joel Kumwenda
Davor Ljubenkov
Nidhi Sinha
Dr. Grace Thomson

Class of 2021

Temofe Akaba
Selim Alan
Bridget Boakye
Lyantoniette Chua
Stephanie Cairns
Peter Furlong
August Gweon
Jason K. Johnson
Roberto Lopez-Davilla
Natalia Menéndez González
Tamra Moore
Oarabile Mudango
Somaeih Nikpoor
Ubongabasi Obot
Khatia Zukhubai

Class of 2020

Giuliano Borter
Regina Kronan
Rebecca Leeper
Paula Soumaya Doumit
Larissa Zutter

www.ingramcontent.com/pod-product-compliance
Lightning Source LLC
Chambersburg PA
CBHW081757200326
41597CB00023B/4053